ANNALS *of* THE NEW YORK ACADEMY OF SCIENCES

.UME
1279

ISBN-10: 1-57331-874-4 ; **ISBN-13:** 978-1-57331-874-7

ISSUE

Neurons, Circuitry, and Plasticity in the Spinal Cord and Brainstem

ISSUE EDITORS

Lea Ziskind-Conhaim,[a] Amy B. MacDermott,[b] Francisco J. Alvarez,[c] John D. Houle,[d] and Shawn Hochman[c]

[a]University of Wisconsin, [b]Columbia University, [c]Emory University School of Medicine, and [d]Drexel University College of Medicine

TABLE OF CONTENTS

Annals of the New York Academy of Sciences (ISSN: 0077-8923 [print]; ISSN: 1749-6632 [online]) is published 30 times a year on behalf of the New York Academy of Sciences by Wiley Subscription Services, Inc., a Wiley Company, 111 River Street, Hoboken, NJ 07030-5774.

Mailing: *Annals of the New York Academy of Sciences* is mailed standard rate.

Postmaster: Send all address changes to ANNALS OF THE NEW YORK ACADEMY OF SCIENCES, Journal Customer Services, John Wiley & Sons Inc., 350 Main Street, Malden, MA 02148-5020.

Disclaimer: The publisher, the New York Academy of Sciences, and the editors cannot be held responsible for errors or any consequences arising from the use of information contained in this publication; the views and opinions expressed do not necessarily reflect those of the publisher, the New York Academy of Sciences, and editors, neither does the publication of advertisements constitute any endorsement by the publisher, the New York Academy of Sciences and editors of the products advertised.

Publisher: *Annals of the New York Academy of Sciences* is published by Wiley Periodicals, Inc., Commerce Place, 350 Main Street, Malden, MA 02148; Telephone: 781 388 8200; Fax: 781 388 8210.

Journal Customer Services: For ordering information, claims, and any inquiry concerning your subscription, please go to www.wileycustomerhelp.com/ask or contact your nearest office. *Americas:* Email: cs-journals@wiley.com; Tel:+1 781 388 8598 or 1 800 835 6770 (Toll free in the USA & Canada). *Europe, Middle East, Asia:* Email: cs-journals@wiley. com; Tel: +44 (0) 1865 778315. *Asia Pacific:* Email: cs-journals@wiley.com; Tel: +65 6511 8000. *Japan:* For Japanese speaking support, Email: cs-japan@wiley.com; Tel: +65 6511 8010 or Tel (toll-free): 005 316 50 480. Visit our Online Customer Get-Help available in 6 languages at www.wileycustomerhelp.com.

Information for Subscribers: *Annals of the New York Academy of Sciences* is published in 30 volumes per year. Subscription prices for 2013 are: Print & Online: US$6,053 (US), US$6,589 (Rest of World), €4,269 (Europe), £3,364 (UK). Prices are exclusive of tax. Australian GST, Canadian GST, and European VAT will be applied at the appropriate rates. For more information on current tax rates, please go to www.wileyonlinelibrary.com/tax-vat. The price includes online access to the current and all online back files to January 1, 2009, where available. For other pricing options, including access information and terms and conditions, please visit www.wileyonlinelibrary.com/access.

Delivery Terms and Legal Title: Where the subscription price includes print volumes and delivery is to the recipient's address, delivery terms are Delivered at Place (DAP); the recipient is responsible for paying any import duty or taxes. Title to all volumes transfers FOB our shipping point, freight prepaid. We will endeavour to fulfill claims for missing or damaged copies within six months of publication, within our reasonable discretion and subject to availability.

Back issues: Recent single volumes are available to institutions at the current single volume price from cs-journals@wiley.com. Earlier volumes may be obtained from Periodicals Service Company, 11 Main Street, Germantown, NY 12526, USA. Tel: +1 518 537 4700, Fax: +1 518 537 5899, Email: psc@periodicals.com. For submission instructions, subscription, and all other information visit: www.wileyonlinelibrary.com/journal/nyas.

Production Editors: Kelly McSweeney and Allie Struzik (email: nyas@wiley.com).

Commercial Reprints: Dan Nicholas (email: dnicholas@wiley.com).

Membership information: Members may order copies of *Annals* volumes directly from the Academy by visiting www. nyas.org/annals, emailing customerservice@nyas.org, faxing +1 212 298 3650, or calling 1 800 843 6927 (toll free in the USA), or +1 212 298 8640. For more information on becoming a member of the New York Academy of Sciences, please visit www.nyas.org/membership. Claims and inquiries on member orders should be directed to the Academy at email: membership@nyas.org or Tel: 1 800 843 6927 (toll free in the USA) or +1 212 298 8640.

Printed in the USA by The Sheridan Group.

View *Annals* online at www.wileyonlinelibrary.com/journal/nyas.

Abstracting and Indexing Services: *Annals of the New York Academy of Sciences* is indexed by MEDLINE, Science Citation Index, and SCOPUS. For a complete list of A&I services, please visit the journal homepage at www. wileyonlinelibrary.com/journal/nyas.

Access to *Annals* is available free online within institutions in the developing world through the AGORA initiative with the FAO, the HINARI initiative with the WHO, and the OARE initiative with UNEP. For information, visit www. aginternetwork.org, www.healthinternetwork.org, www.oarescience.org.

Annals of the New York Academy of Sciences accepts articles for Open Access publication. Please visit http://olabout.wiley.com/WileyCDA/Section/id-406241.html for further information about OnlineOpen.

Wiley's Corporate Citizenship initiative seeks to address the environmental, social, economic, and ethical challenges faced in our business and which are important to our diverse stakeholder groups. Since launching the initiative, we have focused on sharing our content with those in need, enhancing community philanthropy, reducing our carbon impact, creating global guidelines and best practices for paper use, establishing a vendor code of ethics, and engaging our colleagues and other stakeholders in our efforts. Follow our progress at www.wiley.com/go/citizenship.

Ann. N.Y. Acad. Sci. ISSN 0077-8923

ANNALS OF THE NEW YORK ACADEMY OF SCIENCES
Issue: *Neurons, Circuitry, and Plasticity in the Spinal Cord and Brainstem*

Molecular, genetic, cellular, and network functions in the spinal cord and brainstem

Paul S.G. Stein

Biology Department, Washington University, St. Louis, Missouri

Address for correspondence: Paul S.G. Stein, Biology Department, Campus Box 1137, One Brookings Drive, Washington University, St. Louis, MO 63130. stein@wustl.edu

Studies of the model systems of spinal cord and brainstem reveal molecular, genetic, and cellular mechanisms that are critical for network and behavioral functions in the nervous system. Recent experiments establish the importance of neurogenetics in revealing cellular and network properties. Breakthroughs that utilize direct visualization of neuronal activity and network structure provide new insights. Major discoveries of plasticity in the spinal cord and brainstem contribute to basic neuroscience and, in addition, have promising therapeutic implications.

Keywords: spinal cord; brainstem; neurogenetics; neuronal circuits; plasticity

Introduction

Advances in the important model systems of spinal cord and brainstem were highlighted during the four Madison meetings entitled "Cellular and Network Functions in the Spinal Cord" organized by Lea Ziskind-Conhaim.[1–4] The 2012 Madison conference[4] featured talks from 40 speakers[5–44] as well as oral presentations from four trainees[45–48] selected to give platform talks. Two of the invited speakers with abstracts listed in the oral presentations section of the conference program[4] were unable to attend. In addition to the abstracts, the conference program[4] also includes 47 abstracts presented at the poster session.

Participants at the 2005 and the 2007 Madison conferences[1,2] did not submit papers for formal publication. However, speakers at the 2009 conference[3] received invitations to write contributions for publication in *Annals of the New York Academy of Sciences*, titled *Neurons and Networks in the Spinal Cord*.[49] Similarly, speakers at the 2012 conference[4] were invited to contribute to the current conference volume.[50]

The goal of this report is to highlight four spinal cord and brainstem themes discussed by speakers at the 2012 conference and authors of the present volume: neurogenetics, circuits, plasticity, and translational impacts. This report will discuss the contents of the summaries of the oral presentations[5–44] of the conference[4] and published data presented by speakers during the oral sessions.[51–67] Space limitations prevent detailed descriptions of every important result, however.

Neurogenetics of the spinal cord and brainstem

An exciting theme that permeated much of the 2012 conference was the increasing use of neurogenetic techniques in the study of rodents, especially mice, and zebrafish to characterize and manipulate neuronal types in the spinal cord and brainstem.[5,8–13,17–24,26,27,34,36] Several of the speakers, Arber,[68] Fetcho,[69] Goulding,[70] Gray,[71,72] Kiehn,[73,74] Kullander,[53] Ma,[71,75] and Pfaff,[76,77] have reviewed neurogenetic developments in the last decade applied to these model systems.

In the last century, identification of cell types in the nervous system relied mainly on neuroanatomical, electrophysiological, and behavioral techniques. In the last two decades, identification of specific combinations of transcription factors expressed in distinct neuronal populations and the genetic manipulation of these transcription factors have been added to the techniques available for understanding the structure and function of neuronal networks. Of special elegance are genetic techniques[70] that (1) allow the disabling of specific populations of

doi: 10.1111/nyas.12083

neurons at specific times in the animal's life (e.g., only in adults), and/or only in specific locations of the nervous system (e.g., only in the caudal part of the spinal cord); (2) allow the disabling of neurotransmitter function by mutations of specific vesicular transporters (e.g., Vglut2, in specific cell types); (3) label a cell type with a visible protein (e.g., GFP) that serves as an anatomical marker for that specific cell type; and/or (4) add a spanning membrane protein controllable by light or a chemical whose activation will excite (e.g., ChR2) or inhibit (e.g., allostatin receptor) a specific population of neurons.

Transcription factors in the spinal cord and brainstem

Spinal cord sensory processing. Transcription factors are important in the development of cells responsible for sensory transduction and processing.[75] Lumpkin[36,51,78] described the critical role of Merkel cell–neurite complexes in the sensory transduction of fine touch, including such object features as shape, edges, and curvature. These complexes consist of a peripheral Merkel cell that interacts synaptically with a slowly adapting type I (SAI) sensory afferent. SAI afferents can represent the finest spatial details, such as that encoded in Braille-like characters. Conditional deletion of Atoh1 in the body skin results in (1) a failure of development of Merkel cells; (2) a complete loss of SAI afferent responses; and (3) the behavioral loss of the ability to display preferences for specific textures. Many open questions remain, however, concerning the site or sites of the mechanical transduction and the specific mechanisms for the synapse between the Merkel cell and its associated SAI sensory afferent.[51]

Ma[34] focused on the effects of conditional knockouts of Tlx3 in Lbx1 lineage interneurons in the superficial laminae of the dorsal spinal cord and spinotrigeminal nuclei. These knockouts had behavioral deficits when presented with pain and/or itch stimuli but responded well to innocuous cold or warm stimuli. This work extends his prior studies of the role of Tlx3 in primary sensory neurons[75,79] and demonstrates the importance of characterizing the interactions in the interneuronal pathways driven by specific primary sensory neurons.[80]

Brownstone[20] described the dI3 population of dorsally derived interneurons identified by their expression of isl1. Most members of this population are excitatory and glutamatergic. Transmitter outputs of these dI3 interneurons can be disabled by selective elimination of the glutamate transporter Vglut2. This disabling of the outputs of dI3 interneurons has the behavioral consequence of impairing the abilities of the mouse to grasp objects.

Spinal cord interneurons. Goulding[18,70] provided an overview of spinal cord transcription factors. He focused on genetic manipulations of distinct populations of spinal cord inhibitory interneurons characterized by specific transcription factors and involved in processing hindlimb motor outputs, the V1 and V2b interneurons. These interneuronal classes assist in the organization of flexor/extensor rhythmic alternation during locomotion. The results of focused removal of these specific subsets of interneurons indicated that continued usage of these techniques will assist in understanding the structure of the spinal cord locomotor central pattern generator.

Alvarez[23,52] described a specific type of ventral horn V1 interneuron, the Renshaw cell. Alvarez traced the history of Renshaw cell research and highlighted the abilities and limitations of tools available to each experimentalist in each specific era of investigation. The transcription-factor pedigree of this cell type is well characterized as is the fundamental role of transcription factors in setting up many aspects of this cell type's identity, including its cell body location, its neurotransmitter, its input connections, and its output connections. Some features of the Renshaw cell are present before the development of its synaptic inputs and outputs. At present, much is known about this cell type; it will be interesting to see what future tools will be developed that reveal even more.

Kullander described the contribution of the transcription factor Dmrt3 to left–right coordination in both horses and mice.[53,81] Breeds of horses with a mutation in the *Dmrt3* gene have the ability to produce alternative gaits not produced by wild-type horses and tend to use these alternative gaits rather than gallop at high speeds. Mice with a *Dmrt3* mutation have difficulty with high-speed locomotion. In addition, left–right coordination during fictive locomotion in neonatal mutant mice is impaired compared to wild-type mice. Dmrt3 is expressed in the dI6 population of interneurons that express the Viaat marker for inhibitory interneurons; some dI6 subpopulations have axons that cross the midline. Thus interneurons expressing Dmrt3 assist in left–right coordination of locomotion in both mice and horses.

Spinal cord motor neuron identity. Transcription factors play a fundamental role in motor neuron identity.[76] Pfaff[8] described the importance of Lhx3 levels in motor neuron development. Lowering Lhx3 assisted in the production of limb-innervating motor neurons in the lateral motor column (LMC). Elevating Lhx3 assisted in the production of axial-innervating motor neurons in the medial motor column (MMC). Keeping Lhx3 levels high results in the conversion of some LMC motor neurons into MMC motor neurons. Such a conversion changes the rostral–caudal pattern of locomotor output while leaving rhythmicity and left–right coordination unchanged.

Ribera[9] examined the participation of Islet2 in the development of specific populations of secondary motor neurons (SMNs) in zebrafish. Interestingly, she found that signaling in the Notch pathway contributed to how Islet2 specifies the development of specific subtypes of SMNs.

Brainstem chemosensitive neurons. Guyénet[27,82] described the history of a well-studied population of glutamatergic central chemosensitive neurons in the retrotrapezoid nucleus (RTN) on the ventral surface of the brainstem, which express the Phox2b transcription factor. Neurogenetic manipulations of Phox2 revealed the participation of these neurons in the response to hydrogen ion and carbon dioxide levels and in the modulation of the frequency of respiration in mammals. Phox2b mutations in humans cause congenital central hypoventilation syndrome characterized by the lack of central carbon dioxide drive on the rate of ventilation.

Brainstem interneurons. Fetcho focused on the brainstem in zebrafish[13] and described the locations of cells with specific transcription factors and neurotransmitter identities that reside in anatomical stripes in the brainstem.[83] These stripes were revealed by selective labeling with specific fluorescent protein colors based on specific neurotransmitter transporters (e.g., glyt2), or a specific transcription factor (e.g., En1). This ground plan in the fish brainstem for interneuron escape circuitry shares many features with the mouse brainstem ground plan for locomotor circuitry.

Gray[26,72] provided an overview of mouse brainstem transcription factors and focused on genetic manipulations of those transcription factors expressed in brainstem respiratory interneurons. He described the contributions of Atoh1-dependent excitatory interneurons in controlling the relative phasing of inspiratory and expiratory respiratory rhythms. These contributions contrasted with that of Dbx1-dependent excitatory interneurons that participate in both inspiratory and expiratory rhythm generation.

Left–right coordination of locomotion

Kullander reviewed the participation of axon guidance molecules (e.g., the Eph/ephrins and Netrin/DCC) in the proper development of interneurons with commissural axons that assist in left–right coordination during locomotion.[19,53] Mutations in Netrin-1 result in left–right synchrony during fictive locomotion. Null mutations in DCC produce a mouse with no left–right coordination during fictive locomotion.

Kiehn's abstract[17] described experiments that manipulated specific classes of commissural interneurons involved in left–right coordination during locomotion. These experiments extend prior work from the Kiehn laboratory.[84]

Ziskind-Conhaim[22] described the Hb9 population of interneurons and discussed their inputs and outputs during locomotor rhythms. Of special interest are the contralateral dendrites of some Hb9 interneurons and the gap junctions between these dendrites and those of contralateral Hb9 interneurons. In addition, gap junctions are present between ipsilateral pairs of Hb9 interneurons. These electrical synapses may provide an important synchronizing signal for the production of coordinated motor output. She[22,54] also reported on the role of GAD67 commissural inhibitory GABAergic interneurons in contributing to left–right coordination. The contralateral processes of these interneurons synapse upon motor neurons and Renshaw cells, as well as upon contralateral GAD67 neurons. Electrical couplings between GAD67 interneurons may assist in the synchronization of the GABAergic signals that contribute to left–right coordination.

Optogenetic control of spinal cord interneurons

Kiehn's talk included a description of the application of optogenetic techniques to spinal cord excitatory interneurons in neonatal mice.[85] The optically active protein ChR2 was genetically inserted into the plasma membranes of excitatory glutamatergic interneurons expressing the vesicular glutamate transporter Vglut2. Shining light onto these interneurons opens up the ChR2 cation channel, and the resulting net flux of cations through the channel, in turn, activates the Vglut2 interneurons that contain ChR2. When light is shined on the lumbar spinal cord, it produces fictive locomotion. This provides direct support for the participation of glutamatergic interneurons in the activation of locomotor behavior.

Neurogenetics of membrane spanning proteins

Neurogenetics controlling expression of electrical synapses. McCrea[5,86] described the contributions of connexin36 to presynaptic regulation of transmitter release from group I muscle and tendon organ afferents. Connexin36 helps establish some of the electrical synapses between neurons in the central nervous system. In connexin36$^{-/-}$ knockout mice, a variety of techniques demonstrated attenuation of presynaptic inhibition compared to wild-type mice. This work supports the hypothesis that electrical coupling among interneurons involved in presynaptic inhibition helps synchronize this population and assists in the regulation of the magnitude of presynaptic inhibition.

Neurogenetics of ligand-gated receptors and motor system development. Kalb[10,55] discussed the important role of the specific AMPA receptor, GluA1 (also known as GluR1), in regulating the shape of the dendritic tree of motor neurons. This mechanism may serve as a contributing mechanism to experience-based plasticity. The GluA1 knockout mouse demonstrates deficits in interneuronal circuitry and locomotor behavior in addition to showing a diminished dendritic tree. Modifications of specific ligand-gated receptors contribute to the ability of the nervous system to modulate circuit properties in response to changes in the experience of the organism.

Neurogenetics of motor neuron disease

Shneider[11] described work with the FUS gene. Mutations of this gene have been implicated in some families with amyotrophic lateral sclerosis (ALS) and in some individuals with sporadic ALS. He developed a conditional/CRE-dependent *FUS* allele targeted to mouse motor neurons. Human wild-type FUS did not produce a disease phenotype. In contrast, the ALS-associated R521C mutant expressed in mouse motor neurons produced a phenotype with motor neuron loss and associated denervation of limb muscles. This mouse model exhibited other cytological features of human ALS motor neuron disease: it has the potential to assist in revealing some of the genetic mechanisms that produce ALS in humans.

Neuronal circuits in the spinal cord and brainstem

Speakers at the 2012 meeting provided insights into the structure of the neuronal circuits responsible for the production of behavior. In addition to the breakthroughs based on neurogenetics described in the previous section, speakers presented additional new techniques to reveal the locations of well-defined interneuronal populations that synapse upon specific motor neuron pools, as well as to visualize the activities of entire neuronal populations.

Anatomical techniques that reveal interneuron synapses on specific motor neuron pools

Monosynaptically restricted transsynaptic tracing tools. Arber[12,68] described the power of techniques that reveal the locations of interneurons directly presynaptic to the distinct motor neuron pools that innervate specific limb muscles. These transsynaptic virus retrograde-tracing techniques provide the ability to label with high confidence only those interneurons that directly synapse onto a specific motor neuron population. This work demonstrated an anatomical blueprint for motor antagonism by revealing that the cell-body locations of interneurons synapsing upon flexor motor neurons are different from those of interneurons synapsing upon extensor motor neurons. She applied this technique to the C3–C4 propriospinal interneurons that are known to play a critical role in forelimb motor control. Her data expanded prior physiological results with the demonstration of a much wider segmental distribution of these propriospinal interneurons. She also used this technique to study the projections of specific reticular nuclei to the spinal cord. Future applications of this transsynaptic virus have the potential to reveal new characteristics of circuits in the brainstem and spinal cord.

Imaging techniques to visualize activities of neuronal circuits

Techniques to visualize the activities of populations of neurons during the production of behaviors provide insights into the structure and function of neuronal circuits.

Voltage-sensitive dyes. Sato[14,56,87] described spontaneous depolarization waves in mouse embryos detected with a voltage-sensitive dye at E11–E13. These depolarization waves in both the brainstem and spinal cord contribute to the expression of spontaneous embryonic movements. During E11–E12, in the early stages of expression of the depolarization wave, nicotinic acetylcholine receptors are important for excitation. During E13, in the later stages of expression of the depolarization wave, glutamate is the dominant excitatory neurotransmitter. During the early stages of the depolarization wave, the effects of GABA are excitatory because of the chloride equilibrium potential that is more positive than the resting potential. During the later stages of the depolarization wave, the effects of GABA are inhibitory since the chloride equilibrium potential is more negative as a result of developmental changes in the number of specific cation–chloride cotransporters in the plasma membrane.[24,88]

Calcium imaging techniques. Bosma[15,89] used calcium-imaging techniques as well as patch-clamp recordings in the embryonic mouse midbrain to examine membrane mechanisms responsible for spontaneous activity occurring during E11.5–E14.5 in midline serotonergic neurons. These midline neurons are named the *initiation zone* and have features that include a long-duration depolarization event mediated, in part, by a *t*-type calcium channel. Of interest are possible mechanisms responsible for the cessation of spontaneous activity after E14.5, for example, upregulation of a specific potassium channel.

Pfaff[57,90] described the deep-tissue imaging ability of fast-scanning two-photon (2P) microscopy. This technique reports calcium levels in a number of individual V1 interneurons that contain the genetically encoded calcium indicator GCaMP3. Rhythmic oscillations in the activities of several interneurons at a time in a specific volume of the spinal cord were linked to specific motor outputs. The oscillations of some interneurons were in phase with a reference interneuron, while the rhythms of others nearby were out of phase with the reference. This technique has the promise to visualize activities of large populations of specific interneurons during locomotor activity.

O'Donovan's abstract[21] described the use of GCaMP3 expressed throughout the nervous system to image the transverse face of the spinal cord during locomotor-like activity in mice. This technique revealed initial activation of the intermediate zone followed by both dorsal and ventral propagation of activity.

Perreault[43,58] described functional multineuron calcium imaging (fMCI) of motor neurons and of descending commissural interneurons (dCINs) in the spinal cord activated by reticulospinal interneurons in specific locations in the medullary brainstem. In the mouse embryo, this revealed the developmental time course of the reticulospinal system and the subsets of specific populations that are activated by reticulospinal interneurons. Distinct lateral and medial populations of reticulospinal interneurons predominantly activate MMC and LMC motor neurons, respectively. Specific populations of reticulospinal interneurons are selective in activating only a subpopulation of dCINs. Future experiments with this technique will reveal more about how the brainstem activates the spinal cord.

Circuits for processing sensory inputs

Role of inhibition in dorsal horn sensory processing. MacDermott[35,59] discussed the critical importance of inhibition in the processing of sensory signals in lamina II/III in the spinal cord dorsal horn. Glycine receptors are responsible for the majority of postsynaptic fast inhibitory mechanisms. GABA receptors mainly serve to mediate presynaptic inhibition of primary afferent inputs. These presynaptic GABA receptors assist in the control of sensory pathway activation, for example, tactile or nociceptive, in response to a sensory stimulus. In particular, enhancing $GABA_A$ receptor function lowers noxious stimulus–evoked activity in spinal neurons. A possible mechanism for the generation and maintenance of chronic pain is loss of inhibition in the dorsal horn. Work that reveals the mechanisms of dorsal horn inhibitory processing has important therapeutic implications.

Fitzgerald[33,60,91] described critical periods in the development of sensory processing pathways. In the first postnatal week, rat pups are very sensitive to tactile stimulation. At this age, A fiber myelinated afferents from the skin, particularly low-threshold Aβ fibers, are the dominant input. In contrast, nociceptive C fiber inputs are very sparse at this time. Over several postnatal weeks, C fiber inputs increase. The inputs from the C fibers are critical to the development of functional glycinergic inhibition in sensory pathways activated by A fibers. Preventing activity of C fibers postnatally delays the development of these inhibitory connections. Important interactions between genetic mechanisms and activity-dependent mechanisms in the spinal cord help establish mature sensory signaling pathways in the spinal dorsal horn circuits.

Role of delta opioid receptors in the processing of touch. Scherrer[37,92] described the role of delta opioid receptors (DOR) in the processing of information by touch-sensing low-threshold mechanoreceptors (LTMRs). He demonstrated that DOR activation depresses synaptic release of glutamate from LTMRs onto postsynaptic neurons in lamina III–V. This observation indicates the possible therapeutic use of DOR agonists in the management of touch-evoked neuropathic pain.

Presynaptic inhibition of sensory signals. Hochman[42,61,93] focused on presynaptic inhibition (PSI) recorded as a primary afferent depolarization (PAD) in the neonatal rat spinal cord with hindlimb attached. In this preparation that expresses actual locomotor movements of its hindlimbs, it is possible to monitor (1) the PAD with suction electrodes attached to dorsal roots and (2) motor neuron action potentials with suction electrodes attached to ventral roots. The PSI produced during PAD serves to modify sensory processing. Prior work with PAD focused on ipsilateral effects. Hochman reported regarding a strong influence of contralateral limb stance-phase force on the magnitude of the ipsilateral PAD, a new feature of sensory control of motor output.

Circuits for generating motor output patterns
Several speakers emphasized the contributions of central pattern generator (CPG) circuits in the production of motor output patterns.[5,8,12,16–19,21,22,26,28,31,41,44,44] Specific aspects of CPG function are discussed in other sections of this report. This section describes the roles played by spinal cord interneurons with descending axons in the excitation of hindlimb CPGs and the importance of techniques that reveal the modular structure of CPG circuits.

Propriospinal interneurons. Schmidt[44,94] presented an overview of thoracic descending propriospinal interneurons and their transmission of descending excitation of locomotion. The contributions of these interneurons was demonstrated following the transection of all direct bulbospinal pathways with staggered left and right hemisections of the thoracic spinal cord. In a small percentage of *in vitro* neonatal preparations, electrical stimulation of sites in the brainstem evoked locomotor output to the hindlimbs. Specific substances were applied in the bath to spinal segments located in between the rostral hemisected segment and the caudal hemisected segment. Substances that increased the excitability of propriospinal interneurons in these segments also increased the percentage of preparations that expressed hindlimb locomotor rhythms in response to brainstem stimulation. These experiments are a basic neuroscience result and a result with clinical implications since they identify excitatory descending propriospinal interneurons as a target of therapeutic intervention in humans with spinal cord injury.

Modular structure for generating motor output patterns. Giszter[41,62] described the evidence that spinal cord circuits have a modular structure that contributes to motor pattern generation. He presented several different points of view that each supported motor modularity with a focus on motor primitives. These primitives provide an important and useful framework in which to analyze normal coordination patterns during locomotion as well as the disrupted coordination patterns following nervous system injury.

Plasticity of spinal cord and brainstem neuronal circuits

In the last century, traditional views of spinal cord and brainstem neuronal networks held that these circuits were relatively hardwired and served to produce mainly stereotyped behaviors. Toward the end of the last century and in this century, these points of view have been challenged by a number of important researchers examining the effects of injury, patterns of sensory or motor stimulation, and neuromodulatory agents.

Plasticity in response to injury

A critical issue in developing strategies for therapeutic interventions following spinal cord injury is understanding the nature and magnitude of neural plasticity following spinal cord injury. The hope is that there is sufficient neural plasticity to serve as a contributor to circuit repair when used in combination with other therapeutic strategies.

Dual-lesion techniques to reveal spinal cord plasticity. Rossignol[31,63] developed a protocol to reveal spinal cord plasticity following spinal cord injury that starts out by examining the control situation of a complete transection of the adult cat spinal cord and the time course of recovery of locomotion. The experimental situation consists of first performing a lateral hemisection of the cat spinal cord and characterizing the recovery of locomotion following the hemisection. Once that recovery is stabilized, then the second manipulation in the experimental condition is complete transection of the spinal cord below the initial hemisection. Recovery of locomotion following the complete transection in the experimental condition occurs more rapidly than in the control condition. This demonstrates plasticity of spinal cord circuits after the initial lateral hemisection in the experimental situation.

Downregulation of cotransporters following spinal cord injury. Vinay[24,95] described the contributions of the intracellular concentration of chloride ions and the chloride equilibrium potential in the regulation of the effects of GABA and glycine on spinal cord neurons. Adequate expression of the potassium-chloride cotransporter 2 (KCC2) in the plasma membranes of spinal neurons is required for the production of more negative cell membrane voltages in response to GABA and glycine binding to $GABA_A$ and glycine receptors, respectively. Spasticity occurring after spinal injury is, in part, due to the downregulation of the amount of KCC2 in the plasma membrane that results in hyperexcitable spinal neurons contributing to spasticity. Pharmacological techniques that can upregulate KCC2 following spinal cord injury are of interest as a possible mechanism to reduce spasticity.

Plasticity in response to neuromodulatory agents

Emerging concepts of the past two decades are the contributions of neuromodulatory agents in modifying cellular characteristics that alter neuronal network system properties. Future understandings of the abilities of molecules to modify spinal and brainstem circuits that need repair have translational potential.

Ramirez[25,96] presented an overview of neuromodulatory influences on the pre-Bötzinger complex and the effects of these modulators on respiratory behaviors. He highlighted the abilities of neuromodulators to reconfigure neuronal networks. The outcomes of some types of reconfiguration could be beneficial; other types of reconfiguration could lead to detrimental outcomes.

Funk[28,97] focused on the modulatory contributions of glia in the pre-Bötzinger complex. In particular, during hypoxic conditions the transmitter ATP is released in the brainstem ventral respiratory column, including the pre-Bötzinger complex. The binding of ATP to $P2Y_1$ receptors on glial cells leads to the release of glutamate from these cells. The binding of this glutamate to glutamate receptors on neuronal cells in the pre-Bötzinger complex leads to an increase in the frequency of respiratory inspiration.

Wenner[16,64] described the complex effects of endocannabinoids in the developing spinal cord. Blockade of the CB1 receptor, the presynaptic G protein–coupled cannabinoid receptor, leads to a tonic suppression of spontaneous miniature glutamatergic postsynaptic potentials as well as an increase in the frequency of spontaneous network activity in the embryonic chick spinal cord. Blockade of CB1 receptors did not result in a modulation of evoked glutamatergic responses, however.

Plasticity in response to patterns of sensory inputs

An avenue into the injured nervous system that can assist in plasticity and therapeutic interventions is via sensory inputs. Some specific patterns of sensory input may have a greater therapeutic potential than other input patterns. Of interest was the phenomenon of the pattern-sensitive changes in respiratory neuronal circuits in response to various regimes of intermittent hypoxia.

Ramirez[25] described the effects of exposing the respiratory network to (1) acute intermittent hypoxia (AIH), that leads to enhancement of synaptic inhibition and that changes the response to norepinephrine; and (2) chronic intermittent hypoxia

(CIH), which leads to a decrease of high-voltage activated calcium currents and weakening of synaptic excitation.

Kunze[7] focused on the voltage-clamp studies of the synapse in the nucleus of the solitary tract (nTS) for integration of visceral sensory information from the carotid body arterial chemosensory afferent fiber and the second order neuron in the nTS. In response to CIH, this synapse becomes hyperexcitable over the time course of 10 days and can be reversed after 30 days in ambient oxygen. Several mechanisms are responsible for these plastic changes—each with a different time course.

Mitchell[38,65] presented an overview of multiple molecular and cellular mechanisms responsible for phrenic long-term facilitation (pLTF) in response to AIH. He described the "Q" signaling cascade with G_q protein–coupled receptors activated by mild to moderate AIH and the "S" signaling cascade with G_s protein–coupled receptors activated by severe AIH. He presented evidence for crosstalk inhibition between these pathways. Recent work[98] examined motor recovery in rats with a C2 spinal hemisection. In these rats, daily treatment with AIH induced plasticity and improved function in both respiratory and limb motor circuitry. This experimental paradigm presents the opportunity of exploring the therapeutic potential that results from molecular understandings of spinal cord plasticity.

Translational impacts of spinal cord research findings

An important development arising from basic spinal cord research is the possibility that new therapeutic strategies can be developed to assist those with spinal cord injuries. Research from many different areas of neuroscience can contribute.

Regeneration of injured spinal cord pathways

Houle[29,66] emphasized that combinations of therapeutic interventions are providing the greatest chance of getting functional recoveries from spinal cord injuries. His combination strategy includes (1) peripheral nerve grafts (PNGs) placed into the damaged spinal cord region, (2) modulation of the extracellular matrix to degrade substances in the glial scar, and (3) exercise. PNGs support regeneration of axons via the peripheral nerve. The glial scar that forms in the injured region opposes the growth of axons. The application of chondroitinase (ChABC) to this region assists in the breakup of the glial scar

and promotes axonal growth out of the PNG and into the adjacent regions of the spinal cord. Exercise plays a key role in enhancing the effects of the PNG and the ChABC. The molecular consequences of this combination strategy can be directly assayed by measuring levels of neurotrophic factors in the cord. In addition, exogenous application of neurotrophic factors can also enhance the benefits of the combination therapies.

Silver[30,99] described the successes of ChABC therapy combined with PNG in a cervical model of spinal cord injury that used electromyographic recordings from the diaphragm to assay the functional success of the treatment. After successful treatment with this combined therapy, an important additional manipulation in the work was subsequent transection of the nerve graft that leads to a loss of the recovery.

Reier[39,100] used an alternate strategy, that of interneuronal progenitor-enriched transplants into a region of injured cervical spinal cord, to demonstrate recovery of the rat diaphragm in response to respiratory challenges. Transsynaptic retrograde tracing with pseudorabies virus revealed some labeling in donor neurons, suggesting that some of these donors were integrated into the respiratory interneuronal circuitry. Interneuronal circuitry may provide an excellent substrate for therapeutic interventions and functional recoveries from spinal injury.

Jung[32,67] emphasized another strategy to assist in functional recovery following spinal cord injury, neuromuscular electrical stimulation (NMES) therapy. NMES has an advantage in that it can provide patterned muscular activity before return of voluntary muscle activity. Several measures of locomotor behaviors displayed improvements following the NMES treatments.

Neuroprosthesis via direct electrical stimulation of the spinal cord

Mushahwar[40,101] described intraspinal microstimulation (ISMS) in the cat lumbosacral spinal cord as a strategy to restore standing and walking following spinal cord injury. This technique has been refined so that weight-bearing, propulsive overground walking can be achieved with ISMS. These results support the suggestion that the stimulation activates networks within the spinal cord that generate the normal patterns of locomotion.

Perlmutter[6] described a strategy for activation of the ISMS electrodes in the monkey cervical spinal cord. The firing of a single corticomotoneuronal (CM) cell provided an input to an autonomous, brain–computer interface that delivered spike-triggered electrical stimuli via the implanted electrodes. The interface allowed adjustments of the delay between the CM cell firing and the ISMS stimulation. Different amounts of delay resulted in different strengths of CM spike–triggered rectified electromyographic activity. These results indicate that spinal stimulation triggered by single-unit motor cortex activity may have therapeutic effects for those with spinal injury.

Future directions of spinal cord and brainstem research

What can we look forward to in the next decades of spinal cord and brainstem research? The current explosion of new knowledge will continue. Neurogenetics will transform our basic science understandings of neuronal circuits and will contribute to the development of therapeutic strategies to assist those with spinal cord injury. We are at the beginning of a paradigm shift with exciting work to follow. I am confident that spinal cord and brainstem researchers look forward to future conferences with foci on these important model systems.

Acknowledgments

Dr. Lea Ziskind-Conhaim was the driving force responsible for the successes of the four Cellular and Network Functions in the Spinal Cord meetings. I attended all four meetings; my conversations with colleagues at the meetings were uniform in their praise of Dr. Ziskind-Conhaim's outstanding work in all aspects. The 2012 meeting was sponsored by the Society for Spinal Cord Research in Madison, Wisconsin and several groups at the University of Wisconsin–Madison: the School of Medicine and Public Health, the Department of Physiology/Neuroscience, the Department of Anesthesiology, the Center for Stem Cell Research, the Department of Anatomy/Neuroscience, and the Department of Neurological Surgery.

Conflicts of interest

The author declares no conflicts of interest.

References

1. Ziskind-Conhaim, L. 2005. *Cellular and Network Functions in the Spinal Cord-Brainstem*. Madison, WI. Cited February 24, 2013. http://conferencing.uwex.edu/conferences/spinalconference2012/previoussymposia.cfm.
2. Ziskind-Conhaim, L. 2007. *Cellular and Network Functions in the Spinal Cord*. Madison, WI. Cited February 24, 2013. http://conferencing.uwex.edu/conferences/spinalconference2012/previoussymposia.cfm.
3. Ziskind-Conhaim, L. 2009. *Cellular and Network Functions in the Spinal Cord*. Madison, WI. Cited February 24, 2013. http://conferencing.uwex.edu/conferences/spinalconference2009/.
4. Ziskind-Conhaim, L. 2012. *Cellular and Network Functions in the Spinal Cord*. Madison, WI. Cited February 24, 2013. http://conferencing.uwex.edu/conferences/spinalconference2012/.
5. McCrea, D.A., W. Bautista & J. Nagy. 2012. "Critical involvement of connexin 36 gap junctions in presynaptic inhibition in the mouse spinal cord." Summary of oral presentation. http://conferencing.uwex.edu/conferences/spinalconference2012/.
6. Perlmutter, S., Y. Nishimura, R.W. Eaton, *et al.* 2012. "Plasticity of corticospinal synapses produced by activity-dependent spinal stimulation in behaving monkeys." Summary of oral presentation. http://conferencing.uwex.edu/conferences/spinalconference2012/.
7. Kunze, D.L., S. Wang & D. Kline. 2012. "Plasticity of cardiorespiratory reflexes initiated by increased sensory input." Summary of oral presentation. http://conferencing.uwex.edu/conferences/spinalconference2012/.
8. Alaynick, W.A., C.A. Hinckley, B.W. Gallarda, *et al.* 2012. "Motor neuron subtype identity influences CPG activity." Summary of oral presentation. http://conferencing.uwex.edu/conferences/spinalconference2012/.
9. Moreno, R.L. & A.B. Ribera. 2012. "Islet2 and Notch signaling interact to regulate specification of a subset of motor neurons." Summary of oral presentation. http://conferencing.uwex.edu/conferences/spinalconference2012/.
10. Kalb, R. 2012. "Role of GluA1 glutamate receptor subunit in postnatal motor system development." Summary of oral presentation. http://conferencing.uwex.edu/conferences/spinalconference2012/.
11. Sharma, A., A. Lyashchenko & N. Shneider. 2012. "Cell autonomous effects of ALS mutant FUS in motor neurons." Summary of oral presentation. http://conferencing.uwex.edu/conferences/spinalconference2012/.
12. Arber, S. 2012. "Organizational principles of antagonistic motor circuits." Summary of oral presentation. http://conferencing.uwex.edu/conferences/spinalconference2012/.
13. Fetcho, J., F. Minale, M. Koyama, *et al.* 2012. "Femtosecond laser dissection of a circuit for a simple behavioral choice." Summary of oral presentation. http://conferencing.uwex.edu/conferences/spinalconference2012/.
14. Sato, K. & Y. Momose-Sato. 2012. "Voltage-sensitive dye imaging of the spontaneous depolarization wave in the embryonic mouse CNS: origins and pharmacological

natures." Summary of oral presentation. http://conferencing.uwex.edu/conferences/spinalconference2012/.

15. Bosma, M. 2012. "Mechanisms of age-dependent retraction and cessation of spontaneous activity in embryonic mouse hindbrain." Summary of oral presentation. http://conferencing.uwex.edu/conferences/spinalconference2012/.

16. Wenner, P. & C. Gonzalez-Islas. 2012. "Endocannabinoid signaling in the developing motor network." Summary of oral presentation. http://conferencing.uwex.edu/conferences/spinalconference2012/.

17. Kiehn, O. 2012. "Dual organization of commissural pathways involved in left-right alternation during locomotion." Summary of oral presentation. http://conferencing.uwex.edu/conferences/spinalconference2012/.

18. Goulding, M. 2012. "Misbehaving mice: new insights into the organization of the flexor-extensor system." Summary of oral presentation. http://conferencing.uwex.edu/conferences/spinalconference2012/.

19. Memic, F., N. Rabe, H. Gezelius, et al. 2012. "DCC mediated axon guidance of spinal interneurons is essential for normal locomotor central pattern generator function." Summary of oral presentation. http://conferencing.uwex.edu/conferences/spinalconference2012/.

20. Bui, T.V. & R.M. Brownstone. 2012. "Craving the rose: a step towards grasping a thorny spinal microcircuit." Summary of oral presentation. http://conferencing.uwex.edu/conferences/spinalconference2012/.

21. Puhl, J.G. & M.J. O'Donovan. 2012. "Imaging the spatiotemporal organization and propagation of activity at the onset of locomotor-like activity in the neonatal mouse spinal cord." Summary of oral presentation. http://conferencing.uwex.edu/conferences/spinalconference2012/.

22. Ziskind-Conhaim, L., Y. Verbny, T. Mavlyutov, et al. 2012. "Communicating outside the box: far-projecting excitatory and inhibitory interneurons that are part of the resilient locomotor networks." Summary of oral presentation. http://conferencing.uwex.edu/conferences/spinalconference2012/.

23. Alvarez, F.J. 2012. "Renshaw 101: what we know about the development of the recurrent inhibitory circuit." Summary of oral presentation. http://conferencing.uwex.edu/conferences/spinalconference2012/.

24. Vinay, L., R. Bos, P. Boulenguez, et al. 2012. "New perspectives for the treatment of spasticity after spinal cord injury." Summary of oral presentation. http://conferencing.uwex.edu/conferences/spinalconference2012/.

25. Ramirez, J.-M., A.J. Garcia III, F.P. Elsen, et al. 2012. "Network and behavioral consequences of sparse connectivity in the respiratory network." Summary of oral presentation. http://conferencing.uwex.edu/conferences/spinalconference2012/.

26. Tupal, S., H. Wei-Hsiang, H.Y. Zoghbi, et al. 2012. "Neural network underlying complex respiratory behaviors." Summary of oral presentation. http://conferencing.uwex.edu/conferences/spinalconference2012/.

27. Guyénet, P.G. 2012. "The retrotrapezoid nucleus and breathing." Summary of oral presentation. http://conferencing.uwex.edu/conferences/spinalconference2012/.

28. Funk, G.D. 2012. "Purinergic modulation of the preBötzinger Complex inspiratory rhythm generating network." Summary of oral presentation. http://conferencing.uwex.edu/conferences/spinalconference2012/.

29. Houle, J., M.-P. Côté, V. Tom, et al. 2012. "Activity dependent plasticity and axon regeneration in the injured spinal cord." Summary of oral presentation. http://conferencing.uwex.edu/conferences/spinalconference2012/.

30. Silver, J. 2012. "Functional regeneration into and well beyond the glial scar." Summary of oral presentation. http://conferencing.uwex.edu/conferences/spinalconference2012/.

31. Rossignol, S. & M. Martinez. 2012. "A dual spinal cord lesion paradigm to study spinal locomotor plasticity in the cat." Summary of oral presentation. http://conferencing.uwex.edu/conferences/spinalconference2012/.

32. Jung, R., B.K. Hillen, M. Fairchild, et al. 2012. "Accelerating locomotor recovery after spinal contusion." Summary of oral presentation. http://conferencing.uwex.edu/conferences/spinalconference2012/.

33. Fitzgerald, M. 2012. "Critical periods in the development of spinal tactile and nociceptive circuits." Summary of oral presentation. http://conferencing.uwex.edu/conferences/spinalconference2012/.

34. Xu, Y., C. Lopes, C. Birchmeier, et al. 2012. "Ontogeny of spinal neurons required to sense pain and itch." Summary of oral presentation. http://conferencing.uwex.edu/conferences/spinalconference2012/.

35. Bardoni, R., T. Takazawa, C.-K. Tong, et al. 2012. "GABAergic and glycinergic inhibitory control at the boundary between lamina II and III in the dorsal horn." Summary of oral presentation. http://conferencing.uwex.edu/conferences/spinalconference2012/.

36. Lumpkin, E.A. 2012. "Mechanisms of sensory coding in a mammalian touch receptor." Summary of oral presentation. http://conferencing.uwex.edu/conferences/spinalconference2012/.

37. Scherrer, G., R. Bardoni, P. Choudhury, et al. 2012. "Functional organization of opioid receptors in somatosensory neurons." Summary of oral presentation. http://conferencing.uwex.edu/conferences/spinalconference2012/.

38. Mitchell, G.S. 2012. "Multiple pathways to spinal respiratory motor facilitation: functional implications." Summary of oral presentation. http://conferencing.uwex.edu/conferences/spinalconference2012/.

39. Reier, P.J., D.D. Fuller, K.-Z. Lee, et al. 2012. "Spinal respiratory interneurons: plasticity and repair following cervical spinal cord injury (cSCI)." Summary of oral presentation. http://conferencing.uwex.edu/conferences/spinalconference2012/.

40. Mushahwar, V.K., B. Holinski, D. Everaert, et al. 2012. "Overground walking with intraspinal microstimulation in adult cats." Summary of oral presentation. http://conferencing.uwex.edu/conferences/spinalconference2012/.

41. Giszter, S., C. Oza, U.I. Udoekwere, *et al.* 2012. "Spinal cord mechanisms for recovery of function and modularity in the rat." Summary of oral presentation. http://conferencing. uwex.edu/conferences/spinalconference2012/.

42. Hayes, H.B., Y.-H. Chang & S. Hochman. 2012. "Stance-phase force on the opposite limb dictates strength of sensory transmission to the swing limb during locomotion." Summary of oral presentation. http://conferencing. uwex.edu/conferences/spinalconference2012/.

43. Perreault, M.-C., J.C. Glover, K. Szokol, *et al.* 2012. "Specificity of functional connections between descending glutamatergic neurons and commissural interneurons." Summary of oral presentation. http://conferencing.uwex. edu/conferences/spinalconference2012/.

44. Schmidt, B., K. Cowley & B. MacNeil. 2012. "Facilitation of hindlimb stepping through neurochemical excitation of thoracic propriospinal neurons." Summary of oral presentation. http://conferencing. uwex.edu/conferences/spinalconference2012/.

45. Strey, K.A. & T.L. Baker-Herman. 2012. "Inactivity-induced phrenic motor facilitation following reductions in synaptic inputs to phrenic motor neurons requires Tumor Necrosis Factor alpha and atypical Protein Kinase C activity." Summary of oral presentation. http://conferencing. uwex.edu/conferences/spinalconference2012/.

46. Wang, X., J.A. Hayes & C.A. Del Negro. 2012. "Cumulative single-cell laser ablation of functionally identified rhythmic neurons interrogates network properties in mammalian breathing-related circuits *in vitro.*" Summary of oral presentation. http://conferencing.uwex. edu/conferences/spinalconference2012/.

47. Devinney, M.J., N.L. Nichols & G.S. Mitchell. 2012. "Serotonin/adenosine interactions impart pattern sensitivity to phrenic long-term facilitation following moderate hypoxia." Summary of oral presentation. http://conferencing. uwex.edu/conferences/spinalconference2012/.

48. Jirjis, M.B., S.N. Kurpad & B.D. Schmit. 2012. "Diffusion tensor imaging parameters are sensitive to severity of a caudal spinal cord injury." Summary of oral presentation. http://conferencing.uwex.edu/conferences/ spinalconference2012/.

49. Ziskind-Conhaim, L., J.R. Fetcho, S. Hochman, *et al.*, Eds. 2010. "Neurons and Networks in the Spinal Cord." Issue of, *Ann. N.Y. Acad. Sci.* **1198:** 1–293.

50. Ziskind-Conhaim, L., A.B. MacDermott, F.J. Alvarez, J.D. Houle & S. Hochman, Eds. 2013. "Neurons, Circuitry, and Plasticity in the Spinal Cord and Brainstem." Issue of, *Ann. N.Y. Acad. Sci.* **1279:** 1–174. This volume.

51. Maksimovic, S., Y. Baba & E.A. Lumpkin. 2013. Neurotransmitters and synaptic components in the Merkel cell–neurite complex, a gentle-touch receptor. *Ann. N.Y. Acad. Sci.* **1279:** 13–21. This volume.

52. Alvarez, F.J., A. Benito-Gonzalez & V.C. Siembab. 2013. Principles of interneuron development learned from Renshaw cells and the motoneuron recurrent inhibitory circuit. *Ann. N.Y. Acad. Sci.* **1279:** 22–31. This volume.

53. Vallstedt, A. & K. Kullander. 2013. Dorsally derived spinal interneurons in locomotor circuits. *Ann. N.Y. Acad. Sci.* **1279:** 32–42. This volume.

54. Ziskind-Conhaim, L. 2013. Neuronal correlates of the dominant role of GABAergic transmission in the developing mouse locomotor circuitry. *Ann. N.Y. Acad. Sci.* **1279:** 43–53. This volume.

55. Jablonski, A.M. & R.G. Kalb. 2013. GluA1 promotes the activity-dependent development of motor circuitry in the developing segmental spinal cord. *Ann. N.Y. Acad. Sci.* **1279:** 54–59. This volume.

56. Momose-Sato, Y. & K. Sato. 2013. Optical imaging of the spontaneous depolarization wave in the mouse embryo: origins and pharmacological nature. *Ann. N.Y. Acad. Sci.* **1279:** 60–70. This volume.

57. Hinckley, C.A. & S.L. Pfaff. 2013. Imaging spinal neuron ensembles active during locomotion with genetically encoded calcium indicators. *Ann. N.Y. Acad. Sci.* **1279:** 71–79. This volume.

58. Perreault, M-C. & J.C. Glover. 2013. Glutamatergic reticulospinal neurons in the mouse: developmental origins, axon projections, and functional connectivity. *Ann. N.Y. Acad. Sci.* **1279:** 80–89. This volume.

59. Bardoni, R., T. Takazawa, C-K. Tong, *et al.* 2013. Pre-and postsynaptic inhibitory control in the spinal cord dorsal horn. *Ann. N.Y. Acad. Sci.* **1279:** 90–96. This volume.

60. Koch, S.C. & M. Fitzgerald. 2013. Activity-dependent development of tactile and nociceptive spinal cord circuits. *Ann. N.Y. Acad. Sci.* **1279:** 97–102. This volume.

61. Hochman, S., H.B. Hayes, I. Speigel & Y-H. Chang. 2013. Force-sensitive afferents recruited during stance encode sensory depression in the contralateral swinging limb during locomotion. *Ann. N.Y. Acad. Sci.* **1279:** 103–113. This volume.

62. Giszter, S.F. & C.B. Hart. 2013. Motor primitives and synergies in the spinal cord and after injury—the current state of play. *Ann. N.Y. Acad. Sci.* **1279:** 114–126. This volume.

63. Martinez, M. & S. Rossignol. 2013. A dual spinal cord lesion paradigm to study spinal locomotor plasticity in the cat. *Ann. N.Y. Acad. Sci.* **1279:** 127–134. This volume.

64. Wenner, P. 2013. The effects of endocannabinoid signaling on network activity in developing and motor circuits. *Ann. N.Y. Acad. Sci.* **1279:** 135–142. This volume.

65. Devinney, M.J., A.G. Huxtable, N.L. Nichols & G.S. Mitchell. 2013. Hypoxia-induced phrenic long-term facilitation: emergent properties. *Ann. N.Y. Acad. Sci.* **1279:** 143–153. This volume.

66. Houle, J.D. & M-P. Côté. 2013. Axon regeneration and exercise-dependent plasticity after spinal cord injury. *Ann. N.Y. Acad. Sci.* **1279:** 154–163. This volume.

67. Hillen, B.K., J.J. Abbas & R. Jung. 2013. Accelerating locomotor recovery after incomplete spinal injury. *Ann. N.Y. Acad. Sci.* **1279:** 164–174. This volume.

68. Arber, S. 2012. Motor circuits in action: specification, connectivity, and function. *Neuron* **74:** 975–989.

69. McLean, D.L. & J.R. Fetcho. 2011. Movement, technology and discovery in the zebrafish. *Curr. Opin. Neurobiol.* **21:** 110–115.

70. Grossmann, K.S., A. Giraudin, O. Britz, *et al.* 2010. Genetic dissection of rhythmic motor networks in mice. *Prog. Brain Res.* **187:** 19–37.

71. Gray, P.A., H. Fu, P. Luo, *et al.* 2004. Mouse brain organization revealed through direct genome-scale TF expression analysis. *Science* **306:** 2255–2257.

72. Gray, P.A. 2008. Transcription factors and the genetic organization of brain stem respiratory neurons. *J. Appl. Physiol.* **104:** 1513–1521.

73. Kiehn, O. 2006. Locomotor circuits in the mammalian spinal cord. *Annu. Rev. Neurosci.* **29:** 279–306.

74. Kiehn, O. 2011. Development and functional organization of spinal locomotor circuits. *Curr. Opin. Neurobiol.* **21:** 100–109.

75. Liu, Y. & Q. Ma. 2011. Generation of somatic sensory neuron diversity and implications on sensory coding. *Curr. Opin. Neurobiol.* **21:** 52–60.

76. Alaynick, W.A., T.M. Jessell & S.L. Pfaff. 2011. SnapShot: spinal cord development. *Cell* **146:** 178–178 e1.

77. Levine, A.J., K.A. Lewallen & S.L. Pfaff. 2012. Spatial organization of cortical and spinal neurons controlling motor behavior. *Curr. Opin. Neurobiol.* **22:** 812–821.

78. Maricich, S.M., S.A. Wellnitz, A.M. Nelson, *et al.* 2009. Merkel cells are essential for light-touch responses. *Science* **324:** 1580–1582.

79. Lopes, C., Z. Liu, Y. Xu, *et al.* 2012. Tlx3 and runx1 act in combination to coordinate the development of a cohort of nociceptors, thermoceptors, and pruriceptors. *J. Neurosci.* **32:** 9706–9715.

80. Ma, Q. 2012. Population coding of somatic sensations. *Neurosci. Bull.* **28:** 91–99.

81. Andersson, L.S., M. Larhammar, F. Memic, *et al.* 2012. Mutations in DMRT3 affect locomotion in horses and spinal circuit function in mice. *Nature* **488:** 642–646.

82. Guyénet, P.G., R.L. Stornetta & D.A. Bayliss. 2010. Central respiratory chemoreception. *J. Comp. Neurol.* **518:** 3883–3906.

83. Kinkhabwala, A., M. Riley, M. Koyama, *et al.* 2011. A structural and functional ground plan for neurons in the hindbrain of zebrafish. *Proc. Natl. Acad. Sci. U.S.A.* **108:** 1164–1169.

84. Restrepo, C.E., G. Margaryan, L. Borgius, *et al.* 2011. Change in the balance of excitatory and inhibitory midline fiber crossing as an explanation for the hopping phenotype in EphA4 knockout mice. *Eur. J. Neurosci.* **34:** 1102–1112.

85. Hagglund, M., L. Borgius, K.J. Dougherty, *et al.* 2010. Activation of groups of excitatory neurons in the mammalian spinal cord or hindbrain evokes locomotion. *Nat. Neurosci.* **13:** 246–252.

86. Bautista, W., J.I. Nagy, Y. Dai, *et al.* 2012. Requirement of neuronal connexin36 in pathways mediating presynaptic inhibition of primary afferents in functionally mature mouse spinal cord. *J. Physiol. (London)* **590:** 3821–3839.

87. Momose-Sato, Y., T. Nakamori & K. Sato. 2012. Pharmacological mechanisms underlying switching from the large-scale depolarization wave to segregated activity in the mouse central nervous system. *Eur. J. Neurosci.* **35:** 1242–1252.

88. Ben-Ari, Y., M.A. Woodin, E. Sernagor, *et al.* 2012. Refuting the challenges of the developmental shift of polarity of GABA actions: GABA more exciting than ever! *Front Cell Neurosci.* **6:** 35.

89. Bosma, M.M. 2010. Timing and mechanism of a window of spontaneous activity in embryonic mouse hindbrain development. *Ann. N.Y. Acad. Sci.* **1198:** 182–191.

90. Hinckley, C.A. & S.L. Pfaff. 2012. "Spatiotemporal activity patterns of V1 spinal interneurons during fictive locomotion." Summary of poster presentation. http://conferencing.uwex.edu/conferences/spinalconference2012/.

91. Koch, S.C., K.K. Tochiki, S. Hirschberg, *et al.* 2012. C-fiber activity-dependent maturation of glycinergic inhibition in the spinal dorsal horn of the postnatal rat. *Proc. Natl. Acad. Sci. U.S.A.* **109:** 12201–12206.

92. Scherrer, G., N. Imamachi, Y.-Q. Cao, *et al.* 2009. Dissociation of the opioid receptor mechanisms that control mechanical and heat pain. *Cell* **137:** 1148–1159.

93. Hayes, H.B., Y.H. Chang & S. Hochman. 2012. Stance-phase force on the opposite limb dictates swing-phase afferent presynaptic inhibition during locomotion. *J. Neurophysiol.* **107:** 3168–3180.

94. Zaporozhets, E., K.C. Cowley & B.J. Schmidt. 2011. Neurochemical excitation of propriospinal neurons facilitates locomotor command signal transmission in the lesioned spinal cord. *J. Neurophysiol.* **105:** 2818–2829.

95. Boulenguez, P., S. Liabeuf, R. Bos, *et al.* 2010. Downregulation of the potassium-chloride cotransporter KCC2 contributes to spasticity after spinal cord injury. *Nat. Med.* **16:** 302–307.

96. Garcia, A.J., 3rd, S. Zanella, H. Koch, *et al.* 2011. Chapter 3–networks within networks: the neuronal control of breathing. *Prog. Brain. Res.* **188:** 31–50.

97. Huxtable, A.G., J.D. Zwicker, T.S. Alvares, *et al.* 2010. Glia contribute to the purinergic modulation of inspiratory rhythm-generating networks. *J. Neurosci.* **30:** 3947–3958.

98. Lovett-Barr, M.R., I. Satriotomo, G.D. Muir, *et al.* 2012. Repetitive intermittent hypoxia induces respiratory and somatic motor recovery after chronic cervical spinal injury. *J. Neurosci.* **32:** 3591–3600.

99. Alilain, W.J., K.P. Horn, H. Hu, *et al.* 2011. Functional regeneration of respiratory pathways after spinal cord injury. *Nature* **475:** 196–200.

100. White, T.E., M.A. Lane, M.S. Sandhu, *et al.* 2010. Neuronal progenitor transplantation and respiratory outcomes following upper cervical spinal cord injury in adult rats. *Exp. Neurol.* **225:** 231–236.

101. Bamford, J.A. & V.K. Mushahwar. 2011. Intraspinal microstimulation for the recovery of function following spinal cord injury. *Prog. Brain. Res.* **194:** 227–239.

Ann. N.Y. Acad. Sci. ISSN 0077-8923

ANNALS OF THE NEW YORK ACADEMY OF SCIENCES
Issue: *Neurons, Circuitry, and Plasticity in the Spinal Cord and Brainstem*

Neurotransmitters and synaptic components in the Merkel cell–neurite complex, a gentle-touch receptor

Srdjan Maksimovic,[1] Yoshichika Baba,[1] and Ellen A. Lumpkin[1,2]

[1]Department of Dermatology, [2]Department of Physiology and Cellular Biophysics, Columbia University College of Physicians and Surgeons, New York, New York

Address for correspondence: Ellen A. Lumpkin, Ph.D., College of Physicians and Surgeons, Columbia University, Russ Berrie Medical Science Pavilion, 1150 St. Nicholas Avenue, Room 302B, New York, NY 10032. eal2166@columbia.edu

Merkel cells are an enigmatic group of rare cells found in the skin of vertebrates. Most make contacts with somatosensory afferents to form Merkel cell–neurite complexes, which are gentle-touch receptors that initiate slowly adapting type I responses. The function of Merkel cells within the complex remains debated despite decades of research. Numerous anatomical studies demonstrate that Merkel cells form synaptic-like contacts with sensory afferent terminals. Moreover, recent molecular analysis reveals that Merkel cells express dozens of presynaptic molecules that are essential for synaptic vesicle release in neurons. Merkel cells also produce a host of neuroactive substances that can act as fast excitatory neurotransmitters or neuromodulators. Here, we review the major neurotransmitters found in Merkel cells and discuss these findings in relation to the potential function of Merkel cells in touch reception.

Keywords: Merkel cell; touch; mechanotransduction; neurotransmitter; neuromodulator; somatosensory

The somatosensory neurons that innervate our skin constantly update our brains about the objects and environmental factors that surround us. A remarkable feature of our skin's nervous system is that it encodes a diversity of chemicals, temperatures, and physical forces into membrane potential changes that trigger discrete neural signals. These signals are processed by circuitry within the central nervous system to produce distinct percepts such as touch, pain, warmth, cooling, and itch. We rely on this information to navigate our environment and avoid harm. For example, our sense of discriminative touch allows us to perform countless essential behaviors, including feeding and clothing ourselves.

To initiate a range of sensations, cutaneous sensory neurons display an array of anatomical specializations and physiological properties. They can be classified as Aβ, Aδ, or C fibers based on conduction velocity and degree of myelination.[1–3] Aβ afferents are the fastest (∼35–75 m/s in humans and ∼10–25 m/s in mice) owing to their large diameters and thick myelin sheets. C fibers, which have thin, unmyelinated afferents, are the slowest (∼0.5–2 m/s in humans and ≤ 1 m/s in mice). Thinly myelinated Aδ fibers, which have fine axonal diameters compared to Aβ afferents, fall between, with conduction velocities of ∼5–30 m/s in humans and ∼4–10 m/s in mice. Sensory neurons can be further designated as mechanoreceptors, thermoreceptors, and nociceptors, depending on their modality or the sensory stimuli to which they respond.[2–4] Most high-threshold mechanoreceptors, thermoreceptors, and nociceptors fall into Aδ or C fiber classes. These are thought to terminate in free nerve endings innervating the epidermis, dermis, or hair follicles. Most tactile afferents, or low-threshold mechanoreceptors, are classified as Aβ or Aδ afferents. The reader is cautioned that there are numerous exceptions to these general guidelines. For example, a population of unmyelinated low-threshold mechanoreceptors (C-tactile, CT), with conduction velocities of ∼1 m/s in humans and mice, abundantly innervate hairy skin.[5,6]

Tactile afferents terminate in morphologically specialized end organs that govern their mechanosensory responses and allow them to extract distinct features of a complex tactile stimulus.[2,3,7] For example, rapidly adapting afferents that

doi: 10.1111/nyas.12057

encode vibration have encapsulated endings called
Meissner's corpuscles and Pacinian corpuscles. A
surprising variety of rapidly adapting afferents also
innervate hair follicles to signal hair movements.[6]
Slowly adapting afferents produce sustained discharges throughout mechanical stimulation. The
best characterized of these are slowly adapting type I
(SAI) afferents, which are the Aβ afferents that form
complexes with epidermal Merkel cells (Fig. 1).

Merkel cell–neurite complexes are required for SAI responses

Based on their distribution and response properties,
SAI afferents are thought to encode object features,
such as shape, edges, and curvature.[7] These afferents make contacts with Merkel cells, which cluster
in skin regions that are specialized for high tactile acuity. These include fingertips, whisker follicles
(Fig. 1A), and touch domes, which are high-
sensitivity areas of hairy skin (Fig. 1B). Importantly,
SAI afferents have the highest spatial resolution
among mammalian touch receptors and they represent fine spatial details, such as Braille-like characters, with fidelity.[7]

Though it is well established that Merkel cell–
neurite complexes are gentle-touch receptors that
mediate SAI responses, the function of Merkel cells
in discriminative touch is still a mystery. Based
on similarities to inner-ear hair cells, Merkel cells
have been proposed to be mechanosensory cells
that transduce touch and activate afferent neurons
by neurotransmitter release.[8,9] Parallels between
mechanosensory hair cells and Merkel cells are notable. Merkel cells have elongated microvilli sugges-
tive of the hair cell's mechanosensitive stereocilia.[10]
These cell types express the same developmental
transcription factors, including mammalian atonal
homolog 1 (*Atoh1*), growth factor independent 1,
and Pou4F2.[11–15] Moreover, *Atoh1* is absolutely essential for Merkel-cell development, as it is for hair
cells.[16–18]

To test whether Merkel cells are required for touch
sensation, the Cre-loxP system was used to con-
ditionally delete *Atoh1* in the body skin of mice.
Merkel cells failed to develop in these mice; however, touch domes were still innervated by myelinated Aβ afferents.[16] A survey of touch-sensitive afferents in *ex vivo* skin–nerve preparations revealed
the selective and complete loss of SAI responses in
these mice. Maricich *et al.* subsequently reported

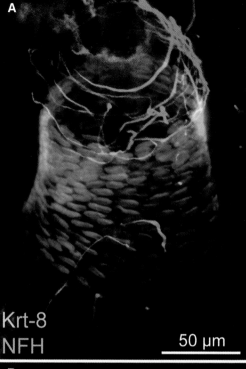

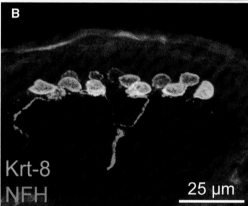

Figure 1. Merkel cell–neurite complexes in mouse whisker follicles (A) and touch domes (B) from the mouse hairy skin. Merkel
cells, marked by keratin 8 (Krt-8; *green*), are in intimate contact
with myelinated sensory afferents, visualized with antibodies
against neurofilament heavy chain (NFH; *red*).

that Merkel-cell knockout mice show a loss of texture preference in behavioral assays.[19] This finding
is exciting because it provides the first behavioral
evidence that animals rely on SAI responses for tex-
tural information. Together, these results indicate
that epidermal Merkel cells play an integral role in
touch-evoked SAI responses;[16] however, they do not

distinguish between a developmental requirement, a mechanosensory function, or an accessory role.

Studies that disrupted Merkel cell–neurite complexes postnatally have yielded conflicting results.[20,21] Removing Merkel cells from the epidermis by photoablation or enzymatic treatment abolished slowly adapting responses in some studies but not in others.[22–26] Mice lacking *p75* neurotrophin receptor, which lose most Merkel cells postnatally, display slowly adapting responses comparable to those of wild-type mice.[27] This disparity from the *Atoh1* phenotype may reflect methodological differences between studies, a developmental requirement for Merkel cells, or a postnatal requirement for *p75* in SAI afferents. Additional studies are needed to distinguish between these possibilities.

A second model for Merkel-cell function posits that Merkel cells are accessory cells rather than sensory receptor cells.[9,28] For example, Merkel cells might release modulatory neurotransmitters that shape the sensitivity of mechanosensitive afferents. Some investigators have argued that the SAI afferent must be the site of mechanotransduction because response latencies at touch onset (∼200 μs) are too short to include synaptic transmission from Merkel cells.[29] A two-receptor-site model, postulating that both Merkel cells and afferent terminals contain mechanotransduction channels, reconciles these short latencies with the Merkel cell's sensory features.[30] In this model, the SAI afferent transduces the phasic component of touch, as do rapidly adapting afferents. It has been proposed that Merkel cells mediate the tonic component of the SAI response. This hypothesis has some support from electrophysiological evidence but remains to be thoroughly tested.[31–33]

The Merkel cell's synaptic-like contacts

The presence of synaptic-like contacts between Merkel cells and sensory afferents suggests that Merkel cells are presynaptic cells. A substantial body of molecular evidence now supports this hypothesis.[13,34,35] Microarray analysis of purified mouse Merkel cells has identified a number of synaptic molecules that are preferentially expressed in Merkel cells.[13] These include essential presynaptic components such as active-zone scaffolding proteins, SNARE complex genes, voltage-activated calcium channels, calcium sensors such as synaptotagmins 1 and 7, and neurotransmitter transporters. En-

richment at the protein level has been verified for many of the key synaptic components.[13,34,35] Furthermore, gene ontology (GO) analysis has revealed that transcripts involved in synaptic transmission are overrepresented in Merkel cells (Fig. 2). It is worth noting that a few studies suggest that Merkel cells form reciprocal synapses with sensory afferents.[36]

Although there is no evidence of clear-core vesicles in Merkel cells, they do contain small, dense-core vesicles, which cluster near synaptic-like densities that mark the junctions with the afferent's membrane.[8,36,37] These vesicles show immunoreactivity for numerous neurotransmitters that vary across species, including adenosine triphosphate (ATP), serotonin (5-HT), vasoactive intestinal polypeptide (VIP), calcitonin gene–related peptide (CGRP), substance P, met-enkephalin, and cholecystokinin octapeptide (CCK8).[9,13,38–47] Glutamate has also been implicated as a potential neurotransmitter at the Merkel cell–neurite complex.[13,31,32,34,48]

This impressive array of neuroactive molecules poses an obvious question: what is the nature of the signal transmitted at the Merkel cell–neurite complex? If Merkel cells act as sensory receptor cells, sensory afferent terminals should contain ionotropic receptors to specific excitatory neurotransmitters released by Merkel cells. On the other hand, if Merkel cells modulate the SAI afferent's response properties, sensory terminals might instead express metabotropic receptors. Of the neurotransmitters that have been localized to Merkel cells, three are classical, small-molecule neurotransmitters that have fast ionotropic receptors: glutamate, serotonin, and ATP. Most other neurotransmitters identified in Merkel cells are neuropeptides that, due to their slower mode of action, are more likely to serve neuromodulatory roles than to mediate fast SAI responses.[49] Functional studies that have tested the involvement of synaptic transmission in Merkel cell–neurite signaling have provided conflicting evidence regarding sensory or modulatory roles.[21,31,32,38,50,51] As a result, the functional significance of the Merkel cell's synapse is still a mystery.

Classical neurotransmitters

Glutamate is the major excitatory neurotransmitter in vertebrates. The expression of the vesicular glutamate transporter 1 (VGLUT1) or 2 (VGLUT2) is the hallmark of glutamatergic synapses. Expression of

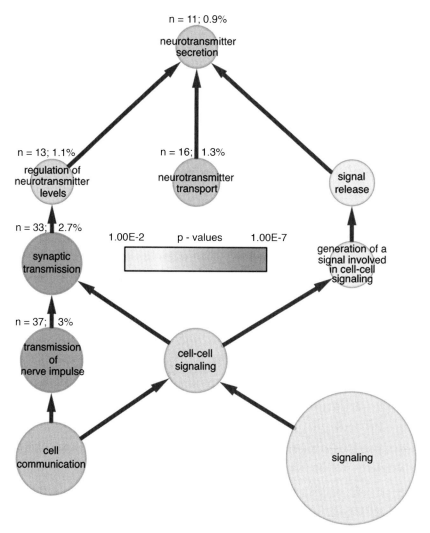

Figure 2. Gene transcripts involved in synaptic transmission and neurotransmitter release are enriched in Merkel cells. This network is a product of a gene ontology (GO) analysis of transcripts enriched in mouse Merkel cells.[13] Compared with the mouse genome overall, genes associated with synaptic transmission and neurotransmitter release are significantly overrepresented in Merkel cells. Node size indicates the number of genes associated with a particular GO term (function). For reference, the absolute (*n*) and relative (%) sizes of nodes associated with synaptic transmission and neurotransmitter release are indicated. Node color indicates the significance level (*P* value) of enrichment. Arrows show parent–child node relationships.

VGLUT1, 2, and 3 in Merkel cells has been reported by several groups.[13,34,35] Even before the discovery of VGLUTs in Merkel cells, Fagan and Cahusac demonstrated by electrophysiology that kynurenate, a broad-spectrum antagonist of ionotropic glutamate receptors, attenuated SAI responses in rat whisker follicles.[31] Interestingly, kynurenate preferentially inhibited the tonic response but had little effect on the phasic response, which lends support to the two-receptor-site model. In the search for ionotropic glutamate receptors in the Merkel cell–neurite complex, one was found: the NMDA receptor, which is a heterotetramer comprising multiple NR1 subunits and at least one NR2 subunit.[52] In the rat whisker follicle, Cahusac *et al.* localized NR1 to Merkel cells and sensory afferents.[50] Surprisingly, they localized the NR2 subunit to Merkel cells rather than to their associated sensory afferents. Thus, their data suggest that it is the Merkel cells and not the sensory afferents that express functional

NMDA receptors.[50] Tachibana *et al.* described an additional glutamate receptor expressed by Merkel cells: metabotropic mGluR5.[38] Electrophysiological experiments with metabotropic glutamate receptor antagonists have yielded confusing results, with some potentiating the SAI response and others suppressing its activity.[32,53] Though these findings might indicate that Merkel-cell glutamate receptors act as autoreceptors that govern Merkel-cell output via a feedback mechanism, Cahusac and Mavulati questioned the selectivity of glutamate receptor antagonists used in their studies.[32] They argue that these compounds might interfere with SAI signaling, not by binding to glutamate receptors, but by inhibiting mechanotransduction channels (or other targets associated with them). Thus, although several lines of evidence support glutamatergic signaling in Merkel cells, how glutamate shapes the SAI response remains an open question.

5-HT has been considered one of the major candidates for a neurotransmitter at the Merkel cell–neurite complex since the discovery of serotonin-like immunoreactivity in Merkel cells of various organisms.[9,40,41,54,55] Like glutamate, 5-HT has both ionotropic and metabotropic receptors. Out of seven 5-HT receptors (5-HT$_1$–5-HT$_7$), only 5-HT$_3$ is ionotropic.[56] Using 5-HT$_2$ and 5-HT$_3$ antagonists, He *et al.* were able to suppress, but not eliminate, rat SAI responses, indicating that 5-HT signaling modulates SAI responses.[57] Though they reported that MDL72222, an ionotropic 5-HT$_3$ receptor antagonist, reduced rat SAI responses, 5-HT$_3$ receptors have yet to be located in the Merkel cell–neurite complex. Furthermore, metabotropic 5-HT$_1$ (though not 5-HT$_2$) receptors have been localized on SAI afferent terminals.[58] Together with the presence of 5-HT transporters in Merkel cells, these data support the modulatory role of 5-HT in the Merkel cell–neurite complex.[58] Combining the same 5-HT$_3$ antagonist, MDL7222, with various levels of displacement, Press *et al.* were able to completely abolish frog SAI responses, providing the first evidence for a fast excitatory serotonergic transmission in the Merkel cell–neurite complex.[33] Thus it is possible that 5-HT can have a dual role in the Merkel cell–neurite complex: as a classical excitatory neurotransmitter and as a neuromodulator of the SAI response.[33]

Though glutamate and 5-HT have been discussed separately throughout this review, it is possible that they are coreleased from Merkel cells during mechanical stimulation. Merkel cells express VG-LUT3, which is also expressed in central serotonergic neurons.[59–61] Thus, as proposed by Nunzi *et al.*, VGLUT3$^+$ Merkel cells might corelease glutamate and serotonin (or any other neurotransmitter for that matter) during tactile stimulation.[35] In support of this hypothesis, Press *et al.* found that in frog Merkel cell–neurite complexes SAI responses are shaped by both 5-HT and glutamate.[33]

ATP has been a neurotransmitter candidate since Toyoshima and Shimamura confirmed its presence in finch Merkel-cell vesicles with a uranaffin reaction.[39] In frogs, large-diameter DRG neurons, which include SAI afferents, express metabotropic P2Y1 ATP receptors, and application of ATP increases their activity.[62] In the rat sinus hair follicle, Tachibana *et al.* located another metabotropic receptor, P2Y2, on Merkel cells but not on sensory affcrents.[38] Although a P2Y1 antibody did not show positive reaction, further histological and physiological experiments are needed before conclusions can be made about the role of ATP in Merkel cell–neurite signaling.

Neuropeptides

Merkel cells have been widely reported to produce neuropeptides, though their identities vary from species to species. Even within a single species, Merkel cells show heterogeneous patterns of neuropeptide expression.[44,63,64] The most common neuropeptides reported in Merkel cells are VIP, CGRP, substance P, met-enkephalin, somatostatin, and CCK8.[13,42–47,63] Boulais *et al.* demonstrated VIP release from cultured Merkel cells.[65] Most of these neuropeptides can modulate neuronal activity through metabotropic receptors coupled to heterotrimeric G proteins.[66] Tachibana *et al.* found G$_o$ and G$_i$ immunoreactivity in SAI afferent terminals of rat and monkey Merkel cell–neurite complexes.[67] Because mostly inhibitory neuropeptide receptors are G$_i$/G$_o$-coupled, it is possible that some of the neuropeptides released by Merkel cells reduce SAI firing rates, though electrophysiological data are still lacking.[51,68] Interestingly, Tachibana and Nawa localized receptors for met-enkephalin, VIP, substance P, and CGRP to Merkel cells rather than sensory terminals, suggesting autocrine or paracrine action of these neuropeptides on Merkel cells.[51]

Some of the neuropeptides found in Merkel cells might serve neuroendocrine rather than sensory functions. Indeed, a subpopulation of noninnervated Merkel cells, which exists in mucosal tissues and hair follicles, have been proposed to be neuroendocrine cells.[21,64,69] Merkel cells have been classified as a part of the amine precursor uptake and decarboxylation (APUD) system.[21,70,71] As other cells of the APUD system, Merkel cells have dense-core vesicles and are positive for biogenic amines and neuropeptides. Recent reports posit that the neuroendocrine function of Merkel cells might affect skin disorders.[65,72–74] It is possible that distinct classes of Merkel cells serve different functions in the skin. Indeed, recent reports propose two secretory pathways in rat Merkel cells: a Ca^+-dependent pathway that serves mechanosensory function and neurotransmitter release, and a Ca^{2+}-independent pathway that serves neuroendocrine functions and neuropeptide release.[65,72] Functional studies, including the analysis of transgenic mouse models, are needed to test this hypothesis.

Summary and open questions

Merkel cells are one of four conserved cell types in the vertebrate epidermis and yet much about their biological function still remains unclear more than a century after Merkel's initial description.[21,75] At this point the role for Merkel cells in touch sensation seems assured: the majority of Merkel cells throughout the vertebrate skin contact sensory terminals and they localize to highly touch-sensitive areas. Moreover, Merkel-cell knockout mice lack SAI responses and display impaired texture-driven behaviors. Electrophysiological data support the two-receptor–site hypothesis that initial, phasic responses are mediated by SAI afferents, while slower, tonic responses are mediated by Merkel cells.[32,33] In the last decade, glutamate has emerged as the most likely neurotransmitter in the Merkel cell–neurite complex, but confirming this will have to wait until additional experiments are performed. Further complicating this problem is accumulating evidence suggesting that Merkel cells cosecrete multiple neuroactive substances that could be involved not only in mechanoreception (transduction and modulation), but also in neuroendocrine roles unrelated to mechanoreception.[33,65,69,72]

The morphology of Merkel cell–neurite contacts, which contains features of both fast synaptic transmission and peptide-based neuromodulation, provides many intriguing questions. Physiological and genetic experiments are needed to provide concrete answers. Because both Merkel cells and SAI afferents are $VGLUT2^+$, cell-type–specific ablation of glutamate signaling is needed to resolve the role of this neurotransmitter in touch-evoked responses.[13] To determine whether Merkel cells are required developmentally or postnatally for SAI responses, rapid and selective Merkel-cell ablation is needed. For example, targeted expression of diphtheria toxin receptor in adult Merkel cells could be achieved with Cre-loxP technology and an appropriate *Cre* driver. Optogenetic tools provide an opportunity to selectively excite and/or inhibit Merkel cells while recording SAI activity. If SAI activity can be provoked by exciting only Merkel cells with light, then Merkel cells indeed release excitatory neurotransmitters. If Merkel-cell inhibition with light completely abolishes touch-evoked SAI responses, then Merkel cells not only release excitatory neurotransmitters but also transduce mechanical stimuli. Given the recent explosion of genetic tools for selectively manipulating the excitability of cell populations, we can expect to have a better understanding of the intriguing Merkel cell and its functions in the near future.

Acknowledgments

The work was supported by NIAMS R21 AR062307 (to E.A.L.), NINDS R01 NS073119 (to E.A.L. and Gregory J. Gerling), and NHLB T32HL087745 (to S.M.). We thank Dr. David Ginty, Dr. Michael Rutlin, and Ms. Kara Marshall for sharing whole-mount skin staining protocols before publication. Confocal microscopy was performed in the Columbia University Skin Disease Research Center Advanced Imaging Core (NIAMS P30AR044535).

Conflicts of interest

The authors declare no conflicts of interest.

References

1. Brown, A.G. & A. Iggo. 1967. A quantitative study of cutaneous receptors and afferent fibres in the cat and rabbit. *J. Physiol.* **193:** 707–733.
2. Gardner, E.P., J.H. Martin & T.M. Jessell. 2000. The bodily senses. In *Principles of Neuro Science*. E.R. Kandel, J.H. Schwartz, & T.M. Jessell, Eds.: 430–449. McGraw-Hill, New York.

3. Rice, F.L. & P.J. Albrecht. 2008. Cutaneous mechanisms of tactile perception: morphological and chemical organization of the innervation to the skin. In *The Senses: A Comprehensive Reference*. D. Smith, S. Firestein & G. Beauchamp, Eds.: 1–31. Academic Press, San Diego.

4. Lumpkin, E.A. & M.J. Caterina. 2007. Mechanisms of sensory transduction in the skin. *Nature* **445:** 858–865, doi:10.1038/nature05662.

5. Olausson, H., J. Wessberg, I. Morrison, *et al.* 2010. The neurophysiology of unmyelinated tactile afferents. *Neurosci. Biobehav. Rev.* **34:** 185–191. Epub 2008 Oct 2008.

6. Li, L. *et al.* 2011. The functional organization of cutaneous low-threshold mechanosensory neurons. *Cell* **147:** 1615–1627.

7. Johnson, K.O. 2001. The roles and functions of cutaneous mechanoreceptors. *Curr. Opin. Neurobiol.* **11:** 455–461.

8. Iggo, A. & A.R. Muir. 1969. The structure and function of a slowly adapting touch corpuscle in hairy skin. *J. Physiol.* **200:** 763–796.

9. Tachibana, T. & T. Nawa. 2002. Recent progress in studies on Merkel cell biology. *Anat. Sci. Int.* **77:** 26–33.

10. Toyoshima, K., Y. Seta, S. Takeda & H. Harada. 1998. Identification of Merkel cells by an antibody to villin. *J. Histochem. Cytochem.* **46:** 1329–1334.

11. Ben-Arie, N. *et al.* 2000. Functional conservation of atonal and Math1 in the CNS and PNS. *Development* **127:** 1039–1048.

12. Wallis, D. *et al.* 2003. The zinc finger transcription factor Gfi1, implicated in lymphomagenesis, is required for inner ear hair cell differentiation and survival. *Development* **130:** 221–232.

13. Haeberle, H. *et al.* 2004. Molecular profiling reveals synaptic release machinery in Merkel cells. *Proc. Natl. Acad. Sci. U.S.A.* **101:** 14503–14508, doi:10.1073/pnas.0406308101.

14. Leonard, J.H. *et al.* 2002. Proneural and proneuroendocrine transcription factor expression in cutaneous mechanoreceptor (Merkel) cells and Merkel cell carcinoma. *Int. J. Cancer* **101:** 103–110, doi:10.1002/ijc.10554.

15. Xiang, M. *et al.* 1997. Role of the Brn-3 family of POU-domain genes in the development of the auditory/vestibular, somatosensory, and visual systems. *Cold Spring Harb. Symp. Quant. Biol.* **62:** 325–336.

16. Maricich, S.M. *et al.* 2009. Merkel cells are essential for light-touch responses. *Science* **324:** 1580–1582, doi:10.1126/science.1172890.

17. Van Keymeulen, A. *et al.* 2009. Epidermal progenitors give rise to Merkel cells during embryonic development and adult homeostasis. *J. Cell Biol.* **187:** 91–100. Epub 2009 Sep 2028.

18. Bermingham, N.A. *et al.* 1999. Math1: an essential gene for the generation of inner ear hair cells. *Science* **284:** 1837–1841.

19. Maricich, S.M., K.M. Morrison, E.L. Mathes & B.M. Brewer. 2012. Rodents rely on Merkel cells for texture discrimination tasks. *J. Neurosci.* **32:** 3296–3300.

20. Ogawa, H. 1996. The Merkel cell as a possible mechanoreceptor cell. *Prog. Neurobiol.* **49:** 317–334.

21. Halata, Z., M. Grim & K.I. Bauman. 2003. Friedrich Sigmund Merkel and his "Merkel cell," morphology, development, and physiology: review and new results. *Anat. Rec. A Discov. Mol. Cell Evol. Biol.* **271:** 225–239, doi:10.1002/ar.a.10029.

22. Ikeda, I., Y. Yamashita, T. Ono & H. Ogawa. 1994. Selective phototoxic destruction of rat Merkel cells abolishes responses of slowly adapting type I mechanoreceptor units. *J. Physiol.* **479**(Pt 1)**:** 247–256.

23. Senok, S.S., K.I. Baumann & Z. Halata. 1996. Selective phototoxic destruction of quinacrine-loaded Merkel cells is neither selective nor complete. *Exp. Brain Res.* **110:** 325–334.

24. Senok, S.S., Z. Halata & K.I. Baumann. 1996. Chloroquine specifically impairs Merkel cell mechanoreceptor function in isolated rat sinus hairs. *Neurosc. Lett.* **214:** 167–170.

25. Diamond, J., L.R. Mills & K.M. Mearow. 1988. Evidence that the Merkel cell is not the transducer in the mechanosensory Merkel cell-neurite complex. *Prog. Brain Res.* **74:** 51–56.

26. Mills, L.R. & J. Diamond. 1995. Merkel cells are not the mechanosensory transducers in the touch dome of the rat. *J. Neurocytol.* **24:** 117–134.

27. Kinkelin, I., C.L. Stucky & M. Koltzenburg. 1999. Postnatal loss of Merkel cells, but not of slowly adapting mechanoreceptors in mice lacking the neurotrophin receptor p75. *Eur. J. Neurosci.* **11:** 3963–3969.

28. Pasche, F., Y. Merot, P. Carraux & J.H. Saurat. 1990. Relationship between Merkel cells and nerve endings during embryogenesis in the mouse epidermis. *J. Invest. Dermatol.* **95:** 247–251.

29. Gottschaldt, K.M. & C. Vahle-Hinz. 1981. Merkel cell receptors: structure and transducer function. *Science* **214:** 183–186.

30. Yamashita, Y. & H. Ogawa. 1991. Slowly adapting cutaneous mechanoreceptor afferent units associated with Merkel cells in frogs and effects of direct currents. *Somatosens. Motor Res.* **8:** 87–95.

31. Fagan, B.M. & P.M. Cahusac. 2001. Evidence for glutamate receptor mediated transmission at mechanoreceptors in the skin. *Neuroreport* **12:** 341–347.

32. Cahusac, P.M. & S.C. Mavulati. 2009. Non-competitive metabotropic glutamate 1 receptor antagonists block activity of slowly adapting type I mechanoreceptor units in the rat sinus hair follicle. *Neuroscience* **163:** 933–941, doi:10.1016/j.neuroscience.2009.07.015.

33. Press, D., S. Mutlu & B. Guclu. 2010. Evidence of fast serotonin transmission in frog slowly adapting type 1 responses. *Somatosens. Mot. Res.* **27:** 174–185, doi:10.3109/08990220.2010.516670.

34. Hitchcock, I.S., P.G. Genever & P.M. Cahusac 2004. Essential components for a glutamatergic synapse between Merkel cell and nerve terminal in rats. *Neurosci. Lett.* **362:** 196–199, doi:10.1016/j.neulet.2004.02.071.

35. Nunzi, M.G., A. Pisarek & E. Mugnaini. 2004. Merkel cells, corpuscular nerve endings and free nerve endings in the mouse palatine mucosa express three subtypes of vesicular glutamate transporters. *J. Neurocytol.* **33:** 359–376.

36. Mihara, M., K. Hashimoto, K. Ueda & M. Kumakiri. 1979. The specialized junctions between Merkel cell and neurite: an electron microscopic study. *J. Invest. Dermatol.* **73:** 325–334.

37. Hartschuh, W. & E. Weihe. 1980. Fine structural analysis of the synaptic junction of Merkel cell-axon-complexes. *J. Invest. Dermatol.* **75:** 159–165.

38. Tachibana, T., M. Endoh, R. Kumakami & T. Nawa. 2003. Immunohistochemical expressions of mGluR5, P2Y2 receptor, PLC-beta1, and IP3R-I and -II in Merkel cells in rat sinus hair follicles. *Histochem. Cell Biol.* **120:** 13–21, doi:10.1007/s00418-003-0540-5.

39. Toyoshima, K. & A. Shimamura. 1991. Uranaffin reaction of Merkel corpuscles in the lingual mucosa of the finch, Lonchula striata var. domestica. *J. Anat.* **179:** 197–201.

40. Garcia-Caballero, T. *et al.* 1989. Localization of serotonin-like immunoreactivity in the Merkel cells of pig snout skin. *Anatom. Record* **225:** 267–271, doi:10.1002/ar.1092250402.

41. English, K.B. *et al.* 1992. Serotonin-like immunoreactivity in Merkel cells and their afferent neurons in touch domes from the hairy skin of rats. *Anat. Rec.* **232:** 112–120, doi:10.1002/ar.1092320112.

42. Alvarez, F.J. *et al.* 1988. Presence of calcitonin gene-related peptide (CGRP) and substance P (SP) immunoreactivity in intraepidermal free nerve endings of cat skin. *Brain Res.* **442:** 391–395.

43. Hartschuh, W., E. Weihe, N. Yanaihara & M. Reinecke. 1983. Immunohistochemical localization of vasoactive intestinal polypeptide (VIP) in Merkel cells of various mammals: evidence for a neuromodulator function of the Merkel cell. *J. Invest. Dermatol.* **81:** 361–364.

44. Garcia-Caballero, T., R. Gallego, E. Roson *et al.* 1989. Calcitonin gene-related peptide (CGRP) immunoreactivity in the neuroendocrine Merkel cells and nerve fibres of pig and human skin. *Histochemistry* **92:** 127–132

45. Gauweiler, B., E. Weihe, W. Hartschuh & N. Yanaihara. 1988. Presence and coexistence of chromogranin A and multiple neuropeptides in Merkel cells of mammalian oral mucosa. *Neurosci. Lett.* **89:** 121–126.

46. Hartschuh, W. & E. Weihe. 1988. Multiple messenger candidates and marker substance in the mammalian Merkel cell-axon complex: a light and electron microscopic immunohistochemical study. *Prog. Brain Res.* **74:** 181–187.

47. Hartschuh, W. *et al.* 1979. Met enkephalin-like immunoreactivity in Merkel cells. *Cell Tissue Res.* **201:** 343–348.

48. Morimoto, R. *et al.* 2003. Co-expression of vesicular glutamate transporters (VGLUT1 and VGLUT2) and their association with synaptic-like microvesicles in rat pinealocytes. *J. Neurochem.* **84:** 382–391.

49. Wellnitz, S.A., D.R. Lesniak, G.J. Gerling & E.A. Lumpkin. 2010. The regularity of sustained firing reveals two populations of slowly adapting touch receptors in mouse hairy skin. *J. Neurophysiol.* **103:** 3378–3388, doi:10.1152/jn.00810.2009.

50. Cahusac, P.M., S.S. Senok, I.S. Hitchcock, *et al.* 2005. Are unconventional NMDA receptors involved in slowly adapting type I mechanoreceptor responses? *Neuroscience* **133:** 763–773.

51. Tachibana, T. & T. Nawa. 2005. Immunohistochemical reactions of receptors to met-enkephalin, VIP, substance P, and CGRP located on Merkel cells in the rat sinus hair follicle. *Arch. Histol. Cytol.* **68:** 383–391.

52. Cull-Candy, S., S. Brickley & M. Farrant. 2001. NMDA receptor subunits: diversity, development and disease. *Curr. Opin. Neurobiol.* **11:** 327–335.

53. Cahusac, P.M. & S.S. Senok. 2006. Metabotropic glutamate receptor antagonists selectively enhance responses of slowly adapting type I mechanoreceptors. *Synapse* **59:** 235–242.

54. Zaccone, G. 1986. Neuron-specific enolase and serotonin in the Merkel cells of conger-eel (Conger conger) epidermis. An immunohistochemical study. *Histochemistry* **85:** 29–34.

55. Toyoshima, K. & A. Shimamura. 1987. Monoamine-containing basal cells in the taste buds of the newt Triturus pyrrhogaster. *Arch. Oral Biol.* **32:** 619–621.

56. Hoyer, D., J.P. Hannon & G.R. Martin. 2002. Molecular, pharmacological and functional diversity of 5-HT receptors. *Pharmacol. Biochem. Behav.* **71:** 533–554.

57. He, L., R.P. Tuckett & K.B. English. 2003. 5-HT2 and 3 receptor antagonists suppress the response of rat type I slowly adapting mechanoreceptor: an in vitro study. *Brain Res.* **969:** 230–236.

58. Tachibana, T., M. Endoh, N. Fujiwara & T. Nawa. 2005. Receptors and transporter for serotonin in Merkel cell-nerve endings in the rat sinus hair follicle. An immunohistochemical study. *Arch. Histol. Cytol.* **68:** 19–28.

59. Fremeau, R.T., Jr. *et al.* 2002. The identification of vesicular glutamate transporter 3 suggests novel modes of signaling by glutamate. *Proc. Natl. Acad. Sci. U.S.A.* **99:** 14488–14493, doi:10.1073/pnas.222546799.

60. Schafer, M.K., H. Varoqui, N. Defamie, *et al.* 2002. Molecular cloning and functional identification of mouse vesicular glutamate transporter 3 and its expression in subsets of novel excitatory neurons. *J. Biol. Chem.* **277:** 50734–50748, doi:10.1074/jbc.M206738200.

61. Gras, C. *et al.* 2002. A third vesicular glutamate transporter expressed by cholinergic and serotoninergic neurons. *J. Neurosci.* **22:** 5442–5451.

62. Nakamura, F. & S.M. Strittmatter. 1996. P2Y1 purinergic receptors in sensory neurons: contribution to touch-induced impulse generation. *Proc. Natl. Acad. Sci. U.S.A.* **93:** 10465–10470.

63. Fantini, F. & O. Johansson. 1995. Neurochemical markers in human cutaneous Merkel cells. An immunohistochemical investigation. *Exp. Dermatol.* **4:** 365–371.

64. Moll, I. *et al.* 2005. Human Merkel cells–aspects of cell biology, distribution and functions. *Eur. J. Cell Biol.* **84:** 259–271.

65. Boulais, N. *et al.* 2009. Merkel cells as putative regulatory cells in skin disorders: an in vitro study. *PLoS One* **4:** e6528.

66. Merighi, A., C. Salio, F. Ferrini & L. Lossi. 2011. Neuromodulatory function of neuropeptides in the normal CNS. *J. Chem. Neuroanat.* **42:** 276–287, Epub 2011 Mar 2016.

67. Tachibana, T., M. Endoh & T. Nawa. 2001. Immunohistochemical expression of G protein alpha-subunit isoforms in

rat and monkey Merkel cell-neurite complexes. *Histochem. Cell Biol.* **116:** 205–213, doi:10.1007/s004180100318.

68. Tallent, M.K. 2008. Presynaptic inhibition of glutamate release by neuropeptides: use-dependent synaptic modification. *Results Probl. Cell Differ.* **44:** 177–200.

69. Boulais, N. & L. Misery. 2007. Merkel cells. *J. Am. Acad. Dermatol.* **57:** 147–165, Epub 2007 Apr. 2006.

70. Pearse, A.G. 1968. Common cytochemical and ultrastructural characteristics of cells producing polypeptide hormones (the APUD series) and their relevance to thyroid and ultimobranchial C cells and calcitonin. *Proc. R. Soc. Lond. B Biol. Sci.* **170:** 71–80.

71. Winkelmann, R.K. 1977. The Merkel cell system and a comparison between it and the neurosecretory or APUD cell system. *J. Invest. Dermatol.* **69:** 41–46.

72. Boulais, N. *et al.* 2009. Rat Merkel cells are mechanoreceptors and osmoreceptors. *PLoS One* **4:** e7759, doi:10.1371/journal.pone.0007759.

73. Liu, T. *et al.* 2011. Neuroendocrine regulatory role of Merkel cells in the pathogenesis of psoriasis. *Scientific Res. Essays* **6:** 4526–4531, doi:10.5897/SRE11.624.

74. Righi, A. *et al.* 2006. Merkel cells in the oral mucosa. *Int. J. Surg. Pathol.* **14:** 206–211.

75. Merkel, F. 1985. Tastzellen und Tastkörperchen bei den Haustehieren und beim Menschen. *Arch. Mikrosk. Anat.* **11:** 636–652.

Ann. N.Y. Acad. Sci. ISSN 0077-8923

ANNALS OF THE NEW YORK ACADEMY OF SCIENCES
Issue: *Neurons, Circuitry, and Plasticity in the Spinal Cord and Brainstem*

Principles of interneuron development learned from Renshaw cells and the motoneuron recurrent inhibitory circuit

Francisco J. Alvarez,[1] Ana Benito-Gonzalez,[1,2] and Valerie C. Siembab[2]

[1]Department of Physiology, Emory University, Atlanta, Georgia. [2]Department of Neurosciences, Cell Biology and Physiology, Wright State University, Dayton, Ohio

Address for correspondence: Francisco J. Alvarez, Department of Physiology, Emory University, Whitehead Research Building, Room 642, 615 Michael Street, Atlanta, GA 30322-3110. francisco.j.alvarez@emory.edu

Renshaw cells provide a convenient model to study spinal circuit development during the emergence of motor behaviors with the goal of capturing principles of interneuron specification and circuit construction. This work is facilitated by a long history of research that generated essential knowledge about the characteristics that define Renshaw cells and the recurrent inhibitory circuit they form with motoneurons. In this review, we summarize recent data on the specification of Renshaw cells and their connections. A major insight from these studies is that the basic Renshaw cell phenotype is specified before circuit assembly, a result of their early neurogenesis and migration. Connectivity is later added, constrained by their placement in the spinal cord. Finally, different rates of synapse proliferation alter the relative weights of different inputs on postnatal Renshaw cells. Based on this work some general principles on the integration of spinal interneurons in developing motor circuits are derived.

Keywords: Interneuron; spinal cord; neurogenesis; synaptogenesis; differentiation

Introduction: current context on interneuronal diversity and spinal circuit development

Spinal cord motor function depends on the wiring and properties of the interneurons that modulate motoneuron firing and motor output. A long-term objective is to gain a better understanding of the development and integration of interneurons in spinal circuits during maturation of motor function. One major challenge for understanding the principles of spinal cord circuit organization and construction is the large diversity of spinal interneurons.[1] Motoneuron to interneuron ratios in the lumbar 5 spinal segment of the newborn mouse are around 1:15 (Ref. 2), representing an intricacy in local circuits that is orders of magnitude higher than those in other parts of the central nervous system; in the cortex, for example, the average ratio of pyramidal cells to interneurons is 10:1.[3,4]

Basic questions then are, How many different types of interneurons exist, and what principal categories explain the function and development of spinal motor circuits. We cannot yet answer these questions, but an overarching organizing principle is that all spinal interneurons throughout the vertebrate phylogeny derive from just 10 progenitor domains,[5,6] each generating an independent interneuron lineage (dl1–dl6 dorsally and V0–V3 ventrally). Each lineage diversifies into multiple subtypes but shares fundamental properties like laminar location, mediolateral positioning, and the primary direction of the axon (ascending, descending, ipsilateral, or contralateral). Mechanisms that create diversity within each canonical group are only now beginning to be understood. As reviewed later on, they start at the progenitor level; thus, adult interneuron phenotypes are specified before circuit assembly, challenging present definitions of interneuron classes based on adult connectivity and circuit function.[7]

doi: 10.1111/nyas.12084

Ann. N.Y. Acad. Sci. 1279 (2013) 22–31 © 2013 New York Academy of Sciences.

Functional subtypes sometimes appear to be derived from a single lineage (e.g., Renshaw cells[8]), but this is not necessarily true for other functional classes, such as Ia inhibitory interneurons (IaINs) that have diverse origins.[9] Thus, it is important to emphasize that physiological and molecular/developmental classifications group interneurons are based on different perspectives and distinguishing criteria. While physiology-based classifications take into account the inputs received by the interneuron and its action on target motoneurons, classifications based on molecular/genetic characteristics consider interneuronal derivation from particular progenitor domains and unique sets of gene coexpression directed by specific transcription factor combinations. These patterns of specific gene expression direct fundamental aspects of cellular development related to migration and placement, axonal projection type (e.g., commissural, ipsilateral, ascending, descending, short, long), and neurotransmitter phenotype. Because these features sometimes define the functional properties of mature neurons, we could expect, in some situations, a close relationship between a unique genetic/developmental subclass and a functional subgroup (e.g., Renshaw cells or cholinergic partition cells); however, in other situations, the correspondence might be not as clear. For example, a basic property that varies among different genetically defined groups is the orientation of the axon, implying that some functions, like reciprocal inhibition, may recruit interneurons from various classes according to the axon trajectories necessary to establish the required synaptic connections. To comprehend the fundamental categories of spinal interneurons, it is therefore necessary to understand the development of their basic properties and connections. A conceptual framework was laid out by the discovery of a few discrete progenitor domains; further work is now needed to understand the process of differentiating adult interneurons from these domains.

Another important difficulty is the changing nature of spinal interneurons, and their inputs, during embryonic and postnatal development. In mammalian species, major spinal inputs and outputs develop sequentially:[10,11] motor axons are first developed and connected to muscle around embryonic day 14 (E14);[12] proprioceptive afferents establish connections in the ventral spinal cord around E16;[13] and finally, descending inputs arrive during late embryonic and early postnatal development (these dates always refer to development in the mouse lumbar spinal cord). The corticospinal tract is one of the last projection systems to form synapses in the lumbar cord, around postnatal day 6 (P6).[14] In contrast, neurogenesis, differentiation, and migration of ventral interneurons are completed by E12, an age at which morphological or functional evidence for synapses is scarce.[15–17] Thus, ventral interneurons are specified, and start differentiation in a largely synapse-free environment. Lack of synapses does not mean that there is no active neurochemical communication among the cellular elements of the earliest spinal network. In fact, E11–E12 mouse spinal cords display spontaneous activity driven by cholinergic, GABAergic, and glycinergic signaling,[18,19] which could rely on the few synapses found at this age, but are more likely paracrine in nature[17] and/or based on electrical coupling. Within this environment of relatively nonspecific neurochemical interactions, synaptic inputs are sequentially added onto specific cellular elements, and this process likely changes the character and function of developing interneurons from birth to circuit maturation.

Our approach is to study a few spinal circuits and interneurons whose adult characteristics and variety are well known. We focus on the final spinal cord output module that consists of motoneurons synaptically coupled to Renshaw cells and IaINs, which respectively exert recurrent and reciprocal inhibition of motoneurons (Fig. 1). Here, we review what we know (or don't know) about Renshaw cells and the recurrent inhibitory circuit they form, making some comparisons to the sparse information available on IaIN development.

Basic properties of recurrent inhibition and Renshaw cells

Renshaw cells and motoneurons form a recurrent inhibitory circuit that controls motor output. Individual Renshaw cells receive inputs from particular motor pools and spread their inhibitory output to the same motoneurons, their synergists (i.e., motor pools exerting a similar action on the same joint), and sometimes selected motor pools across joints.[20–23] Renshaw cells are located in the most ventral regions of LVII and LIX, defined since 1965 as the *Renshaw cell area*,[24] and send axons into the adjacent ventral funiculus. In the mature spinal

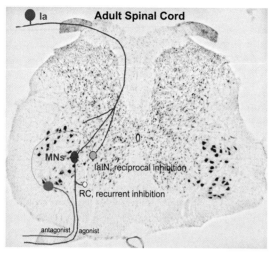

Figure 1. Diagram of basic connectivity in the recurrent and reciprocal inhibitory circuit controlling motor output, superimposed on a Nissl stained section of the lumbar spinal cord. Motoneurons are arranged in pools that innervate different muscles. While Renshaw cells receive inputs from certain pools and provide feedback inhibition to the same motoneurons and its synergists, Ia inhibitory interneurons mediate reciprocal inhibition, such that they inhibit motor pools with antagonist actions to the muscle of origin of the Ia afferent, thus permitting smooth flexor extension around individual joints. Ia inhibitory interneurons and motoneurons receiving common Ia inputs also receive similar excitatory drive from other systems, such that activation of motor pools is always coupled with relaxation of antagonists. However, Ia inhibitory interneuron activation is tightly controlled, in part, by the Renshaw cells themselves, so that the amount of relaxation or cocontraction of antagonist muscles is finely modulated. Excitatory drive onto motoneurons acts on this last order circuit involved in the last step of motor control. Each of the two interneurons display specific placement in the ventral horn and in connectivity with target motoneurons and incoming Ia afferents, as depicted in the diagram.

cord, Renshaw axons bifurcate, sending both ascending and descending branches, from which collaterals enter lamina IX and primarily target motoneuron proximal dendrites.[25] Renshaw cells are thus designed to monitor motor output and exert feedback modulation of excitatory inputs, such as Ia afferent inputs, to motoneuron dendrites, as first reported by Renshaw in 1941.[20]

The principal distinguishing feature of Renshaw cells is that they are the main intraspinal target of motor axons. In addition, Renshaw cells exhibit other properties, like calbindin immunoreactivity, that distinguish them from other interneurons (Table 1; see also Ref. 27 for more detailed descrip-

tions and references). Together, they construct interneurons that, in the adult, integrate incoming motor axon synaptic inputs to their dendrites[27] and are strongly modulated by a high density of large inhibitory synapses located more proximally.[28–30] The excitatory postsynaptic potentials (EPSPs) evoked by motor axons on Renshaw cells are of longer duration than on muscle, in part because of the incorporation of slow NMDA components in the Renshaw cell response in addition to the faster nicotinic responses.[31–34] Renshaw cell output is characterized by burst firing that outlasts the synaptic input, at least in the typical recording conditions of classical experiments (with pentobarbital anesthetized cats),[21–23] and in more recent *in vitro* preparations, like the isolated spinal cord and transverse spinal cord slices from mice. Their inhibitory synapses on motoneurons are designed to further lengthen inhibitory synaptic action by addition of GABAergic components that have slower time courses than pure inhibitory glycinergic neurotransmission[35] (see also Ref. 36 for more recent data). Temporal summation of successive slow IPSPs during burst firing explains the typical compound recurrent IPSP recorded after a ventral root volley, and consisting of a slow "Renshaw ripple" in the rising phase and long decays.[22] The combination of these cellular properties produces recurrent IPSPs that are of longer duration than the more phasic reciprocal inhibition elicited by IaINs.

Expression of calbindin, a calcium-buffering protein, is a distinguishing feature of Renshaw cells that is abundant in their axons and dendrites.[26] Its function has not been investigated directly, but the BAPTA-like calcium-buffering properties of calbindin, and its presence throughout the Renshaw cell axon and synaptic terminals, could facilitate summation of successive release events in a synaptic train, as occurs in other calbindin-expressing interneurons.[37] On the other hand, dendritic calbindin is well positioned to modulate calcium transients induced by opening calcium-permeable nicotinic and NMDA receptors associated with motor axon synapses.[33] In summary, Renshaw cells are endowed with cellular properties (Table 1) that give them quite distinct functional characteristics. An important question then is, How do these properties develop to construct the mature Renshaw phenotype.

Table 1. Major features that characterize the Renshaw cell phenotype

Property	Distinguishing features
Cell body location	Ventral most LVII and LIX in front of ventral root exit
Axon	Ipsilateral, ventral funiculus
Characteristic input	Motor axons/nicotinic cholinoreceptive neurons
Characteristic output	Homonymous motoneurons and synergists
Firing properties	Burst firing to motor axon inputs in anesthetized animals
Neurotransmitter phenotype	Glycine and GABA
Calcium regulation	Strong expression of calbindin calcium-buffering protein
Synaptic architecture	Large inhibitory synapses on cell body and proximal dendrites, and segregation of excitatory synapses to more distal dendrites

Lineage origins of Renshaw cells and temporal control of cell phenotype

It has been known for some time that Renshaw cells originate from p1 progenitors and therefore belong to the V1 interneuron subclass.[8] They share with other V1s the early expression of the transcription factor, engrailed-1, an inhibitory phenotype, and extension of ipsilateral ascending axons. However, within this class they constitute just under 10% of all V1s in the mouse lumbar cord.[38] How is their phenotype specified within the V1 class?

Several mechanisms can create interneuron diversity from single progenitor domains. In some instances, daughter cells are differentially fated through asymmetric last divisions, as for example the diversification of p2-derived interneurons into excitatory V2a and inhibitory V2b cells.[39–41] In other cases, different lineages are created through spatial and/or temporal heterogeneity within each progenitor domain, as occurs during specification of subclasses of V0 commissural interneurons from distinct groups of p0 progenitors.[42,43] Remarkably, these mechanisms are conserved from zebrafish to mice, but the phenotypic variation generated from each domain is much larger in mice, as should be expected, given the larger complexities of limbed terrestrial motor function. While, in zebrafish and tadpoles, p1 progenitors generate one type of interneuron of well-defined morphology and function during swimming,[44,45] a recent preliminary report divides V1 interneurons in the neonatal mouse lumbar cord into over half a dozen populations, each defined by overlapping expression of 18 transcription factors.[46]

Renshaw cell specification from the V1 lineage is based on temporal control of their neurogenesis.[47]

V1 interneurons can be divided into early- and late-generated groups defined by the time they exit the progenitor cell cycle and start differentiation. These groups are further distinguished by expression of the transcription factor, Forkhead box protein P2 (FoxP2). Early-born V1s start differentiation between E9.5 and E10.5, and lack FoxP2 expression; many upregulate calbindin soon after being generated. Within this group, Renshaw cells constitute a homogenous cohort generated in a narrow temporal window at the beginning of V1 neurogenesis.[47,48] Thus, in the E10–10.5 lumbar cord, most differentiating V1s are calbindin Renshaw cell precursors. In contrast, late-born V1s exit the progenitor cell cycle between E11 and E12, rapidly upregulate FoxP2, and do not express calbindin. Interestingly, some FoxP2 V1s display synaptic connectivity in the postnatal spinal cord that resembles that of IaINs, and some upregulate a different calcium binding protein (parvalbumin) in the second postnatal week.[9,47] Temporal control of neurogenesis is currently considered a primary mechanism, inducing ventral interneuron diversity,[49] as is the case for the V1 lineage.

A yet unresolved question is whether the early generation of Renshaw cells is due to derivation from an early p1 pool of progenitor cells that then becomes exhausted or if, on the contrary, Renshaw cells are clonally related to later V1s. A clonal analysis of V1 phenotypes derived from single p1 progenitors will be necessary to resolve this issue. The first possibility might advocate for a certain genetic uniqueness of Renshaw cells. In any case, early genesis confers Renshaw cells with properties that influence their differentiation and integration into specific synaptic circuitry. First, early generation is related to a specific transcriptional code, inducing a *Renshaw cell genetic program*.[48] Second, early genesis permits Renshaw precursors to enter a migratory pathway that places them in a close relationship with motor axons,[47] and finally, early V1s are specifically associated with calbindin expression.[47]

A Renshaw cell–specific transcriptional code and calbindin expression

Recent work has described several transcription factors involved in Renshaw cell differentiation[48] and whose expression is specifically associated with the earliest interneurons generated from p1 progenitors. Two key transcription factors are Forkhead box protein d3 (Foxd3) and v-maf musculoaponeurotic fibrosarcoma oncogene homolog B (MafB). Both are also expressed by other V1 and non-V1 interneurons, being that the timing and length of expression are important factors in Renshaw cell specification. Foxd3 is broadly, and transiently, expressed by V1s just after birth, while Renshaw cells maintain Foxd3 expression for a longer period. Deletion of Foxd3 prevents Renshaw cells from expressing calbindin and MafB, and from forming a recurrent inhibitory circuit. MafB is expressed later and its deletion results in calbindin downregulation in differentiated Renshaw cells.[48] Foxd3 and MafB are therefore related to expression of calbindin in Renshaw cells. Other early-generated V1s also transiently express calbindin,[47] but while calbindin is downregulated in most V1s during development, mature Renshaw cells retain strong expression.[9] The previously mentioned results can thus be interpreted as Foxd3 being involved in induction of a calbindin V1 phenotype (and perhaps also other Renshaw properties), and its prolonged expression, together with MafB, acting to consolidate calbindin expression and the Renshaw phenotype.

Renshaw cell circumferential migration and placement at the ventral root exit

At the time when Renshaw cells are born, there are few cellular elements in the ventral horn other than motoneurons, which are generated at times overlapping with Renshaw cells. This early specification promotes a close relationship between motor axons and Renshaw cells because of the availability of a migratory pathway toward the ventral root that is taken by the early cohort of V1s (Fig. 2). The scarcity of cells in the mantle layer

during the time Renshaw cell precursors originate implies that these precursors are rapidly placed at the lateral border of the embryonic cord, above motor pools. Soon after, they extend ventrally oriented neurites that surround the motor pools circling the lateral edge of the embryonic ventral horn, and end in a growth cone that stops at the ventral root exit. Later, their cell bodies translocate following the path established by these neurites. The end result is that differentiating Renshaw cells are distributed among exiting motor axons. Later generated V1s do not display this circumferential migration and accumulate medially to motoneurons, thus distal to motor axons.[47]

These findings suggest that placement among exiting motor axons early in development is a critical step that defines the preferential targeting of Renshaw cells by motoneurons. For this to occur, two mechanisms are necessary: first, Renshaw cells need to migrate toward the ventral root exit; second, motor axons must be prevented from spreading outside the ventral root area. Although the signals that control these processes are unknown, clearly both mechanisms occur (Fig. 2) and there are interesting parallels in other spinal systems. For example, placement was identified as the major principle predicting connectivity among interneurons and motoneurons in tadpole spinal cords,[50,51] and recent elegant studies showed that positional cues are a major influence restricting monosynaptic Ia afferent connections to the correct motor pools.[52,53] Recent molecular studies also suggest that this process is based on constraining the trajectories of different inputs while placing specific cellular targets in their paths. Thus, Ia afferent axon trajectories are controlled by semaphorin/plexin interactions,[54] while motor pool placement is influenced by the transcription factor, FoxP1, perhaps by regulating patterns of cadherin-catenin signaling, which directly control topographic order and motor pool positions.[53,55]

In conclusion, the mechanisms that define early cellular placement seem to have an important influence on the major inputs that different cells will receive later. This is because, as shown for Ia afferent trajectories and the extension of motor axon collaterals in embryos, different axonal systems are allowed to course only through specific regions of the ventral horn, each containing specific neural targets. Such topographic relationships establish an early connectivity map that can

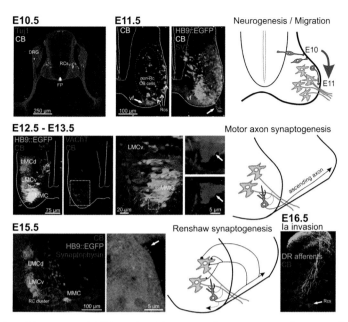

Figure 2. Early formation of the recurrent inhibitory circuit in mouse embryos. In E10.5 embryos, Renshaw cell precursors (RCs) are identified as the first calbindin (CB)-IR interneurons generated in the spinal cord. Other CB-expressing cells are present in the floor plate (FP) and dorsal root ganglion (DRG). This E10.5 section was counterstained with Tuj1 antibodies to depict the location of differentiating neurons and their axons. At this age, ventral roots are already formed and motor axons are entering the limb buds. CB-IR Renshaw cells are exiting the progenitor zone (negative for Tuj1 immunoreactivity) and position themselves laterally. By E11.5, two populations of CB-IR interneurons are present in the ventral horn. Most CB-IR interneurons in central regions are not RCs; they correspond with interneurons that, in their majority, will downregulate CB during embryonic and early postnatal development. RC precursors are located at the lateral edge and actively migrate toward the ventral root. This pathway surrounds the motoneurons (labeled by EGFP, driven by the Hb9 promoter), which have not yet started to separate into discrete columns (although they already express different EGFP levels). At this time, the ventral funiculus is formed (axons heavily immunoreactive for SV2) and CB-IR RC axons appear for the first time at this location. Between E12 and E13, the major motor columns of the lumbar spinal cord (MMC, LMCd, LMCv) can be distinguished by different levels of EGFP expression and they start to segregate spatially. During this time, we find the first evidence of immunoreactivity against the vesicular acetylcholine transporter (VAChT). VAChT-IR axons are restricted to locations containing motor pools, motor axons, and RCs. The first evidence of EGFP and VAChT-IR processes in contact with RCs (arrows) is also apparent at this time, suggesting the possibility of early motor axon synaptic interactions. RCs emit an ascending axon located in the ventral funiculus, but this has no synaptic collaterals entering the spinal cord. A large invasion of the motor pools by synaptophysin-containing axons does not occur until E15. Before that time, all synaptic markers are restricted to the developing white matter. The presence of synaptophysin processes coincides with the presence of CB-IR axons putatively originating from RCs around motoneurons. It is also at this time that the spatial relationships of different motor columns mature and a tight cluster of RCs is formed between the LMC and MMC in front of the ventral root, now placed more ventrally instead of more laterally, due to spinal cord morphogenetic maturation. Some CB-IR processes are in contact with EGFP-labeled motoneurons and contain synaptophysin, suggesting the presence of synapses. Finally, other inputs, like Ia afferents, are added later. In the image, dorsal root (DR) axons were anterogradely labeled by filling the dorsal roots with FITC-conjugated dextrans in a section that was also immunostained for CB. CB-IR RCs are at the very bottom of the ventral horn, which explains the relatively late formation of Ia afferent synapses on them.

be later refined through other mechanisms, including activity-dependent synaptic strengthening, or weakening, of specific inputs.

Early synaptogenesis and formation of the recurrent inhibitory circuit

Most pre- and postsynaptic markers of excitatory and inhibitory synapses are upregulated in the mouse spinal cord at E13 or later, and this corresponds with a large increase in the number of synapses detected by electron microscopy.[15, 16] Before synaptogenesis, synaptic proteins, including those related to synaptic vesicles, are diffusely distributed in axons, but at around E13, vesicular acetylcholine transporter (VAChT) immunoreactive motor axons establish a close relationship with calbindin-IR Renshaw cell bodies and dendrites (Fig. 2), providing the first evidence of putative synapses in the ventral horn. Interestingly, at E13, there is no anatomical evidence for synaptic connections from Renshaw cells onto motoneurons. Renshaw cell axons are first found in the ventral funiculus at E11, immediately after Renshaw cell migration is completed. From E11 to E13, they extend for a few segments rostrally without forming collateral branches entering lamina IX. The earliest evidence for calbindin synaptic collaterals innervating lumbar motoneurons was found at E15, and these synapses increase in frequency by E17 (Fig. 2). By P4, a secondary descending branch has formed that innervates motoneurons caudal to the Renshaw cell body location.[56] It is noteworthy that, in the chick embryonic spinal cord, the spread of recurrent inhibition has been shown to undergo refinement during embryonic development; however, the exact mechanisms have not yet been directly investigated.[57]

Although many details still need elucidation, one conclusion is clear: the recurrent inhibitory circuit is formed in two steps. First, motor axons synapse onto Renshaw cells and, later, Renshaw axons synapse onto motoneurons. The significance of this chronological order is unknown, but one possibility is that Renshaw axons, similar to thalamocortical axons,[58] require a waiting period until their targets mature certain properties (importantly, their final spatial distributions, see Fig. 2). This sequence of events does not imply that Renshaw differentiation requires early motor axon synaptic input. Rather, Renshaw cells differentiate in normal numbers and locations in choline acetyltransferase knockouts[19] and recurrent inhibitory circuit connections form normally in the absence of synaptic activity.[48] Thus, activity is dispensable during early circuit formation but influences later circuit maturation.

Late addition of Ia afferent inputs

As mentioned previously, different inputs arrive at the spinal cord sequentially, and one set of inputs that reaches the Renshaw area relatively late are Ia afferent axons. Ia afferent synapses are formed on a few Renshaw cells around E16–E17. At birth, most Renshaw cells receive Ia afferent inputs capable of evoke firing. By P10, they have spread and innervate all Renshaw cells in lumbar segments.[56] The presence of these synapses was surprising because adult Ren-

shaw cells are believed to lack Ia afferent input.[7,31] Its functional significance during development is yet unknown, but it does not imply that Renshaw cells share circuit functions with IaINs in newborns. Renshaw cells receive fewer Ia afferent synapses than do IaINs in neonates, and it is unlikely that Ia inputs on Renshaw cells are organized in reciprocal circuits. Following the idea of positional specification of sensory afferent inputs, it follows that the origins and strength of Ia inputs is likely influenced by V1 interneuron location. Thus, IaINs are distributed throughout the dorsoventral extent of LVII and receive a higher density of Ia afferent vesicular glutamate transporter 1 (VGLUT1) synapses than the more ventrally located Renshaw cells.[9] Moreover, pools of IaINs located at different positions receive innervation from Ia afferents originating from different muscles.[59,60] This cannot be the case for Renshaw cells, which are all clustered in the most ventral region of the ventral horn, an area traversed by Ia afferents from just a few muscle groups.[61] Given the large input resistance of Renshaw cells at P4, dorsal root stimulation elicits robust Renshaw responses;[56] however, Ia afferent firing might not be very active in neonates since Ia afferent stretch responses are still maturing.[62] As reviewed subsequently, this input weakens in Renshaw cells during later circuit maturation while being strengthened in IaINs. Thus, all V1s seem permissive for receiving Ia afferent synapses, but the initial Ia input strengths (and maybe their organization) are related to cell location. During later maturation, Ia inputs become strengthened or deselected in different V1s.

Maturation of synaptic numbers and input selection

There is a significant addition of synapses in the mouse spinal cord during the first two postnatal weeks, a period that also coincides with the development of weight-bearing locomotion, reflex maturation, and improvement of motor coordination. Correspondingly, both Ia afferent and motor axon synapses proliferate on Renshaw cells from P0 to P15; however, while motor synaptic density on Renshaw dendrites is maintained after P15, Ia afferent synaptic density decreases.[9,56] This decrease is not the result of active synapse withdrawal, but of arrested proliferation creating mismatches with dendritic growth. Ia synapses on mature Renshaw cells also display decreased synaptic complex sizes[56]

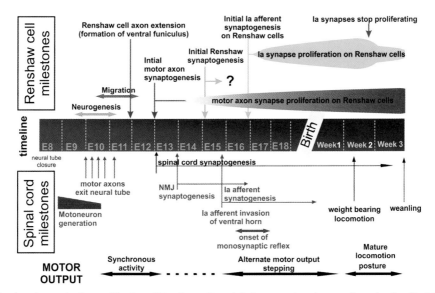

Figure 3. Landmark steps in the specification of Renshaw cells and their connections (events above the time line) in the context of other developmental processes in the spinal cord (events below the time line), and the corresponding motor output. Dates are approximations because of differences associated with spinal cord rostrocaudal level and even small differences within litters, with embryos showing more or less advanced maturation.

and AMPA receptor content (Francisco J. Alvarez, unpublished observations). In contrast, Ia synaptic density is maintained on IaINs during the same period.[9] Interestingly, the reduction of Ia afferent synaptic density on Renshaw cells is prevented in a mouse model with postnatal muscle overexpression of neurotrophin-3 driven by the myosin light chain promoter (see Fig. 3 in Ref. 63), a situation that increases Ia input strength.[64] Therefore, postnatal maturation of synaptic numbers seems to be an activity-dependent process that shifts the relative weights of different inputs on Renshaw firing. Mature Renshaw cells are driven to fire by motor axon synapses and not by Ia afferents. During the same period, inhibitory synaptic strength is adjusted by the amount of excitation that Renshaw cells receive from motor axons,[29] a process that could further decrease the influence of weaker excitatory inputs.

In summary, postnatal regulation of synapse numbers reconfigures Renshaw cell synaptic integrative properties by strengthening some inputs and weakening/silencing others within the connectivity plan laid down in embryo. This tuning is also affected by dendritic growth,[56] developmental regulation of inhibition,[29] alterations in neurotransmitter receptor subunit compositions,[28] and expression of different voltage-dependent channels (for exam-

ple, faster KV3 channels upregulate in late postnatal Renshaw cells).[65] This process implies that perhaps not all anatomical connectivity is functionally expressed in normal conditions in adult spinal networks, an idea that has also recently been proposed in the tadpole spinal cord.[51]

Conclusions

Although many details are still missing, the developmental model of Renshaw cells provides, for the first time, a view of interneuronal development that is sufficient for a preliminary description of several discrete phases (Fig. 3). First, interneuron specification occurs during neurogenesis on the basis of the induction of specific genetic programs at the progenitor level. These programs control axonal trajectories, neurochemical phenotypes, migration and spinal cord placement, and are determinants of later circuit assembly. Second, a connectivity blueprint is established on the basis of positional information, such that growing axons are channeled along pathways that cross locations containing specific interneuron targets. Third, initial synaptogenesis occurs among neighboring axons and dendrites through mechanisms that are independent of activity. Given the high excitability of embryonic interneurons, most synapses are initially capable of evoking postsynaptic firing in embryonic and early postnatal interneurons. Fourth, later postnatal synapse proliferation and maturation is dependent on activity, and occurs in conjunction with maturation of other postsynaptic properties that influence the synaptic integrative capabilities and excitability of interneurons. Postnatal maturation selects from the previously formed connectome the inputs that will exert major influence on firing activity of adult interneurons. This final phase results in mature circuits capable of producing adult coordinated motor output.

Conflicts of Interest

The authors declare no conflicts of interest.

References

1. Brownstone, R.M. & T.V. Bui. 2010. Spinal interneurons providing input to the final common path during locomotion. *Prog. Brain Res.* **187:** 81–95.
2. Chen, W.V. *et al.* 2012. Functional significance of isoform diversification in the protocadherin gamma gene cluster. *Neuron* **75:** 402–409.
3. Olbrich, H.G. & H. Braak. 1985. Ratio of pyramidal cells versus non-pyramidal cells in sector CA1 of the human Ammon's horn. *Anat. Embryol.* **173:** 105–110.
4. Gabbott, P.L. & M.G. Stewart. 1987. Distribution of neurons and glia in the visual cortex (area 17) of the adult albino rat: a quantitative description. *Neuroscience* **21:** 833–845.
5. Goulding, M. 2009. Circuits controlling vertebrate locomotion: moving in a new direction. *Nat. Rev. Neurosci.* **10:** 507–518.
6. Grillner, S. & T.M. Jessell. 2009. Measured motion: searching for simplicity in spinal locomotor networks. *Curr. Opin. Neurobiol.* **19:** 572–586.
7. Jankowska, E. 1992. Interneuronal relay in spinal pathways from proprioceptors. *Prog. Neurobiol.* **38:** 335–378.
8. Sapir, T. *et al.* 2004. Pax6 and engrailed 1 regulate two distinct aspects of Renshaw cell development. *J. Neurosci.* **24:** 1255–1264.
9. Siembab, V.C. *et al.* 2010. Target selection of proprioceptive and motor axon synapses on neonatal V1-derived Ia inhibitory interneurons and Renshaw cells. *J. Comp. Neurol.* **518:** 4675–4701.
10. Altman, J. & S.A. Bayer. 2001. *Development of the Human Spinal Cord.* New York: Oxford University Press.
11. Ladle, D.R., E. Pecho-Vrieseling & S. Arber. 2007. Assembly of motor circuits in the spinal cord: driven to function by genetic and experience-dependent mechanisms. *Neuron* **56:** 270–283.
12. Hippenmeyer, S. *et al.* 2005. A developmental switch in the response of DRG neurons to ETS transcription factor signaling. *PLoS Biol.* **3:** e159.
13. Chen, H.H. & E. Frank. 1999. Development and specification of muscle sensory neurons. *Curr. Opin. Neurobiol.* **9:** 405–409.
14. Bareyre, F.M. *et al.* 2005. Transgenic labeling of the corticospinal tract for monitoring axonal responses to spinal cord injury. *Nat. Med.* **11:** 1355–1360.
15. Vaughn, J.E., C.K. Henrikson & J.A. Grieshaber. 1974. A quantitative study of synapses on motor neuron dendritic growth cones in developing mouse spinal cord. *J. Cell. Biol.* **60:** 664–672.
16. Vaughn, J.E. *et al.* 1975. Genetically-associated variations in the development of reflex movements and synaptic junctions within an early reflex pathway of mouse spinal cord. *J. Comp. Neurol.* **161:** 541–553.
17. Scain, A.L. *et al.* 2010. Glycine release from radial cells modulates the spontaneous activity and its propagation during early spinal cord development. *J. Neurosci.* **30:** 390–403.
18. Hanson, M.G. & L.T. Landmesser. 2003. Characterization of the circuits that generate spontaneous episodes of activity in the early embryonic mouse spinal cord. *J. Neurosci.* **23:** 587–600.
19. Myers, C.P. *et al.* 2005. Cholinergic input is required during embryonic development to mediate proper assembly of spinal locomotor circuits. *Neuron* **46:** 37–49.
20. Renshaw, B. 1941. Influence of discharge of motoneurons upon excitation of neighbouring motoneurons. *J. Neurophysiol.* **4:** 167–183.
21. Renshaw, B. 1946. Central effects of centripetal impulses in axons of spinal ventral roots. *J. Neurophysiol.* **9:** 191–204.
22. Eccles, J.C., P. Fatt & K. Koketsu. 1954. Cholinergic and inhibitory synapses in a pathway from motor-axon collaterals to motoneurones. *J. Physiol.* **126:** 524–562.
23. Eccles, J.C. *et al.* 1961. Distribution of recurrent inhibition among motoneurones. *J. Physiol.* **159:** 479–499.
24. Thomas, R.C. & V.J. Wilson. 1965. Precise localization of Renshaw cells with a new marking technique. *Nature* **206:** 211–213.
25. Fyffe, R.E. 1991. Spatial distribution of recurrent inhibitory synapses on spinal motoneurons in the cat. *J. Neurophysiol.* **65:** 1134–1149.
26. Alvarez, F.J. & R.E. Fyffe. 2007. The continuing case for the Renshaw cell. *J. Physiol.* **584**(Pt 1): 31–45.
27. Alvarez, F.J. *et al.* 1999. Distribution of cholinergic contacts on Renshaw cells in the rat spinal cord: a light microscopic study. *J. Physiol.* **515** (Pt 3): 787–797.
28. Gonzalez-Forero, D. & F.J. Alvarez. 2005. Differential postnatal maturation of GABAA, glycine receptor, and mixed synaptic currents in Renshaw cells and ventral spinal interneurons. *J. Neurosci.* **25:** 2010–2023.
29. Gonzalez-Forero, D. *et al.* 2005. Regulation of gephyrin cluster size and inhibitory synaptic currents on Renshaw cells by motor axon excitatory inputs. *J. Neurosci.* **25:** 417–429.
30. Nishimaru, H. *et al.* 2010. Inhibitory synaptic modulation of renshaw cell activity in the lumbar spinal cord of neonatal mice. *J. Neurophysiol.* **103:** 3437–3447.
31. Mentis, G.Z. *et al.* 2005. Noncholinergic excitatory actions of motoneurons in the neonatal mammalian spinal cord. *Proc. Natl. Acad. Sci U.S.A.* **102:** 7344–7349.
32. Nishimaru, H. *et al.* 2005. Mammalian motor neurons corelease glutamate and acetylcholine at central synapses. *Proc. Natl. Acad. Sci. U.S.A.* **102:** 5245–5249.

33. Lamotte d'Incamps, B. & P. Ascher. 2008. Four excitatory postsynaptic ionotropic receptors coactivated at the motoneuron-Renshaw cell synapse. *J. Neurosci.* **28:** 14121–14131.

34. Lamotte d'Incamps, B., E. Krejci & P. Ascher. 2012. Mechanisms shaping the slow nicotinic synaptic current at the motoneuron-renshaw cell synapse. *J. Neurosci.* **32:** 8413–8423.

35. Cullheim, S. & J.O. Kellerth, 1981. Two kinds of recurrent inhibition of cat spinal alpha-motoneurones as differentiated pharmacologically. *J. Physiol.* **312:** 209–224.

36. Bhumbra, G.S. *et al.* 2012. Co-release of GABA does not occur at glycinergic synapses onto lumbar motoneurons in juvenile mice. *Front. Cell Neurosci.* **6:** 8.

37. Blatow, M. *et al.* 2003. Ca2+ buffer saturation underlies paired pulse facilitation in calbindin-D28k-containing terminals. *Neuron* **38:** 79–88.

38. Alvarez, F.J. *et al.* 2005. Postnatal phenotype and localization of spinal cord V1 derived interneurons. *J. Comp. Neurol.* **493:** 177–192.

39. Del Barrio, M.G. *et al.* 2007. A regulatory network involving Foxn4, Mash1 and delta-like 4/Notch1 generates V2a and V2b spinal interneurons from a common progenitor pool. *Development* **134:** 3427–3436.

40. Peng, C.Y. *et al.* 2007. Notch and MAML signaling drives Scl-dependent interneuron diversity in the spinal cord. *Neuron* **53:** 813–827.

41. Kimura, Y., C. Satou & S. Higashijima. 2008. V2a and V2b neurons are generated by the final divisions of pair-producing progenitors in the zebrafish spinal cord. *Development* **135:** 3001–3005.

42. Lanuza, G.M. *et al.* 2004. Genetic identification of spinal interneurons that coordinate left-right locomotor activity necessary for walking movements. *Neuron* **42:** 375–386.

43. Satou, C., Y. Kimura & S. Higashijima. 2012. Generation of multiple classes of V0 neurons in zebrafish spinal cord: progenitor heterogeneity and temporal control of neuronal diversity. *J. Neurosci.* **32:** 1771–1783.

44. Higashijima, S. *et al.* 2004. Engrailed-1 expression marks a primitive class of inhibitory spinal interneuron. *J. Neurosci.* **24:** 5827–5839.

45. Li, W.C. *et al.* 2004. Primitive roles for inhibitory interneurons in developing frog spinal cord. *J. Neurosci.* **24:** 5840–5848.

46. Bikoff, J.B. *et al.* 2012. *Subtype Diversity of Spinal V1 Interneurons Defined by Transcription Factor Expression in Society for Neuroscience.* New Orleans, LA: Neuroscience Meeting Planner. New Orleans, LA: Society for Neuroscience, Online.

47. Benito-Gonzalez, A. & F.J. Alvarez. 2012. Renshaw cells and Ia inhibitory interneurons are generated at different times from p1 progenitors and differentiate shortly after exiting the cell cycle. *J. Neurosci.* **32:** 1156–1170.

48. Stam, F.J. *et al.* 2012. Renshaw cell interneuron specialization is controlled by a temporally restricted transcription factor program. *Development* **139:** 179–190.

49. Tripodi, M. & S. Arber. 2012. Regulation of motor circuit assembly by spatial and temporal mechanisms. *Curr. Opin. Neurobiol.* **22:** 615–623.

50. Li, W.C. *et al.* 2007. Axon and dendrite geography predict the specificity of synaptic connections in a functioning spinal cord network. *Neural. Dev.* **2:** 17.

51. Roberts, A., W.C. Li & S.R. Soffe. 2011. A functional scaffold of CNS neurons for the vertebrates: the developing Xenopus Laevis spinal cord. *Develop. Neurobiol.* **72:** 575–584.

52. Jessell, T.M., G. Surmeli & J.S. Kelly. 2011. Motor neurons and the sense of place. *Neuron* **72:** 419–424.

53. Surmeli, G. *et al.* 2011. Patterns of spinal sensory-motor connectivity prescribed by a dorsoventral positional template. *Cell* **147:** 653–665.

54. Pecho-Vrieseling, E. *et al.* 2009. Specificity of sensory-motor connections encoded by Sema3e-Plxnd1 recognition. *Nature* **459:** 842–846.

55. Demireva, E.Y. *et al.* 2011. Motor neuron position and topographic order imposed by beta- and gamma-catenin activities. *Cell* **147:** 641–652.

56. Mentis, G.Z. *et al.* 2006. Primary afferent synapses on developing and adult Renshaw cells. *J. Neurosci.* **26:** 13297–13310.

57. Xu, H. *et al.* 2007. Developmental reorganization of the output of a GABAergic interneuronal circuit. *J. Neurophysiol.* **97:** 2769–2779.

58. Ghosh, A. & C.J. Shatz, 1993. A role for subplate neurons in the patterning of connections from thalamus to neocortex. *Development* **117:** 1031–1047.

59. Hultborn, H., E. Jankowska & S. Lindstrom, 1971. Recurrent inhibition of interneurones monosynaptically activated from group Ia afferents. *J. Physiol.* **215:** 613–636.

60. Jankowska, E. & S. Lindstrom. 1972. Morphology of interneurones mediating Ia reciprocal inhibition of motoneurones in the spinal cord of the cat. *J. Physiol.* **226:** 805–823.

61. Ishizuka, N. *et al.* 1979. Trajectory of group Ia afferent fibers stained with horseradish peroxidase in the lumbosacral spinal cord of the cat: three dimensional reconstructions from serial sections. *J. Comp. Neurol.* **186:** 189–211.

62. Vejsada, R. *et al.* 1985. The postnatal functional development of muscle stretch receptors in the rat. *Somatosens. Res.* **2:** 205–222.

63. Mentis, G.Z. *et al.* 2010. Mechanisms regulating the specificity and strength of muscle afferent inputs in the spinal cord. *Ann. NY. Acad. Sci.* **1198:** 220–230.

64. Wang, Z. *et al.* 2007. Prenatal exposure to elevated NT3 disrupts synaptic selectivity in the spinal cord. *J. Neurosci.* **27:** 3686–3694.

65. Song, Z.M. *et al.* 2006. Developmental changes in the expression of calbindin and potassium-channel subunits Kv3.1b and Kv3.2 in mouse Renshaw cells. *Neuroscience* **139:** 531–538.

Ann. N.Y. Acad. Sci. ISSN 0077-8923

ANNALS OF THE NEW YORK ACADEMY OF SCIENCES
Issue: *Neurons, Circuitry, and Plasticity in the Spinal Cord and Brainstem*

Dorsally derived spinal interneurons in locomotor circuits

Anna Vallstedt and Klas Kullander

Unit of Developmental Genetics, Science for Life Laboratory, Department of Neuroscience, Uppsala University, Uppsala, Sweden

Address for correspondence: Klas Kullander, Department of Neuroscience, Uppsala University, Husargatan 3, Box 593, 751 24 Uppsala, Sweden. klas.kullander@neuro.uu.se

During neuronal circuit formation, axons are guided to their targets by the help of axon guidance molecules, which are required for establishing functional circuits. A promising system to dissect the development and functionalities of neuronal circuitry is the spinal cord central pattern generator (CPG) for locomotion, which converts a tonic supraspinal drive to rhythmic and coordinated movements. Here we describe concepts arising from genetic studies of the locomotor network with a focus on the position and roles of commissural interneurons. In particular, this involves studies of several families of axon guidance molecules relevant for midline crossing, the Eph/ephrins and Netrin/DCC. Effects on developing commissural interneurons in mice with aberrant midline axon guidance capabilities suggest that, in addition to ventral populations, dorsal commissural interneurons also play a role in coordinating locomotor circuitry. Recent findings implicate the novel dI6 interneuron marker Dmrt3 in this role. Strikingly, mutations in Dmrt3 result in divergent gait patterns in both mice and horses.

Keywords: neuronal network; spinal cord; central pattern generator; Dmrt3; netrin; Eph

Introduction

Locomotor central pattern generators (CPGs) in the spinal cord are neuronal networks that produce rhythmic activities necessary for coordinated trunk and limb movements.[1–5] The CPGs are mainly responsible for the generation of a stable rhythm while at the same time coordinating flexion extension of limbs and alternation of movement between the right and left sides of the body. Left–right locomotor coordination requires commissural interneurons (CINs) defined by their projections to the contralateral side of the spinal cord. In mammals, many CINs are rhythmically active during locomotor-like activity, and both excitatory and inhibitory CINs are considered to coordinate left–right activities during locomotion.[6–9] Lesion experiments in neonatal spinal cords have shown that the dorsal spinal cord is dispensable for rhythmic and coordinated locomotor activity[10] and that normal left–right alternating locomotion disappears after cutting the ventral commissure. These experiments suggest that basic left–right coordination is mediated by ventrally located CINs, at least in the perinatal rat spinal cord.

Mouse mutants that display aberrant axon guidance over the midline are useful to analyze the contribution of neuronal subpopulations to left–right regulation of locomotor CPG activity. Such studies have raised a possible role for dorsally originating interneurons in locomotor coordination.

Studies aimed at identifying the subcomponents of the locomotor CPG, as well as ancillary components responsible for left–right coordination, are challenging but nevertheless critical to dissecting the development and function of locomotor neuronal circuitry. Subpopulations of spinal cord interneurons can be identified during development by their expression of specific homeodomain transcription factors.[11] Dorsal progenitor cells give rise to six early classes of neurons, dI1 to dI6, and ventral progenitors give rise to motor neurons and four classes of interneurons, V0 to V3. Several of these early classes of neurons have been found to produce CINs.[12–17] Out of the ventral subtypes, the V0 and V3 populations are considered to generate CINs and have been investigated in genetic studies on left–right locomotor coordination. During development some dorsally born neurons migrate ventrally,[12–14,16]

doi: 10.1111/j.1749-6632.2012.06801.x

Ann. N.Y. Acad. Sci. 1279 (2013) 32–42 © 2013 New York Academy of Sciences.

suggesting that neurons originating from the dorsal spinal cord might participate in ventral-located circuitries. Moreover, dorsal interneurons are activated during locomotion as demonstrated by c-Fos activation.[18,19] Indeed, dorsally originating interneuron populations also extend commissural projections,[17,20–22] and project directly to motor neurons[23] and are therefore candidates regulating midline coordination during locomotion.

Netrin-dependent interneurons

During spinal cord development, differentiating commissural interneurons send their axons toward and across the floor plate to form the ventral commissure. A set of axon guidance molecules, the Netrins, acts as diffusible floor plate chemotrophic guidance cues for commissural axons in the mouse spinal cord.[24] Netrins bind to and activate the Netrin receptor DCC, which mediates an attractive effect to the growing axons, and in Netrin-1 knockout mice, the majority of dorsal commissural interneurons no longer find and cross the midline.[25] Given their important role in guiding neurons toward the midline, Netrin-1 and DCC are obvious candidates for providing guidance cues to commissural interneurons involved in CPG left–right coordination (Fig. 1A). Indeed, formation of an alternating locomotor CPG network requires the presence of Netrin protein as loss of Netrin-1 results in complete synchrony between the two sides during fictive locomotion.[17] Therefore, Netrin-1 and DCC mouse mutants were used to address the role of commissural interneuron subpopulations in left–right coordination.

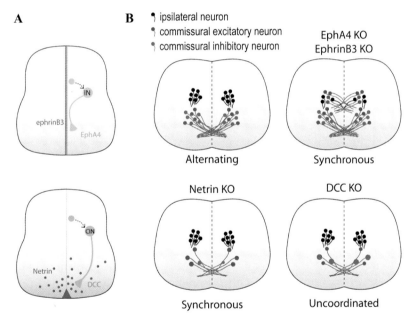

Figure 1. Schematic models of CPG networks in axon guidance mutant animals. (A) EphA4/ephrinB3 and Netrin-1/DCC have mirror roles in the guiding of axons away or toward the midline. While EphA4 prevents axon fibers from crossing the midline and remaining ipsilateral, Netrin attracts fibers to the midline. (B) Consequently, *EphA4/ephrinB3* and *Netrin-1/DCC* mutant mice differ in their shaping of locomotor circuitry. Under normal conditions (upper left), excitatory and inhibitory CINs are balanced to produce alternating left/right CPG activities. In EphA4- and ephrinB3-null mutants, aberrant midline crossing of ipsilateral interneurons shifted the balance between excitation and inhibition, leading to synchronous left/right CPG output. In Netrin-1-null mutants, reduced midline crossing of fibers also led to a similar shift in the balance between excitation and inhibition but was due to reduced number of inhibitory CINs. Again, this resulted in a synchronous output, but in contrast to EphA4/ephrinB3-null mutants, the synchrony is not reversible by pharmacological strengthening of inhibition (not shown). This might be explained by a too limited number of commissural inhibitory fibers left in *Netrin*-1-null mutants in which the drug can be active. We hypothesize that under normal conditions, inhibitory connections dominate with alternation as a result. When excitatory action over the midline is strengthened, as in the *EphA4*-null mice, or when inhibitory action over the midline is weakened, as in the *Netrin-1*-null mutant mice, the excitatory contralateral component takes over, leading to coupling of the two spinal cord half centers and synchrony of the rhythmic output. Finally, in DCC-null mutants, a decreased midline crossing led to a severe reduction of both excitatory and inhibitory fibers, resulting in a complete collapse of left–right coordination.

The number of CINs and fibers crossing the spinal cord midline are strongly reduced in mouse mutants lacking *Netrin-1* (Fig. 1B, Refs. 17 and 25). Since Netrin-1 is a potential attractant for fibers from cells originating from both dorsal and ventral neuroepithelium, it has been unclear which neuronal subpopulations and functions are dependent of Netrin-mediated axon guidance. Commissural interneurons are found within the dI1-dI3, dI5, dI6, V0, and V3 subpopulations of neurons. Of these, dI1–dI3, dI5, dI6, and V0d cells were most severely affected in *Netrin-1* mutant mice, having a 75–80% reduction of commissural traceable axons, indicating a strong dependence on Netrin-1 to properly find the midline. Interestingly, the ventral-most population (V3) was completely unaffected.

There are approximately twice as many inhibitory CIN terminals than excitatory ones in both rat and mouse spinal cords.[17,26] In Netrin-1 mutants, a greater number of inhibitory than excitatory CINs are lost, resulting in the converse situation—a prevailing majority of excitatory CINs. Since the remaining coordination is synchronous, excitatory CIN action is presumably more important for general cross-midline activation than for maintenance of left–right alternation. Furthermore, since mice that, in principle, lack excitatory activity in the spinal cord through loss of VGLUT2 did not show any locomotor rhythm or coordination deficiencies,[27–29] inhibitory action is sufficient to produce normal rhythm and coordination. Inhibitory commissural connections have previously been suggested to be major constituents of left–right phasing during locomotion,[8,14,30–34] and in cats, such interneurons have been implicated in mediation of the crossed reflexes and in the selection of different motor patterns.[35,36] Functional roles for Netrin-1–guided commissural interneurons in the spinal cord are evident, and together with the fact that Netrin-1 preferentially attracts dorsally originating CINs of an inhibitory character to the floorplate, we should consider those as potential populations regulating left/right alternation.

Lack of DCC results in collapsed left–right coordination

Mice that carry a null mutation of DCC display a severe loss of interneuronal subpopulations originating from commissural progenitors as well as matured commissural interneurons.[37] The loss of CINs is accompanied by completely uncoordinated left–right ventral root activities during fictive locomotion (see schematic in Fig. 1B). Thus, DCC plays a crucial role in the formation of spinal neuronal circuitry coordinating left–right activities. Of note, similar to Netrin-1 mutants, flexion-extension ventral root activities remained alternating. The loss of commissural fibers from V3 neurons in mice lacking DCC was the only significant difference compared to Netrin-1 mutant mice, and resulted in a complete loss of coordination between the left and right side, emphasizing the fundamental role of V3 CINs in coordinating synchronous activities over the midline. Taken together, in mice lacking DCC, a limited number of CINs crossed the midline, which functionally resulted in uncoordinated left–right activity during fictive locomotion. This severe reduction of CINs did not shift the balance between excitatory and inhibitory fibers over the midline, likely due to an almost similar loss of CINs originating from different progenitor domains. In summary, while Netrin and DCC mutant mice have a similar severe reduction of CINs, their fictive locomotion phenotypes are markedly different, producing synchronous versus uncoordinated patterns of activation (see schematic in Fig. 1B). Thus, axon guidance mediated by Netrin and DCC is required to form a circuitry capable of coordinating left–right activities.

Lessons from Eph and ephrins

Previous studies have shown that the axon guidance molecules EphA4 and ephrinB3 cooperate to prevent ipsilateral interneurons from crossing the midline in the spinal cord and if either molecule is deleted in mice, this will result in a synchronous hopping gait.[38] To further investigate the origin of these interneurons, we generated a conditional allele targeting the third exon of *EphA4* (EphA4lox; Supporting Fig. S1), which was found to delete the EphA4 protein in a spatially restricted manner.[39] In addition, mice with a complete deletion of *EphA4*, produced by crossing *EphA4lox* mice with a general Cre deleter (PGK-Cre; *EphA4lox/lox)*, displayed all characteristics of regular EphA4-null mutants, including a synchronous gait.[39,40] To achieve restricted deletion of EphA4 in the dorsoventral axis of the lumbar spinal cord (Fig. 2A and B), we crossbred the conditional *EphA4* allele to mice expressing

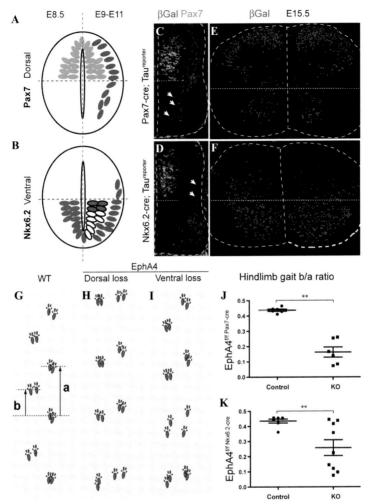

Figure 2. Deletion of EphA4 in dorsal progenitor cells leads to near synchronous gait. (A and B) Schematic outline of Pax7 and Nkx6.2 expression in cells of the developing spinal cord. Pax7-expressing dorsal progenitors in the neural tube[78,79] are labeled in green. At e8.5, Nkx6.2 is expressed in a broad ventral domain of the mouse neural tube (blue). At later stages (e9.5–e11.5), Nkx6.2 expression is largely confined to a narrow domain immediately ventral of the horizontal midline (blue).[70] Note that cells in the entire ventral domain (black circles) have been exposed to Cre at e8.5 and are therefore subsequently detectable when these cells mature into neurons (red). (C–F) Analysis of Pax7-Cre and Nkx6.2-Cre–mediated recombination. Pax7-Cre and Nkx6.2-Cre mice were crossed with Tau^{mEGFP,NlsLacZ} reporter mice and β-Gal expression in e11.5 (C and D) and e15.5 (E and F) embryos was analyzed. At e11.5, costaining of Pax7 and β-Gal in Pax7-Cre/Tau^{mEGFP,NlsLacZ} embryos display predominantly dorsal expression of β-Gal (C). When cells originating from Pax7 progenitors start to differentiate into neurons, Pax7 is downregulated and the expression of β-Gal driven from the promoter of the axonal microtubule-associated protein tau is initiated. This enables detection of neurons during their radial and/or ventral migration from the Pax7 mediodorsal compartment between e9.5 and e11.5 (red in A and B). Some expression in ventrally located cells is detected (arrows), largely corresponding to the earliest born dorsal neurons that migrate ventrally. Nkx6.2-Cre/Tau^{mEGFP,NlsLacZ} embryos at e11.5 show an extensive and strong expression in ventrally located neurons together with a weak expression in a few cells that appear to be migrating ventrally from dorsal positions (arrows, D). At e15.5, the expression in β-Gal in Pax7-Cre/Tau^{mEGFP,NlsLacZ} spinal cords is restricted to dorsal lamina and to scattered ventral and intermediate cells (E). In Nkx6.2-Cre/Tau^{mEGFP,NlsLacZ} spinal cords, the expression of β-Gal is widespread in ventral cells and is present in some cells that are more dorsally located (F). Weak dotted lines indicate the boundary between dorsal and ventral horns. (G–K) Gait analyses of WT, EphA4$^{f/fPax7-cre}$ (H) and EphA4$^{f/fNkx6.2-cre}$ mice (I). EphA4 conditional knockout mice were generated by homologous recombination in embryonic stem cells, essentially as described previously.[40] Forepaws (red) and hindpaws (blue) were painted and the distance between left and right paws (b in G) and distance covered by the same paw (a in G) was determined. Comparison of the b/a ratio of wild-type (n = 6) and EphA4$^{f/fPax7-cre}$ mice (n = 6) show a pronounced decrease in all mutant mice analyzed (E). The b/a ratio of EphA4$^{f/fNkx6.2-cre}$ mice (n = 8) compared to Wt (n = 8) show a variable phenotype with some animals displaying a close to normal gait pattern while some are more affected (F).

Cre in the dorsal (Pax7-Cre[41]) or ventral progenitor domain of the spinal cord (Nkx6.2-Cre[42]). The specificities of the two Cre alleles were validated using *TaumGFP-nlslacZ* reporter mice,[43] followed by histochemical stains with antibodies against β-galactosidase and Pax7. At E11.5, the dorsoventral specificities are clearly recognized, whereas at E15.5, the pattern of lacZ expression suggests that some of the neurons born in the Pax7 expression domain migrate ventrally (Fig. 2C and E). Similarly, but less pronounced, some of the neurons born in the Nkx6.2 domain have a tendency to expand dorsally (Fig. 2D and F). Whereas neonatal fictive locomotion was normal, gait analysis demonstrated that the walking pattern of *Pax7-Cre; EphA4lox/lox* animals markedly differed from control animals (Fig. 2G–K). The b/a ratio, indicating the stride distance between the left and right hindpaw, was close to 0.5 for controls (including *EphA4lox/lox* animals), while it was significantly lower (0.15) for *Pax7-Cre; EphA4lox/lox* mice (Fig. 2J). The b/a ratio in *Nkx6.2-Cre; EphA4lox/lox* mice was also significantly affected, although with an average value of 0.25, suggesting a lesser influence on locomotor coordination from EphA4-dependent neurons originating from the ventral domain (Fig. 2K).

This genetic deletion of *EphA4* in the dorsal or ventral domain of the spinal cord revealed that mice affected in the dorsal subpopulations produce close to a synchronous gait. Mice with an *EphA4* deletion in the ventral spinal cord were also affected, but to a lesser degree. This suggests that dorsally originating ipsilateral interneurons, when misdirected, are interfere with normal locomotion. Thus, subtype-specific deletion of EphA4 in dorsal spinal cord populations demonstrated a stronger influence on left–right coordination compared with deletion in ventral populations. Likewise, studies of DCC and Netrin axon guidance molecules suggest that dorsal populations play a role in coordination of locomotion. Neurons originating from dorsal progenitors are at least partly responsible for the phenotype observed in *EphA4*[−/−] and *Ntn1*[−/−] mice, emphasizing the importance of investigating these populations in relation to locomotor network coordination.

Location and function of locomotor interneurons

V0 interneurons arise from p0 progenitor cells expressing the Dbx1 homeodomain (HD) protein and consist of two populations of cells, one of which expresses the HD protein Evx1.[12,15,16] When leaving the proliferative zone, these cells take on a ventral migratory route to settle in lamina VII/VIII and extend their axons rostrally on the contralateral side of the spinal cord. Dbx1 is expressed in progenitor cells giving rise to both V0V and V0D neurons and the loss of both V0V and V0D in Dbx1-knockout mice led to intermittent episodes of synchrony between left and right ventral roots during fictive locomotion. Sim1-expressing V3 neurons, which arise from Nkx2.2-positive p3 progenitors, are predominantly commissural excitatory and part of this population appears to settle close to their origin in lamina VIII.[44] Silencing of the V3 population using Sim-Cre/TeNT resulted in an irregular and imbalanced motor rhythm.[45]

Although ipsilateral populations of neurons do not directly regulate bilateral coordination, they might provide ipsilateral input to CINs and vice versa, several CINs have been shown to synapse on ipsilateral neurons on the opposite site of the spinal cord.[46,47] For this reason, genetic studies of specific ipsilateral subpopulations with regard to their function during locomotion are informative of left–right activities. V1 progenitors expressing HD proteins Dbx2 and Nkx6.2 give rise to Engrailed 1 (En1)-positive V1 interneurons.[48] These cells migrate to a ventrolateral position in lamina VII where they develop local projections.[49] Studies of En1/2-expressing V1 interneurons have shown that in "simpler" vertebrates, such as the fish[50] and frog,[51] these neurons represent a homogenous cell population of ipsilateral glycinergic inhibitory interneurons that play important roles in motor control and sensory gating during swimming, while they appear to have more heterogeneous functions in higher vertebrates.[52–54] Studies of fictive locomotion in neonatal preparations of En1 null "knock-in" mice and acute silencing of V1 neurons via the allatostatin receptor have shown to slow the speed of locomotion but with remaining normal left–right coordination activities.[55] V2 interneurons develop from Irx3 and Nkx6.1-expressing p2 progenitors into two separate populations of postmitotic cells, one expressing Chx10 and Lhx3 and the other expressing Gata2 and Gata3.[56,57] The V2 interneurons migrate laterally to their location in lamina VII. Ablation of V2a interneurons with a Chx10-DTA strategy results in greater variability in cycle period and amplitude of locomotor bursts[46] during fictive locomotion and locomotion in adult mice show a speed-dependent loss of left–right alternation defined by a transition to synchronous gait at high speed.[58] Thus, ventral-originating populations of neurons are involved in various aspects of locomotion. With regard to left–right coordination, the rhythms of fictive locomotion were irregular with episodes of synchrony and alternation (V0)[49] or drifting in and out of strict alternation (V2a).[14] Consequently, it seems likely that multiple neuronal subtypes originating from several ventral progenitor domains are involved in the different aspects of left–right coordination.

Of the dorsal IN populations, commissural inhibitory dI6 neurons, which settle in laminae VII and VIII, are the most promising candidate neurons for left–right alternating circuitry. This is supported by the idea that inhibitory commissural connections are the major pathways responsible for coordinating the left–right phasing during locomotion.[14,30–33] Moreover, dI6 neurons share some properties with dorsal V0 interneurons in that they originate from progenitors with similar character and migrate along similar routes.[14,17] Interestingly, one subset of dI6 neurons have been demonstrated to phase-lock with ventral root outputs, which is in concert with a role to coordinate motor output during locomotion.[59] In the same study, another subset was demonstrated to oscillate intrinsically when isolated from excitatory input, potentially indicating an involvement in locomotor rhythm generation. Possible roles for dI4 and dI5 interneurons for CPG coordination remains to be determined; notably, however, a cohort of dI4 neurons have been reported to form contacts on Ia afferent terminals near MNs.[60] Interestingly, dI3 neurons have been demonstrated to directly contact MNs by rabies virus tracing[61] and Isl1 positive dI3 cells, which migrate ventrally to lamina VII, receive direct primary afferent inputs as well as rhythmic inputs, suggesting a role for these dI3 neurons in sensorimotor integration.[62] Finally, dI1-dI2 interneurons are less likely to contribute to CPG coordination, since they have been suggested as part of ascending pathways, including the spinocerebellar and the spinothalamic tract.[12,13] In any case, multiple lines of evidence suggest that neurons from the dorsal neuroepithelium are also likely to contribute to locomotor coordination.

A novel dI6 marker having a role for gait

In a search for genes expressed predominantly in cholinergic cells located in the P11 ventral spinal cord, we found *Dmrt3*.[63] Another study using a similar screen strategy also identified Dmrt3 as a marker for a subset of spinal cord neurons.[64] The DMRT family consists of three members (1–3), which carry a DM (dsx and mab-3) DNA-binding domain conferring sequence-specific DNA binding distinct from a classical zinc finger.[65] Dmrt3, the least characterized of the three members, is expressed in neural and germ cells and has been associated with sexual development.[66] At postnatal stages, Dmrt3 is expressed in the ventral spinal cord while at earlier time points, expression was evident in more dorsally located cells, indicative of dorsal to ventral migration.[67] The origin of the Dmrt3 cells was characterized using markers for dorsal and ventral progenitors and found to be a subset of dI6 neurons (Fig. 3A). At E12.5 and E14.5, there was also a partial overlap with neurons positive for the Wilms tumor 1 (WT1) protein, which has been suggested to label a dI6 population.[68] Retrograde tracing experiments demonstrated that Dmrt3 interneurons extend projections both ipsi- and contralateral and transsynaptic pseudorabies virus-based tracing demonstrated direct connections to motor neurons both ipsi- and contralateral to the injected muscle (schematized in Fig. 3B). Finally, Dmrt3 interneurons are VIAAT-positive/VGLUT2-negative suggesting that they are inhibitory.

Independent of these efforts, in a genome-wide association study (GWAS), Leif Andersson and his coworkers identified a region of the chromosome associated with the ability to pace in Icelandic horses. They explored the circumstance that some Icelandic horses are four-gaited, while some are five-gaited having the additional ability to pace at high speed, so called flying pace. The identified region spanned four genes; and when revealed to us, we informed the Andersson group of our cellular data. Thus, even if pure positional cloning would have identified the mutation, it was clear that of the four genes, *Dmrt3* was the obvious candidate. Sequencing of the *Dmrt3* gene in pace and a nonpace Icelandic horses soon thereafter identified a nonsense mutation in the *Dmrt3* gene, verifying the GWAS result. The mutation is a single nucleotide substitution that leads to a premature stop codon and expression of a truncated

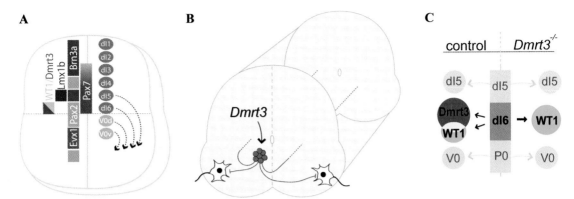

Figure 3. Characteristics of the Dmrt3 spinal cell population. (A) Schematic spinal cord cross section showing progenitor and transcription factor domains. Dmrt3+ cells originate from the ventral-most part of the dorsal domain (border indicated by line). Dmrt3+ cells overlap with the dI4/dI6/V0d marker Pax2 but not with the V0v/V0C/V0G marker Evx1 or the dI5 marker Lmx1b. Dmrt3 and WT1 show a partial overlap. (B) Schematic of position and demonstrated possible projections of Dmrt3 neurons in the mouse spinal cord. (C) Schematic illustration of fate change in the Dmrt3 dI6 population of neurons.

version of Dmrt3.[67] The mutation is present on both alleles in pacing Icelandic horses and has a major effect on the pattern of locomotion in horses. Domestic horses with the ability to perform alternate gaits at intermediate speeds, possessing toelt, classicofino, and fox trot are all homozygous for the Dmrt3 mutation, while nongaited horses are homozygous for the wild-type allele. Moreover, the Dmrt3 mutation has accumulated in harness race horses in which the transition from trot to gallop leads to disqualification from the race.[67]

We analyzed a possible role for Dmrt3 in dI6 population development in *Dmrt3*-null mutant mice. These mice have a shorter lifespan and occasional male sexual developmental abnormalities.[69] In general, loss of transcription factor expression within spinal cord progenitor domains results in specification defects in the produced neurons, presumably by suppression of differentiation programs that operate in adjacent domains.[68] For example, in *Dbx1* mutant mice, progenitors that give rise to V0 interneurons instead generate V1 and dI6 interneurons,[14,16] and loss of the V1-specific *Nkx6.2* homeobox transcription factor results in increased numbers of V0 interneurons at the expense of V1s.[70] The dI6 population, as well as the flanking dI5 and V0d populations, remained of normal sizes in *Dmrt3*−/− mice.[67] In contrast, the number of WT1+ neurons increased, while the number of commissural interneurons decreased, likely explained by an altered fate of the Dmrt3 neuron population (Fig. 3C).

Drug-induced fictive locomotion in Dmrt3 mutant mice displayed a strikingly uncoordinated and irregular firing rhythm both between the left–right and flexion–extension outputs. Moreover, airstepping in 1- and 4-day-old mice was also severely disturbed. Despite this very irregular motor output, walking abnormalities in adult knockout mice were not previously reported.[69] A closer examination on a treadmill revealed an inability of Dmrt3 mutant mice to run at higher velocities. At a speed where gait recordings were possible, significant increases in stance and swing times were apparent.[67] Dmrt3+ cells are inhibitory and can have ipsi- and/or contralateral projections, and since previous studies have suggested that inhibitory commissural connections are major constituents responsible for left–right phasing during locomotion,[71–73] a role for Dmrt3 in left–right coordination was envisaged. Interestingly, in Dmrt3 mutant mice, left–right as well as flexion–extension perinatal locomotion collapses, and of note, this is the first example of a mouse mutant with an ipsilateral (flexion-extension) coordination failure. Yet, the locomotor coordination abnormalities in adults were relatively mild and confined to changes in flexion-extension circuitry, suggesting a remarkable adaptability in locomotor circuit development and function. The observation of an extensive flexibility of spinal cord circuitry has been reported previously.[74] While the most likely explanation for the observed defects lies in the changed fate within the dI6 neuron population, the extent of circuit reorganization is not known and requires further investigation, such as the determination of the physiological character of the Dmrt3 population, the role of the increased WT1 population, possible regulation of other transcription factors, and effects on other possible dI6 subpopulations.

In quadrupedal mammalian locomotion, speed regulation is accompanied by a change of gait. Interneuron populations are recruited to locomotor coordination depending on running speed while in zebrafish, the recruitment of spinal interneurons at higher speeds results in silencing of interneurons recruited at lower speeds.[75] In mice, an increase in running speed leads to a gradual recruitment of interneurons[76] and such involvement of V2a interneurons is required for the maintenance of left–right alternation at escalating velocities.[58] In the same way, Dmrt3 dI6 neurons may have a similar role in the flexion-extension cycle. At lower velocities, the loss of Dmrt3 does not seem to have a large effect, while at intermediate running speeds, *Dmrt3*-null mice increase their stride length, and finally, they refuse to run at higher speeds. This suggests a role of Dmrt3 in maintaining strict and swift flexion-extension alternation.

Despite the differences in speed and size between mice and horses, the phenotypic dependencies downstream of Dmrt3 are relatively comparable. Running at higher speeds is characterized by shortening the extensor phase, while the flexor phase is largely unchanged.[77] An increased stride length, which was observed in *Dmrt3*-null mice, may thus be a selectable advantage for harness racing horses. Interestingly, harness racing horses are not able to keep up with nonharness racing horses during gallop, which complements the finding that Dmrt3-null mice give up at higher treadmill speeds. The truncated Dmrt3 allows horses to move beyond its three natural gaits and to perform alternative gaits. In the mouse, lack of Dmrt3 results in impaired limb coordination in newborns and in adults it affects their ability to run at higher speeds. Thus, the loss of Dmrt3 results in a more permissive and flexible state of the locomotor circuit. This suggests that Dmrt3-positive neurons might be important for maintaining precise flexion-extension alternation. Taken together, a role for Dmrt3 neurons in coordinating both left–right alternation and flexor–extensor motor neuron activity seems plausible.

Our data demonstrate that Dmrt3 cells serve as both contralateral and ipsilateral neurons, and the complete collapse of early CPG coordination opens for speculative thoughts regarding the role of these neurons in organizing spinal output. To us, the most parsimonious explanation positions Dmrt3 neurons as coordinators between left–right and flexion–extension activities. We hypothesize that these neurons are required for the early CPG coordination circuitry, and in adults, they may act as phase-lock neurons to control and secure a robust gait. Interestingly, the observed flexibility of the developing spinal cord circuitry suggests that alternative circuitries are shaped in the absence of Dmrt3. Neurons, which replace the role of Dmrt3-dependent neurons, are clearly able to create left–right as well as flexion–extension activities, however, not with the same restraining action in adults. Presumably, the contra- and ipsilateral projections of the Dmrt3 neurons are unique features among locomotor circuit neurons and cannot be replaced entirely. The fact that Dmrt3 neurons are all inhibitory implies that they are positioned so that flexion activates Dmrt3 neurons, which results in inhibition of ipsilateral extension and contralateral flexion activities.

In conclusion, Dmrt3 is specifically expressed in the dI6 subdivision of spinal cord neurons, takes part in neuronal specification within this subdivision, and is critical for the development of a coordinated locomotor network with emphasis on the regulation of swing and stance phases in the mouse. Dmrt3 interneurons project fibers ipsi- and contralaterally, which together with a significant phenotype in flexor-extensor coordination in fictive locomotion, positions Dmrt3 in a pivotal role for configuring the spinal circuits controlling stride. The discovery of a previously unknown molecule required for shaping a neuronal ensemble coordinating limb movement in at least two species, and with factual relevance for horse locomotion, will have an impact on our understanding of gait control and the underlying flexibility in neuronal circuit hardwiring. Further studies of the Dmrt3 neurons will be necessary to define the exact role of these neurons in the locomotor network. Given the results from studies of axon guidance molecules and the evident important role for dI6 neurons, we conclude that dorsally derived spinal interneurons indeed play a significant role in locomotor circuits.

Acknowledgments

We thank M. Tessier-Lavigne (Netrin-1), E. Andersson and J. Ericson (Nkx6.2-Cre), and M. Capecchi (Pax7-Cre) for providing mice, and Dr. T. Boesl for the generation of EphA4 transgenic mice. Work in the lab is supported by grants from the Swedish Research Council—Medicine and Health, the Swedish Brain Foundation, the Hållsten Foundation, the Knut and Alice Wallenberg Foundation, and Uppsala University. K.K. is a Royal Swedish Academy of Sciences Research Fellow supported by a grant from the Knut and Alice Wallenberg Foundation.

Supporting Information

Additional supporting information may be found in the online version of this paper:

Figure S1. Design and production of an EphA4 conditional allele. (A) Schematic overview of the construct used to target genomic DNA in ES cells. A cDNA encoding EphA4 from exon IV and onward and a polyA (pA) tail was fused to exon III with $5'$ genomic sequence remaining intact. LoxP sites were inserted before and after the exonIII-cDNA. After the second loxP site, exon III with its splice acceptor site was inserted to allow production of an alternative transcript after excision of the EphA4 cDNA fusion. An internal ribosome entry site (IRES) sequence was placed before a lacZ-TTC reporter construct for production of retrogradely transported β-gal protein. Finally, a Frt-flanked Neo cassette was introduced for positive selection of embryonic stem cell clones. (B) Southern blotting experiments verified that out of 200 clones, seven were identified as true homolog recombinants. Detection with a probe for the Neo cassette resulted in a single band, suggesting the absence of extra insertions in the genome. (C) The oligos okk100 (CAGTAATTTTCTTCTTCACTC) and okk101 (ACCTGGTAGGTTCGGATCGGT) were able to detect the wild type and loxP alleles by PCR.

Conflicts of interest

The authors declare no conflicts of interest.

References

1. Goulding, M. *et al.* 2002. The formation of sensorimotor circuits. *Curr. Opin. Neurobiol.* **12:** 508–515.

2. Grillner, S. 2006. Biological pattern generation: the cellular and computational logic of networks in motion. *Neuron* **52:** 751–766.

3. Kiehn, O. & K. Kullander. 2004. Central pattern generators deciphered by molecular genetics. *Neuron* **41:** 317–321.

4. Pearson, K.G. 1993. Common principles of motor control in vertebrates and invertebrates. *Annu. Rev. Neurosci.* **16:** 265–297.

5. Kullander, K. 2005. Genetics moving to neuronal networks. *Trends Neurosci.* **28:** 239–247.

6. Brownstone, R.M. & J.M. Wilson. 2008. Strategies for delineating spinal locomotor rhythm-generating networks and the possible role of Hb9 interneurones in rhythmogenesis. *Brain Res. Rev.* **57:** 64–76.

7. Grillner, S. 2003. The motor infrastructure: from ion channels to neuronal networks. *Nat. Rev. Neurosci.* **4:** 573–586.

8. Kiehn, O. 2006. Locomotor circuits in the mammalian spinal cord. *Annu. Rev. Neurosci.* **29:** 279–306.

9. McCrea, D.A. & I.A. Rybak. 2008. Organization of mammalian locomotor rhythm and pattern generation. *Brain Res. Rev.* **57:** 134–146.

10. Kjaerulff, O. & O. Kiehn. 1996. Distribution of networks generating and coordinating locomotor activity in the neonatal rat spinal cord in vitro: a lesion study. *J. Neurosci.* **16:** 5777–5794.

11. Alaynick, W.A., T.M. Jessell & S.L. Pfaff. 2011. SnapShot: spinal cord development. *Cell* **146:** 178–178 e171.

12. Bermingham, N.A. *et al.* 2001. Proprioceptor pathway development is dependent on Math1. *Neuron* **30:** 411–422.

13. Gross, M.K., M. Dottori & M. Goulding. 2002. Lbx1 specifies somatosensory association interneurons in the dorsal spinal cord. *Neuron* **34:** 535–549.

14. Lanuza, G.M. *et al.* 2004. Genetic identification of spinal interneurons that coordinate left–right locomotor activity necessary for walking movements. *Neuron* **42:** 375–386.

15. Moran-Rivard, L. *et al.* 2001. Evx1 is a postmitotic determinant of v0 interneuron identity in the spinal cord. *Neuron* **29:** 385–399.

16. Pierani, A. *et al.* 2001. Control of interneuron fate in the developing spinal cord by the progenitor homeodomain protein Dbx1. *Neuron* **29:** 367–384.

17. Rabe, N. *et al.* 2009. Netrin-1-dependent spinal interneuron subtypes are required for the formation of left–right alternating locomotor circuitry. *J. Neurosci.* **29:** 15642–15649.

18. Dai, X. *et al.* 2005. Localization of spinal neurons activated during locomotion using the c-fos immunohistochemical method. *J. Neurophysiol.* **93:** 3442–3452.

19. Dai, Y. *et al.* 2009. Electrophysiological and pharmacological properties of locomotor activity-related neurons in cfos-EGFP mice. *J. Neurophysiol.* **102:** 3365–3383.

20. Avraham, O. *et al.* 2009. Transcriptional control of axonal guidance and sorting in dorsal interneurons by the Lim-HD proteins Lhx9 and Lhx1. *Neural. Develop.* **4:** 21.

21. Helms, A.W. & J.E. Johnson. 1998. Progenitors of dorsal commissural interneurons are defined by MATH1 expression. *Development* **125:** 919–928.

22. Müller, T. *et al.* 2002. The homeodomain factor lbx1 distinguishes two major programs of neuronal differentiation in the dorsal spinal cord. *Neuron* **34:** 551–562.

23. Tripodi, M., A.E. Stepien & S. Arber. 2011. Motor antagonism exposed by spatial segregation and timing of neurogenesis. *Nature* **479:** 61–66.

24. Kennedy, T.E. *et al.* 1994. Netrins are diffusible chemotropic factors for commissural axons in the embryonic spinal cord. *Cell* **78:** 425–435.

25. Serafini, T. *et al.* 1996. Netrin-1 is required for commissural axon guidance in the developing vertebrate nervous system. *Cell* **87:** 1001–1014.

26. Wéber, I. *et al.* 2007. Neurotransmitter systems of commissural interneurons in the lumbar spinal cord of neonatal rats. *Brain Res.* **1178:** 65–72.

27. Gezelius, H. *et al.* 2006. Role of glutamate in locomotor rhythm generating neuronal circuitry. *J. Physiol.* **100:** 297–303.

28. Wallén-Mackenzie, A. *et al.* 2006. Vesicular glutamate transporter 2 is required for central respiratory rhythm generation but not for locomotor central pattern generation. *J. Neurosci.* **26:** 12294–12307.

29. Wallen-Mackenzie, A., H. Wootz & H. Englund. 2010. Genetic inactivation of the vesicular glutamate transporter 2 (VGLUT2) in the mouse: what have we learnt about functional glutamatergic neurotransmission? *Upsala J. Med. Sci.* **115:** 11–20.

30. Cohen, A.H. & R.M. Harris-Warrick. 1984. Strychnine eliminates alternating motor output during fictive locomotion in the lamprey. *Brain Res.* **293:** 164–167.

31. Cowley, K.C. & B.J. Schmidt. 1995. Effects of inhibitory amino acid antagonists on reciprocal inhibitory interactions during rhythmic motor activity in the in vitro neonatal rat spinal cord. *J. Neurophysiol.* **74:** 1109–1117.

32. Grillner, S. & P. Wallen. 1980. Does the central pattern generation for locomotion in lamprey depend on glycine inhibition? *Acta Physiol. Scand.* **110:** 103–105.

33. Jankowska, E. & B.R. Noga. 1990. Contralaterally projecting lamina VIII interneurones in middle lumbar segments in the cat. *Brain Res.* **535:** 327–330.

34. Soffe, S.R., J.D. Clarke & A. Roberts. 1984. Activity of commissural interneurons in spinal cord of Xenopus embryos. *J. Neurophysiol.* **51:** 1257–1267.

35. Harrison, P.J., E. Jankowska & D. Zytnicki. 1986. Lamina VIII interneurones interposed in crossed reflex pathways in the cat. *J. Physiol.* **371:** 147–166.

36. Jankowska, E. *et al.* 2005. Functional differentiation and organization of feline midlumbar commissural interneurones. *J. Physiol.* **565:** 645–658.

37. Rabe Bernhardt, N. *et al.* 2012. DCC mediated axon guidance of spinal interneurons is essential for normal locomotor central pattern generator function. *Develop. Biol.* **366:** 279–289.

38. Kullander, K. *et al.* 2003. Role of EphA4 and EphrinB3 in local neuronal circuits that control walking. *Science* **299:** 1889–1892.

39. Filosa, A. *et al.* 2009. Neuron-glia communication via EphA4/ephrin-A3 modulates LTP through glial glutamate transport. *Nature Neurosci.* **12:** 1285–1292.

40. Kullander, K. *et al.* 2001. Kinase-dependent and kinase-independent functions of EphA4 receptors in major axon tract formation in vivo. *Neuron* **29:** 73–84.

41. Keller, C. *et al.* 2004. Pax3:Fkhr interferes with embryonic Pax3 and Pax7 function: implications for alveolar rhabdomyosarcoma cell of origin. *Genes Develop.* **18:** 2608–2613.

42. Baudet, C. *et al.* 2008. Retrograde signaling onto Ret during motor nerve terminal maturation. *J. Neurosci.* **28:** 963–975.

43. Hippenmeyer, S. *et al.* 2005. A developmental switch in the response of DRG neurons to ETS transcription factor signaling. *PLoS Biol.* **3:** e159.

44. Briscoe, J. *et al.* 1999. Homeobox gene Nkx2.2 and specification of neuronal identity by graded Sonic hedgehog signalling. *Nature* **398:** 622–627.

45. Zhang, Y. *et al.* 2008. V3 spinal neurons establish a robust and balanced locomotor rhythm during walking. *Neuron* **60:** 84–96.

46. Crone, S.A. *et al.* 2008. Genetic ablation of V2a ipsilateral interneurons disrupts left–right locomotor coordination in mammalian spinal cord. *Neuron* **60:** 70–83.

47. Quinlan, K.A. & O. Kiehn. 2007. Segmental, synaptic actions of commissural interneurons in the mouse spinal cord. *J. Neurosci.* **27:** 6521–6530.

48. Burrill, J.D. *et al.* 1997. PAX2 is expressed in multiple spinal cord interneurons, including a population of EN1 +interneurons that require PAX6 for their development. *Development* **124:** 4493–4503.

49. Saueressig, H., J. Burrill & M. Goulding. 1999. Engrailed-1 and netrin-1 regulate axon pathfinding by association interneurons that project to motor neurons. *Development* **126:** 4201–4212.

50. Higashijima, S. *et al.* 2004. Engrailed-1 expression marks a primitive class of inhibitory spinal interneuron. *J. Neurosci.* **24:** 5827–5839.

51. Li, W.C. *et al.* 2004. Primitive roles for inhibitory interneurons in developing frog spinal cord. *J. Neurosci.* **24:** 5840–5848.

52. Alvarez, F.J. *et al.* 2005. Postnatal phenotype and localization of spinal cord V1 derived interneurons. *J. Comp. Neurol.* **493:** 177–192.

53. Sapir, T. *et al.* 2004. Pax6 and engrailed 1 regulate two distinct aspects of renshaw cell development. *J. Neurosci.* **24:** 1255–1264.

54. Wenner, P., M.J. O'Donovan & M.P. Matise. 2000. Topographical and physiological characterization of interneurons that express engrailed-1 in the embryonic chick spinal cord. *J. Neurophysiol.* **84:** 2651–2657.

55. Gosgnach, S. *et al.* 2006. V1 spinal neurons regulate the speed of vertebrate locomotor outputs. *Nature* **440:** 215–219.

56. Karunaratne, A. *et al.* 2002. GATA proteins identify a novel ventral interneuron subclass in the developing chick spinal cord. *Develop. Biol.* **249:** 30–43.

57. Smith, E. *et al.* 2002. Coexpression of SCL and GATA3 in the V2 interneurons of the developing mouse spinal cord. *Dev. Dyn.* **224:** 231–237.

58. Crone, S.A. *et al.* 2009. In mice lacking V2a interneurons, gait depends on speed of locomotion. *J. Neurosci.* **29:** 7098–7109.

59. Dyck, J., G.M. Lanuza & S. Gosgnach. 2012. Functional characterization of dI6 interneurons in the neonatal mouse spinal cord. *J. Neurophysiol.* **107:** 3256–3266.

60. Betley, J.N. *et al.* 2009. Stringent specificity in the construction of a GABAergic presynaptic inhibitory circuit. *Cell* **139:** 161–174.

61. Stepien, A.E., M. Tripodi & S. Arber. 2010. Monosynaptic rabies virus reveals premotor network organization and synaptic specificity of cholinergic partition cells. *Neuron* **68:** 456–472.

62. Bui, T.V. & R.M. Brownstone. 2010. dI3 neurons are rhythmically active during fictive locomotion. Presented at Society for Neuroscience, San Diego.

63. Enjin, A. *et al.* 2010. Identification of novel spinal cholinergic genetic subtypes disclose Chodl and Pitx2 as markers for fast motor neurons and partition cells. *J. Comp. Neurol.* **518:** 2284–2304.

64. Zagoraiou, L. *et al.* 2009. A cluster of cholinergic premotor interneurons modulates mouse locomotor activity. *Neuron* **64:** 645–662.

65. Hong, C.S., B.Y. Park & J.P. Saint-Jeannet. 2007. The function of Dmrt genes in vertebrate development: it is not just about sex. *Dev. Biol.* **310:** 1–9.

66. Kim, S. *et al.* 2003. Sexually dimorphic expression of multiple doublesex-related genes in the embryonic mouse gonad. *Gene Exp. Patterns* **3:** 77–82.

67. Andersson, L.S. *et al.* 2012. Mutations in DMRT3 affect locomotion in horses and spinal circuit function in mice. *Nature* **488:** 642–646.

68. Goulding, M. 2009. Circuits controlling vertebrate locomotion: moving in a new direction. *Nature Rev.* **10:** 507–518.

69. Ahituv, N. *et al.* 2007. Deletion of ultraconserved elements yields viable mice. *PLoS Biol.* **5:** e234.

70. Vallstedt, A. *et al.* 2001. Different levels of repressor activity

71. assign redundant and specific roles to Nkx6 genes in motor neuron and interneuron specification. *Neuron* **31:** 743–755.

71. Cowley, K.C. & B.J. Schmidt. 1995. Effects of inhibitory amino acid antagonists on reciprocal inhibitory interactions during rhythmic motor activity in the in vitro neonatal rat spinal cord. *J. Neurophysiol.* **74:** 1109–1117.

72. Jankowska, E. & B.R. Noga. 1990. Contralaterally projecting lamina VIII interneurones in middle lumbar segments in the cat. *Brain Res.* **535:** 327–330.

73. Grillner, S. & P. Wallen. 1980. Does the central pattern generation for locomotion in lamprey depend on glycine inhibition? *Acta Physiol. Scand.* **110:** 103–105.

74. Harris-Warrick, R.M. 2011. Neuromodulation and flexibility in Central Pattern Generator networks. *Curr. Opin. Neurobiol.* **21:** 685–692.

75. McLean, D.L. *et al.* 2007. A topographic map of recruitment in spinal cord. *Nature* **446:** 71–75.

76. Zhong, G., K. Sharma & R.M. Harris-Warrick. 2011. Frequency-dependent recruitment of V2a interneurons during fictive locomotion in the mouse spinal cord. *Nat. Commun.* **2:** 274.

77. Grillner, S. 1975. Locomotion in vertebrates: central mechanisms and reflex interaction. *Physiol. Rev.* **55:** 247–304.

78. Ericson, J. *et al.* 1996. Two critical periods of Sonic Hedgehog signaling required for the specification of motor neuron identity. *Cell* **87:** 661–673.

79. Jostes, B., C. Walther & P. Gruss. 1990. The murine paired box gene, Pax7, is expressed specifically during the development of the nervous and muscular system. *Mech. Dev.* **33:** 27–37.

Ann. N.Y. Acad. Sci. ISSN 0077-8923

ANNALS OF THE NEW YORK ACADEMY OF SCIENCES
Issue: *Neurons, Circuitry, and Plasticity in the Spinal Cord and Brainstem*

Neuronal correlates of the dominant role of GABAergic transmission in the developing mouse locomotor circuitry

Lea Ziskind-Conhaim

Department of Neuroscience, University of Wisconsin-Madison, School of Medicine and Public Health, Madison, Wisconsin

Address for correspondence: Lea Ziskind-Conhaim, Department of Neuroscience, University of Wisconsin, 129 SMI, 1300 University Ave., Madison, WI 53706. lziskind@wisc.edu

GABA and glycine are the primary fast inhibitory neurotransmitters in the mammalian spinal cord, but they differ in their regulatory functions, balancing neuronal excitation in the locomotor circuitry in the mammalian spinal cord. This review focuses on the unique role of GABAergic transmission during the assembly of the locomotor circuitry, from early embryonic stages when $GABA_A$ receptor–activated membrane depolarizations increase network excitation, to the period of early postnatal development, when GABAergic inhibition plays a primary role in coordinating the patterns of locomotor-like motor activity. To gain insight into the mechanisms that underlie the dominant contribution of GABAergic transmission to network activity during that period, we examined the morphological and electrophysiological properties of a subpopulation of GABAergic commissural interneurons that fit well with their putative function as integrated components of the rhythm-coordinating networks in the mouse spinal cord.

Keywords: GABA; commissural interneurons; inhibitory transmission; locomotion; spinal cord

Introduction

The locomotor central pattern generators (CPGs) in the spinal cord of all vertebrates are composed of ipsilateral rhythmogenic interneurons that control the timing of bouts of motor activity and interneuronal populations with complimentary functions that coordinate the patterns of motor outputs to execute a variety of locomotor movements.[1–5] Interactive, heterogeneous populations of excitatory and inhibitory commissural interneurons that project across the midline to innervate contralateral target neurons are responsible for adjusting limb positions during various phases of the step cycle.[6] Alternating left–right steps are executed through reciprocal inhibitory interactions between rhythm generators in the two sides of the spinal cord.[7–10] Multiple groups of GABAergic and glycinergic inhibitory commissural interneurons are involved in controlling motor outputs, but to date little is known about the morphological characteristics and intrinsic properties of identified populations of inhibitory commissural interneurons with known functions in the locomotor CPG.

GABAergic neurons are broadly distributed in all laminae of the mammalian spinal cord[11,12] and presumably modulate the activity of various neuronal populations with diverse sensory and motor functions. Synaptic interactions with numerous neuronal groups might explain the dominant role of GABAergic transmission in coordinating the patterns of locomotor activity during development. Recent studies have analyzed the morphological and electrophysiological properties of identified ipsilateral and commissural GABAergic interneurons that are putative components of the CPG in the mouse spinal cord.[13,14] This review focuses on the correlation between the robust function of GABAergic transmission in the locomotor circuitry and the properties of a subpopulation of GABAergic commissural interneurons in the developing mouse spinal cord.[14]

Double-edged GABAergic inhibition regulates neuronal excitation in sensory–motor networks

The spinal cord receives a continuous stream of sensory information from the body that is crucial for

doi: 10.1111/nyas.12064

adjusting muscle activity during movements. Proprioceptive and cutaneous signals are transmitted to motoneurons either directly or indirectly through premotor excitatory and inhibitory interneurons. GABA and glycine are the primary fast inhibitory neurotransmitters in the spinal cord, but they differ in their cellular function and network activity. One of the principal differences between the two amino acids is the distinct double-edged pre- and postsynaptic inhibitory function of GABAergic transmission in the sensory–motor circuitry. Postsynaptically, GABA$_A$ receptor–gated Cl$^-$ influx generates inhibitory postsynaptic currents (IPSCs) with a characteristic slow time course, resulting in a long-lasting inhibitory action compared to glycine-mediated inhibition. In addition, GABA$_A$-activated Cl$^-$ efflux in presynaptic terminals leads to primary afferent depolarization (PAD), which suppresses glutamate release from nerve terminals of proprioceptive afferent projections.

It should be noted that while antidromically propagating PAD blocks electrical activity generated in the periphery, PADs can reach firing threshold and trigger spontaneous activity in centrally projecting terminals in the neonatal rat spinal cord.[15] It is conceivable that electrical coupling between GABAergic interneurons in laminae IV–VI[16] synchronizes their activity and increases the strength of presynaptic inhibition. The anatomical identification and functional organization of interneurons responsible for presynaptic inhibition have been examined and discussed extensively[17–21] and will not be elaborated on in this short review.

The physiological significance of the double-edge GABAergic inhibition is the robust excitation in response to dorsal root stimulation in the presence of selective GABA$_A$ receptor antagonists. Blocking GABAergic transmission prolongs the duration and increases the spread of network excitation in all regions of the cord to a greater extent than blocking glycinergic inhibition (Fig. 1).[22] It is likely that the longer duration network excitation results from the slow decay time constant of GABA$_A$ receptor–mediated synaptic currents.[23,24] The spread of the robust activity beyond regions affected by glycinergic inhibition might be attributed to the suppression of GABAergic presynaptic inhibition in the intermediate area and increased glutamate release from group I muscle and tendon afferents that synapse onto numerous interneuronal populations in intermediate and ventral laminae.

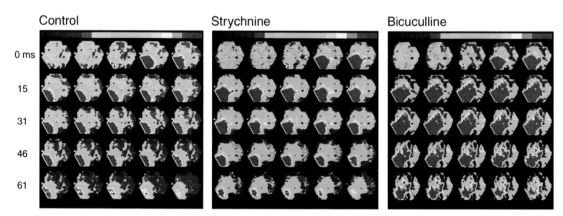

Figure 1. Real time maps of voltage-sensitive dye showing spatiotemporal patterns of dorsal root–evoked optical responses during the inhibition of glycinergic and GABAergic transmission in a transverse slice (400 μm) of a P3 rat spinal cord. The spatiotemporal dynamics of the spread of potential changes in sensory–motor circuits were monitored with a 464 multiple-photodiode device.[22] Inhibitory synaptic transmission was blocked by strychnine (0.5 μM) or bicuculline (5 μM), glycine and GABA$_A$ receptor antagonists, respectively. Dorsal root–evoked optical signals spread along the dorsoventral axis (dorsal is down). The amplitude of membrane depolarizations (red) increased in the presence of strychnine, but it lasted longer and spread to larger areas in the presence of bicuculline. Maps were taken with 10× objective that covers mostly the stimulated side of the cord and are shown at 3.2-ms intervals for a total period of 80 ms. The sequence starts from left to right in the upper row. Reproduced with permission from Ziskind-Conhaim and Redman.[22]

GABAergic transmission regulates the synchronous patterns of spontaneous and induced locomotor activity in the mouse spinal cord

In the developing CNS, GABA and glycine receptor activation triggers membrane depolarizations and neuronal excitation,[25,26] as the result of an early expression of the cotransporter Na–K–Cl isoform 1 (NKCC1), which raises cytosolic Cl^-, and a delayed onset of the cotransporter K–Cl isoform 2 (KCC2), which decreases Cl^- concentrations at later stages of embryonic/postnatal development.[27–30]

Spontaneous motor activity

During the developmental period of strong NKCC1 activity, Cl^- outflux through both GABA- and glycine-gated channels[31] contributes to spontaneous network excitation that is crucial for a wide range of cellular events, including axon outgrowth, maturation of receptors and ion channels, synapse formation, and the development of locomotor-like activity in the spinal cord.[32–40] Exposure of the isolated embryonic rat spinal cord to exogenous GABA and/or glycine triggers GABA$_A$ receptor–activated outward Cl^- currents and membrane depolarizations that are significantly larger than those generated by activation of glycine-gated channels.[41,42] The large GABAergic depolarizations in embryonic spinal cords are at least partially responsible for initiating spontaneous local bursts of motor activity and coordinating the synchronous rhythms between the left and right sides of the embryonic mouse spinal cord.[43] During that phase, cholinergic rather than glutamatergic transmission underlies the rhythmogenic activity, raising the possibility that reciprocal excitation through cholinergic and GABAergic synapses, such as those between motoneurons, Renshaw cells and reciprocal Ia neurons amplifies spontaneous activity in the local circuitry.[43,44] The relative contribution of GABA$_A$ and glycine transmission to spontaneous rhythmic activity in the developing rodent spinal cord remains controversial. On the basis of the effects of GABA$_A$ and glycine receptor antagonists on spontaneous motor activity, it has been suggested that glycinergic rather than GABAergic interneurons are part of the networks generating rhythmic activity in the embryonic rat spinal cord.[45] However, it is possible that the high concentrations of strychnine ($>6\,\mu M$) that was used in that study cross-interacted with GABA$_A$ receptors

and the effect was related to blocking both glycinergic and GABAergic transmission. Strychnine blocks glycine- and GABA-activated currents with IC_{50} of 39 nM and 1.5 μM, respectively.[23]

Neurochemically induced bouts of motor activity

Similar to spontaneous motor activity at early embryonic development, neurochemically induced motor outputs using neurotransmitter agonists such as 5-HT trigger left–right synchronous bouts of motor activity.[46] From the onset of rhythmogenesis, the induced synchronicity between the left and right sides of the spinal cord is mediated primarily via GABAergic commissural interneurons.[47] The role of GABAergic transmission in regulating the rhythmic activity persists when the pattern of alternating left–right motor outputs emerges at late embryonic development.

Neuronal correlates

To increase our understanding of the strong inhibitory function of GABA on the pattern of contralateral motor activity we have examined the morphological and electrophysiological properties of lamina VIII GFP-expressing commissural interneoruns.[14] In the mouse line used in our study (GAD67::GFP), the *GAD1* gene that encodes the enzyme GAD67 promotes the expression of GFP, and based on electrophysiological criteria, our conclusion is that lamina VIII GFP$^+$ neurons comprise a homogenous subpopulation of GABAergic interneurons (GAD67 INs). The number of GABAergic commissural interneurons peaks at around embryonic day 15 (E15),[48] when abundant projections of relatively undifferentiated GFP$^+$ commissural interneurons traverse the midline to terminate in the contralateral side of the spinal cord (Fig. 2A, see also Fig. 1 in Ref. 14). The dense distribution of GAD67 INs in the ventral spinal cord[49] and the extensive GABAergic innervation of embryonic motoneurons are probably responsible for the dominant role of GABA-activated depolarizations in generating spontaneous motor activity during the assembly of the locomotor circuitry. Integrated synaptic outputs from GAD67 INs are probably amplified by the extensive electrical coupling between them and mostly unidentified GFP$^-$ neurons that are assumed to be inhibitory in nature (Fig. 2B).

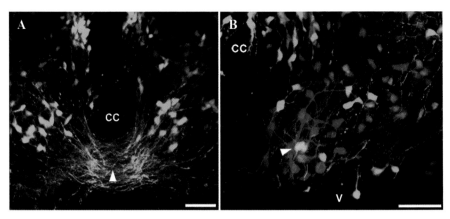

Figure 2. A large number of GABAergic commissural interneurons are distributed in the ventral spinal cord of embryonic spinal cords of GAD67::GFP mice. (A) Numerous lamina VIII GFP[+] interneurons with abundant projections that cross the midline (arrowhead) at E14. The number of these neurons decreases significantly after birth (see Fig. 1 in Ref. 14). (B) The spread of Neurobiotin from an injected GAD67 IN (arrowhead) to mostly unidentified GFP[−] neurons suggests that the GABAergic interneuron is electrically coupled to numerous unidentified neurons. The neuron filled with Neurobiotin (red) was GFP[+] (green), appearing yellow in the overlay. The images are superimpositions of 35–40 confocal optical sections obtained with 0.5-μm Z-steps. Central canal (CC). Scale bars: 50 μm. Modified with permission from Wu *et al.*[14]

The dominant role of GABAergic inhibition in generating left–right alternating bouts of motor activity

The transition from left–right synchronous to alternating bouts of motor activity in the neonatal spinal cord is temporally correlated with a decrease in the activity of the cotransporter NKCC1 and increased activity of the cotransporter KCC2[29,30], which lowers cytosolic Cl[−]. The shift in the activity balance of the cotransporters results in GABA/glycine-activated Cl[−] influx and membrane hyperpolarizations.[24] The change in cytosolic Cl[−] occurs gradually, leading to a period of coexisting depolarizing and hyperpolarizing responses.[42,50] The switch from excitatory to inhibitory commissural drive is associated with: (1) a significant decrease in the number of lamina VIII GABAergic commissural interneurons and a parallel increase in the density of GABAergic neurons in the dorsal horn,[14,49] (2) a substantial reduction in the number of electrically coupled GAD67 INs as demonstrated morphologically by fewer Neurobiotin-filled neurons in the neonatal mouse spinal cord (Fig. 2B; compare Figs. 3 and 10 in Ref. 14), and (3) an increase in the number of ventral glycinergic neurons.[51,52] These changes are correlated with a trend toward stronger glycinergic transmission, as indicated by the shift from primarily GABAergic to glycinergic miniature inhibitory postsynaptic cur-

rents (mIPSCs) in spinal motoneurons.[24] The ratio of GABAergic to glycinergic mIPSCs is 51% versus 24% at E17–E18 and 17% versus 62% at P1–P3, respectively.[24] However, throughout these developmental changes GABAergic interneurons maintain their dominant role in regulating left–right locomotor-like activity.[53–55]

Spontaneous motor activity

Irregular spontaneous motor outputs that occur randomly after birth are often not phase correlated between the two sides of the mouse spinal cord. Similar to spontaneous activity in the embryonic spinal cord, suppressing GABAergic transmission triggers long-duration synchronous bursts of motor activity in the left–right sides of the cord (Fig. 3B and Ref. 53). In contrast, exposure to strychnine at concentrations that do not cross interact with GABA$_A$ receptors (0.5 μM) increases the frequency of random motoneurons firing (Fig. 3A), but it does not initiate synchronous activity between the two sides of the spinal cord. The findings that blocking glycinergic transmission has little effect on spontaneous activity support a previous study showing that at low concentrations (1 μM), strychnine does not trigger coordinated motor activity in the neonatal rat spinal cord.[56]

Induced motor activity

In the neonatal mouse spinal cord, neurochemically induced left–right alternating locomotor-like

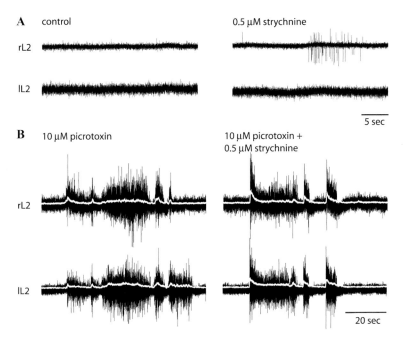

Figure 3. Strychnine and picrotoxin differ in their effects on spontaneous motor activity recorded in the spinal cords of postnatal mice. (A) Strychnine, a specific glycine receptor antagonist at the concentration used, did not trigger bouts of motor activity in the P2 mouse. Control: Typically, spontaneous activity was recorded infrequently, and it was not apparent in left–right L2 ventral root electroneurograms in this cord. Strychnine: Exposure to strychnine initiated only one episode of unilateral burst. (B) Exposure to picrotoxin, a specific GABA$_A$ receptor antagonist at low concentration, generated bilaterally synchronous bouts of motor activity of variable intervals and durations. The electroneurograms were rectified and smoothed (white traces) to show the slow rising initial peaks (1.0 seconds average rise time). Those were followed by multiple, noncorrelated peaks. Subsequent application of strychnine significantly decreased the rise time of the initial peak (average of 0.5 seconds), but it did not alter the pattern of coordinated synchronous activity. Reproduced with permission from Hinckley *et al.*[53]

motor outputs are transformed to robust synchronous bouts of motor activity in the presence of GABA$_A$ but not glycine receptor antagonists.[53] We have proposed that GABAergic neurons control the onset and duration of induced bouts of motor outputs while glycinergic inhibition stabilizes the left–right alternating pattern of motor outputs. Our findings differ from those reported in the developing rat spinal cord. In the brainstem–spinal cord preparation of newborn rats, blocking GABA$_A$ receptors via exposure to low concentrations of bicuculline (2–10 μM, at segments rostral to the lumbar enlargement), strengthened NMA-induced left–right alternating activity recorded in the lumbar spinal cord.[54] In contrast, high concentrations of strychnine (5 μM) synchronized bouts of motor activity induced by the rhythmogenic cocktail NMA/5-HT.[57] Similarly, before birth, 10 μM bicuculline failed to change the alternating phase between the two sides of most spinal cords, but 5 μM strychnine synchronized the left–right bouts

of motor activity.[47] Therefore, it has been suggested that in the rat spinal cord, glycinergic transmission provides the dominant inhibitory drive responsible for the left–right alternating motor outputs. As emphasized in the previous section, some of the ambiguity regarding the contribution of GABA and glycine to induced locomotor-like motor outputs might be rooted in the nonspecific effects of strychnine that at high concentrations blocked both glycine and GABA$_A$ receptors.[23,58–60] However, this will not explain the findings that bicuculline does not affect the alternating rhythms. It is conceivable that the roles of GABAergic and glycinergic transmission in rhythm-coordinating networks differ in the mouse and rat spinal cord. It is also possible that the rhythmogenic substances NMA and/or NMA/5-HT excite different groups of commissural inhibitory interneurons than those activated by a mixture of low concentration of NMA, 5-HT, and dopamine, the rhythmogenic cocktail routinely used in our studies.

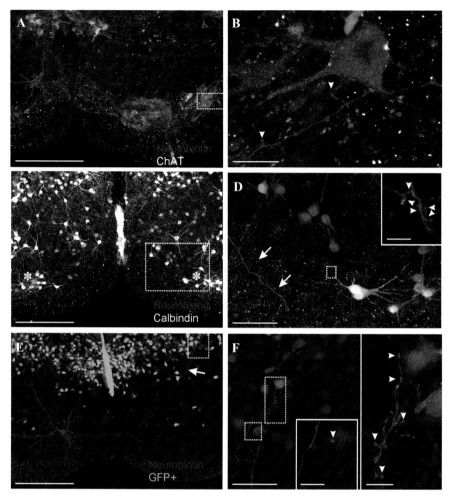

Figure 4. GAD67 INs synapse onto neurons distributed in ventral, intermediate, and dorsal laminae in the contralateral side of the postnatal spinal cord. (A) Confocal image of Neurobiotin-filled (red) GFP+ IN and choline acetyltransferase (ChAT)–expressing motoneurons (white, lamina IX) and cholinergic interneurons located dorsally around the central canal. The region marked by the dashed box is shown at higher magnification in B. (B) Neurobiotin-filled axon collateral (red) synapse onto ChAT-IR motoneuron dendrites (white, arrowheads). (C) Neurobiotin-filled axonal branches projecting toward ventral calbindin-IR (white) Renshaw cell pools that are marked with yellow asterisks. The boxed area is shown at higher magnification in D. (D) Several collateral branches (arrows) from the parent axon projected toward Renshaw cell pools and the cell dendrite. En passant boutons are apparent along their routes (*Inset*: higher magnification of the boxed: Neurobiotin-filled varicosities from the axon (arrowheads) seem to make contacts along the dendritic process of a calbindin+ cell. (E) Neurobiotin-filled GAD67 IN with axonal branch that projected dorsally through contralateral GAD67/GFP neurons (arrow). The region boxed is shown at higher magnification in F. (F) Synaptic boutons and varicosities were located in close apposition to several GFP+ neurons in this area. *Insets*: Higher magnification of the dashed box regions. Putative synaptic contacts (arrowheads) were observed along the somas and dendrites of contralateral GFP+ neurons. Scale bars: 200 μm in A, C, and E; 20 μm in B; 50 μm in D and F (10 μm in D, F insets). Reproduced with permission from Wu et al.[14]

Properties of GABAergic interneurons that are putative components of the locomotor circuitry

The finding that GABAergic transmission exerts a stronger control of the coordinated activity between the left and right sides of the cord is somewhat surprising because the number of glycinergic commissural interneurons is higher than the number of GABAergic commissural interneurons in the ventral spinal cord of newborn mice[61] and rats.[62] Inhibitory commissural interneurons can be

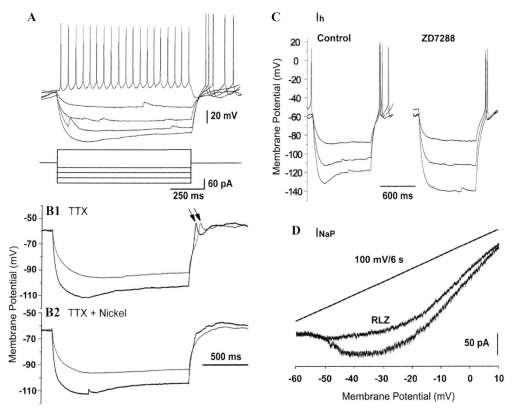

Figure 5. Electrophysiological properties of neonatal GAD67 commissural interneurons. (A) Voltage responses to prolonged positive and negative currents. Positive current injection (80 pA/s) evoked repetitive firing (19 Hz) without spike-frequency adaptation. Negative current injections produced hyperpolarization-activated depolarization sags and rebound action potentials. Resting membrane potential was approximately –55 mV. The input resistance was 604 MΩ. (B) In a different cell, subthreshold postinhibitory rebounds (PIRs) were generated in the presence of TTX (1 μM) at the break from hyperpolarizing currents (arrows). The potentials were blocked by nickel (100 μM). The input resistance was 702 MΩ. (C) Hyperpolarization-activated depolarization sags were generated in response to long hyperpolarization steps (1 second) at potentials more negative than –90 mV. ZD7288 (50 μM) blocked the depolarizations sags. (D) Persistent sodium current (I_{NaP}) was produced in response to a slow depolarization ramp (100 mV/6 seconds) after blocking potassium and calcium currents. The region of negative slope conductance was significantly reduced by riluzole (RLZ, 10 μM). Reproduced with permission from Wu *et al.*[14]

divided into two categories: neurons with short-range segmental projections and long-range intersegmental segmental projections.[63–71] To date, little is known about the morphological and electrophysiological properties of short-range homogenous groups of GABAergic and glycinergic neurons with distinct functions in the locomotor CPG. Identifying such populations is problematic because, except for motoneurons and Hb9-expressing ipsilateral interneurons,[72] most ventral neurons with presumed functions are not clustered in defined nuclei. The ability to express green fluorescent protein (GFP) in specific groups of GABAergic neu-

rons greatly facilitated the characterization of their properties and probable functions in locomotor networks. For example, the gene *GAD2* that encodes the GABA-synthesizing enzyme glutamate decarboxylase65 (GAD65) was used to promote the expression of GFP in ipsilateral interneurons in the dorsomedial spinal cords of the GAD65::GFP mouse line. On the basis of morphological observations, the GFP+ neurons make direct synaptic contacts onto ipsilateral motoneurons and their induced rhythmic firing is out of phase with bouts of motor activity.[13] These findings suggest that the neurons compose part of the circuitry that inhibits ipsilateral motoneurons during locomotor-like activity. To gain an insight into the cellular properties that underlie the strong inhibitory control of GABAergic commissural interneurons on the pattern of contralateral motor activity, we examined the morphological and electrophysiological properties of lamina VIII GAD67 INs.[14]

Morphological properties of GAD67 INs

Axonal projections of GAD67 INs traverse the midline to innervate target neurons in the contralateral side and their induced firing episodes are out of phase with bouts of contralateral motor activity and in phase with ipsilateral motor output. These properties fit the classification of locomotor-related interneurons that are part of commissural networks coordinating the left–right alternating motor activity. The multidirectional, complex dendritic arbors of Neurobiotin-filled GAD67 INs (Fig. 4) suggest that the interneurons receive diverse excitatory and inhibitory synaptic inputs as well as inputs from electrically coupled neurons (see Fig. 8 in Ref. 14). Their elaborated axonal branching in contralateral ventral, intermediate, and dorsal laminae form bouton-like varicosities in close apposition to contralateral motoneurons, Renshaw cells, and unidentified GFP$^+$ neurons in the intermediate and dorsal spinal cord (Fig. 4), suggesting that GAD67 INs synapse onto diverse neuronal populations and modulate the excitability of neuronal groups with different functions in the cord. Blocking GABAergic transmission disinhibits various neuronal groups and can result in a robust excitation and bouts of motor activity that are synchronous between the two sides of the cord (Fig. 3, and Fig. 7 in Ref. 53).

Electrophysiological properties of GAD67 INs

Lamina VIII GAD67 INs share common intrinsic properties that match well their function as inhibitory commissural interneurons in locomotor networks. One of their hallmarks is the spontaneous firing at a relatively constant frequency ≥ 3 Hz at resting membrane potentials (approximately –60 mV). GAD67 INs are highly excitable and are capable of firing at frequencies >30 Hz during intracellular injection of 100–150 pA depolarizing currents.[14] It is conceivable that their persistent spontaneous firing results in tonic release of GABA that suppresses excitatory commissural interneurons responsible for the generation of spontaneous left–right synchronous motor activity when GABAergic transmission is blocked (Fig. 3). Rhythmic increase in excitatory postsynaptic currents at the onset of induced locomotor-like activity initiates voltage oscillations and firing episodes in GAD67 INs. Voltage-gated currents can further amplify locomotor-like membrane oscillations in GAD67 INs. Three types of currents that are part of the ionic repertoire of other rhythmically active commissural neurons[63,73] are expressed in GAD67 INs: persistent sodium current, hyperpolarizing activated inward current, and the T type calcium current (Fig. 5 and Ref. 14). Taken together, the morphological characteristics and intrinsic properties of GAD67 INs are consistent with their probable function in exerting strong inhibitory control of neuronal excitation in rhythm-coordinating circuits in the developing mouse spinal cord.

Conclusion

GABAergic interneurons play an important role in coordinating spontaneous and neurochemically induced locomotor-like activity in the immature mouse spinal cord. To the best of our knowledge the study of GAD67 INs is the first one to provide a detailed characterization of the morphological and electrophysiological properties of an identified subpopulation of GABAergic commissural interneurons that likely regulate the excitability of various neuronal populations and thereby play a key inhibitory function in the locomotor circuitry. Future studies of other groups of glycinergic and GABAergic commissural interneurons will provide much-needed information about the mechanisms that underlie their distinct contributions to the coordinated bilateral motor activity in the mammalian spinal cord.

Conflicts of interest

The author declares no conflicts of interest.

References

1. Grillner, S. & P. Wallen. 1985. Central pattern generators for locomotion, with special reference to vertebrates. *Annu. Rev. Neurosci.* **8:** 233–261.
2. Kiehn, O. & S.J. Butt. 2003. Physiological, anatomical and genetic identification of CPG neurons in the developing mammalian spinal cord. *Prog. Neurobiol.* **70:** 347–361.

3. Grillner, S. 2006. Biological pattern generation: the cellular and computational logic of networks in motion. *Neuron* **52:** 751–766.

4. Endo, T. & O. Kiehn. 2008. Asymmetric operation of the locomotor central pattern generator in the neonatal mouse spinal cord. *J. Neurophysiol.* **100:** 3043–3054.

5. McCrea, D.A. & I.A. Rybak. 2008. Organization of mammalian locomotor rhythm and pattern generation. *Brain Res. Rev.* **57:** 134–146.

6. Jankowska, E. 2008. Spinal interneuronal networks in the cat: elementary components. *Brain Res. Rev.* **57:** 46–55.

7. Jankowska, E., M.G. Jukes, S. Lund & A. Lundberg. 1967. The effect of DOPA on the spinal cord. 6. Half-centre organization of interneurones transmitting effects from the flexor reflex afferents. *Acta Physiol. Scand.* **70:** 389–402.

8. Cowley, K.C. & B.J. Schmidt. 1995. Effects of inhibitory amino acid antagonists on reciprocal inhibitory interactions during rhythmic motor activity in the in vitro neonatal rat spinal cord. *J. Neurophysiol.* **74:** 1109–1117.

9. Bonnot, A., P.J. Whelan, G.Z. Mentis & M.J. O'Donovan. 2002. Locomotor-like activity generated by the neonatal mouse spinal cord. *Brain Res. Brain Res. Rev.* **40:** 141–151.

10. Butt, S.J. & O. Kiehn. 2003. Functional identification of interneurons responsible for left-right coordination of hindlimbs in mammals. *Neuron* **38:** 953–963.

11. Barber, R.P., J.E. Vaughn & E. Roberts. 1982. The cytoarchitecture of GABAergic neurons in rat spinal cord. *Brain Res.* **238:** 305–328.

12. Mackie, M., D.I. Hughes, D.J. Maxwell, *et al.* 2003. Distribution and colocalisation of glutamate decarboxylase isoforms in the rat spinal cord. *Neuroscience* **119:** 461–472.

13. Wilson, J.M., E. Blagovechtchenski & R.M. Brownstone. 2010. Genetically defined inhibitory neurons in the mouse spinal cord dorsal horn: a possible source of rhythmic inhibition of motoneurons during fictive locomotion. *J. Neurosci.* **30:** 1137–1148.

14. Wu, L. *et al.* 2011. Properties of a distinct subpopulation of GABAergic commissural interneurons that are part of the locomotor circuitry in the neonatal spinal cord. *J. Neurosci.* **31:** 4821–4833.

15. Bos, R., F. Brocard & L. Vinay. 2011. Primary afferent terminals acting as excitatory interneurons contribute to spontaneous motor activities in the immature spinal cord. *J. Neurosci.* **31:** 10184–10188.

16. Bautista, W., J.I. Nagy, Y. Dai & D.A. McCrea. 2012. Requirement of neuronal connexin36 in pathways mediating presynaptic inhibition of primary afferents in functionally mature mouse spinal cord. *J. Physiol.* **590:** 3821–3839.

17. Jankowska, E., D. McCrea, P. Rudomin & E. Sykova. 1981. Observations on neuronal pathways subserving primary afferent depolarization. *J. Neurophysiol.* **46:** 506–516.

18. Rudomin, P. & R.F. Schmidt. 1999. Presynaptic inhibition in the vertebrate spinal cord revisited. *Exp. Brain Res.* **129:** 1–37.

19. Rudomin, P. 2009. In search of lost presynaptic inhibition. *Exp. Brain Res.* **196:** 139–151.

20. Shreckengost, J., J. Calvo, J. Quevedo & S. Hochman. 2010. Bicuculline-sensitive primary afferent depolarization re-

mains after greatly restricting synaptic transmission in the mammalian spinal cord. *J. Neurosci.* **30:** 5283–5288.

21. Hochman, S., J. Shreckengost, H. Kimura & J. Quevedo. 2010. Presynaptic inhibition of primary afferents by depolarization: observations supporting nontraditional mechanisms. *Ann. N.Y. Acad. Sci.* **1198:** 140–152.

22. Ziskind-Conhaim, L. & S. Redman. 2005. Spatiotemporal patterns of dorsal root-evoked network activity in the neonatal rat spinal cord: optical and intracellular recordings. *J. Neurophysiol.* **94:** 1952–1961.

23. Jonas, P., J. Bischofberger & J. Sandkuhler. 1998. Corelease of two fast neurotransmitters at a central synapse. *Science* **281:** 419–424.

24. Gao, B.X., C. Stricker & L. Ziskind-Conhaim. 2001. Transition from GABAergic to glycinergic synaptic transmission in newly formed spinal networks. *J. Neurophysiol.* **86:** 492–502.

25. Sibilla, S. & L. Ballerini. 2009. GABAergic and glycinergic interneuron expression during spinal cord development: dynamic interplay between inhibition and excitation in the control of ventral network outputs. *Prog. Neurobiol.* **89:** 46–60.

26. Ben-Ari, Y. *et al.* 2012. Refuting the challenges of the developmental shift of polarity of GABA actions: GABA more exciting than ever! *Front Cell Neurosci.* **6:** 35.

27. Ben-Ari, Y., J.L. Gaiarsa, R. Tyzio & R. Khazipov. 2007. GABA: a pioneer transmitter that excites immature neurons and generates primitive oscillations. *Physiol. Rev.* **87:** 1215–1284.

28. Fiumelli, H. & M.A. Woodin. 2007. Role of activity-dependent regulation of neuronal chloride homeostasis in development. *Curr. Opin. Neurobiol.* **17:** 81–86.

29. Delpy, A., A.E. Allain, P. Meyrand & P. Branchereau. 2008. NKCC1 cotransporter inactivation underlies embryonic development of chloride-mediated inhibition in mouse spinal motoneuron. *J. Physiol.* **586:** 1059–1075.

30. Stil, A. *et al.* 2011. Contribution of the potassium-chloride co-transporter KCC2 to the modulation of lumbar spinal networks in mice. *Eur. J. Neurosci.* **33:** 1212–1222.

31. Reichling, D.B., A. Kyrozis, J. Wang & A.B. MacDermott. 1994. Mechanisms of GABA and glycine depolarization-induced calcium transients in rat dorsal horn neurons. *J. Physiol.* **476:** 411–421.

32. Ziskind-Conhaim, L. 1998. Physiological functions of GABA-induced depolarizations in the developing rat spinal cord. *Perspect. Dev. Neurobiol.* **5:** 279–287.

33. Moody, W.J. & M.M. Bosma. 2005. Ion channel development, spontaneous activity, and activity-dependent development in nerve and muscle cells. *Physiol. Rev.* **85:** 883–941.

34. Hanson, M.G. & L.T. Landmesser. 2004. Normal patterns of spontaneous activity are required for correct motor axon guidance and the expression of specific guidance molecules. *Neuron* **43:** 687–701.

35. Myers, C.P. *et al.* 2005. Cholinergic input is required during embryonic development to mediate proper assembly of spinal locomotor circuits. *Neuron* **46:** 37–49.

36. Akerman, C.J. & H.T. Cline. 2006. Depolarizing GABAergic conductances regulate the balance of excitation to inhibition in the developing retinotectal circuit in vivo. *J. Neurosci.* **26:** 5117–5130.

37. Gonzalez-Islas, C. & P. Wenner. 2006. Spontaneous network activity in the embryonic spinal cord regulates AMPAergic and GABAergic synaptic strength. *Neuron* **49:** 563–575.

38. Wang, D.D., A.R. Kriegstein & Y. Ben-Ari. 2008. GABA regulates stem cell proliferation before nervous system formation. *Epilepsy Curr.* **8:** 137–139.

39. Ben-Ari, Y. & N.C. Spitzer. 2010. Phenotypic checkpoints regulate neuronal development. *Trends Neurosci.* **33:** 485–492.

40. Blankenship, A.G. & M.B. Feller. 2010. Mechanisms underlying spontaneous patterned activity in developing neural circuits. *Nat. Rev. Neurosci.* **11:** 18–29.

41. Wu, W.L., L. Ziskind-Conhaim & M.A. Sweet. 1992. Early development of glycine- and GABA-mediated synapses in rat spinal cord. *J. Neurosci.* **12:** 3935–3945.

42. Gao, B.X. & L. Ziskind-Conhaim. 1995. Development of glycine- and GABA-gated currents in rat spinal motoneurons. *J. Neurophysiol.* **74:** 113–121.

43. Hanson, M.G. & L.T. Landmesser. 2003. Characterization of the circuits that generate spontaneous episodes of activity in the early embryonic mouse spinal cord. *J. Neurosci.* **23:** 587–600.

44. Alvarez, F.J. & R.E. Fyffe. 2007. The continuing case for the Renshaw cell. *J. Physiol.* **584:** 31–45.

45. Ren, J. & J.J. Greer. 2003. Ontogeny of rhythmic motor patterns generated in the embryonic rat spinal cord. *J. Neurophysiol.* **89:** 1187–1195.

46. Ozaki, S., T. Yamada, M. Iizuka, *et al.* 1996. Development of locomotor activity induced by NMDA receptor activation in the lumbar spinal cord of the rat fetus studied in vitro. *Brain Res. Dev. Brain Res.* **97:** 118–125.

47. Nakayama, K., H. Nishimaru & N. Kudo. 2002. Basis of changes in left-right coordination of rhythmic motor activity during development in the rat spinal cord. *J. Neurosci.* **22:** 10388–10398.

48. Allain, A.E., A. Bairi, P. Meyrand & P. Branchereau. 2004. Ontogenic changes of the GABAergic system in the embryonic mouse spinal cord. *Brain Res.* **1000:** 134–147.

49. Phelps, P.E., A. Alijani & T.S. Tran. 1999. Ventrally located commissural neurons express the GABAergic phenotype in developing rat spinal cord. *J. Comp. Neurol.* **409:** 285–298.

50. Ballerini, L. & M. Galante. 1998. Network bursting by organotypic spinal slice cultures in the presence of bicuculline and/or strychnine is developmentally regulated. *Eur. J. Neurosci.* **10:** 2871–2879.

51. Allain, A.E., A. Bairi, P. Meyrand & P. Branchereau. 2006. Expression of the glycinergic system during the course of embryonic development in the mouse spinal cord and its co-localization with GABA immunoreactivity. *J. Comp. Neurol.* **496:** 832–846.

52. Allain, A.E., L. Segu, P. Meyrand & P. Branchereau. 2010. Serotonin controls the maturation of the GABA phenotype in the ventral spinal cord via 5-HT1b receptors. *Ann. N.Y. Acad. Sci.* **1198:** 208–219.

53. Hinckley, C., B. Seebach & L. Ziskind-Conhaim. 2005. Distinct roles of glycinergic and GABAergic inhibition in co-ordinating locomotor-like rhythms in the neonatal mouse spinal cord. *Neuroscience* **131:** 745–758.

54. Cazalets, J.R., Y. Sqalli-Houssaini & F. Clarac. 1994. GABAergic inactivation of the central pattern generators for locomotion in isolated neonatal rat spinal cord. *J. Physiol.* **474:** 173–181.

55. Bertrand, S. & J.R. Cazalets. 1999. Presynaptic GABAergic control of the locomotor drive in the isolated spinal cord of neonatal rats. *Eur. J. Neurosci.* **11:** 583–592.

56. Bracci, E., L. Ballerini & A. Nistri. 1996. Spontaneous rhythmic bursts induced by pharmacological block of inhibition in lumbar motoneurons of the neonatal rat spinal cord. *J. Neurophysiol.* **75:** 640–647.

57. Cazalets, J.R., M. Borde & F. Clarac. 1996. The synaptic drive from the spinal locomotor network to motoneurons in the newborn rat. *J. Neurosci.* **16:** 298–306.

58. Yoon, K.W., V.E. Wotring & T. Fuse. 1998. Multiple picrotoxinin effect on glycine channels in rat hippocampal neurons. *Neuroscience* **87:** 807–815.

59. Tapia, J.C. & L.G. Aguayo. 1998. Changes in the properties of developing glycine receptors in cultured mouse spinal neurons. *Synapse* **28:** 185–194.

60. Chattipakorn, S.C. & L.L. McMahon. 2002. Pharmacological characterization of glycine-gated chloride currents recorded in rat hippocampal slices. *J. Neurophysiol.* **87:** 1515–1525.

61. Restrepo, C.E. *et al.* 2009. Transmitter-phenotypes of commissural interneurons in the lumbar spinal cord of newborn mice. *J. Comp. Neurol.* **517:** 177–192.

62. Weber, I., G. Veress, P. Szucs, *et al.* 2007. Neurotransmitter systems of commissural interneurons in the lumbar spinal cord of neonatal rats. *Brain Res.* **1178:** 65–72.

63. Butt, S.J., R.M. Harris-Warrick & O. Kiehn. 2002. Firing properties of identified interneuron populations in the mammalian hindlimb central pattern generator. *J. Neurosci.* **22:** 9961–9971.

64. Stokke, M.F., U.V. Nissen, J.C. Glover & O. Kiehn. 2002. Projection patterns of commissural interneurons in the lumbar spinal cord of the neonatal rat. *J. Comp. Neurol.* **446:** 349–359.

65. Birinyi, A., K. Viszokay, I. Weber, *et al.* 2003. Synaptic targets of commissural interneurons in the lumbar spinal cord of neonatal rats. *J. Comp. Neurol.* **461:** 429–440.

66. Jankowska, E., I. Hammar, U. Slawinska, *et al.* 2003. Neuronal basis of crossed actions from the reticular formation on feline hindlimb motoneurons. *J. Neurosci.* **23:** 1867–1878.

67. Jankowska, E., S.A. Edgley, P. Krutki & I. Hammar. 2005. Functional differentiation and organization of feline midlumbar commissural interneurones. *J. Physiol.* **565:** 645–658.

68. Hammar, I., B.A. Bannatyne, D.J. Maxwell, *et al.* 2004. The actions of monoamines and distribution of noradrenergic and serotoninergic contacts on different subpopulations of commissural interneurons in the cat spinal cord. *Eur. J. Neurosci.* **19:** 1305–1316.

69. Matsuyama, K., S. Kobayashi & M. Aoki. 2006. Projection patterns of lamina VIII commissural neurons in the lumbar spinal cord of the adult cat: an anterograde neural tracing study. *Neuroscience* **140:** 203–218.

70. Carlin, K.P., Y. Dai & L.M. Jordan. 2006. Cholinergic and serotonergic excitation of ascending commissural neurons

in the thoraco-lumbar spinal cord of the neonatal mouse. *J. Neurophysiol.* **95:** 1278–1284.

71. Quinlan, K.A. & O. Kiehn. 2007. Segmental, synaptic actions of commissural interneurons in the mouse spinal cord. *J. Neurosci.* **27:** 6521–6530.

72. Hinckley, C.A., R. Hartley, L. Wu, *et al.* 2005. Locomotor-like rhythms in a genetically distinct cluster of interneu-rons in the mammalian spinal cord. *J. Neurophysiol.* **93:** 1439–1449.

73. Zhong, G., M. Diaz-Rios & R.M. Harris-Warrick. 2006. Intrinsic and functional differences among commissural interneurons during fictive locomotion and serotoner-gic modulation in the neonatal mouse. *J. Neurosci.* **26:** 6509–6517.

Ann. N.Y. Acad. Sci. ISSN 0077-8923

ANNALS OF THE NEW YORK ACADEMY OF SCIENCES
Issue: *Neurons, Circuitry, and Plasticity in the Spinal Cord and Brainstem*

GluA1 promotes the activity-dependent development of motor circuitry in the developing segmental spinal cord

Angela M. Jablonski[1] and Robert G. Kalb[2,3]

[1]Neuroscience Graduate Group, Department of Neuroscience, Perelman School of Medicine, University of Pennsylvania, Philadelphia, Pennsylvania. [2]Department of Pediatrics, Division of Neurology, Research Institute, Children's Hospital of Philadelphia, Philadelphia, Pennsylvania. [3]Department of Neurology, Perelman School of Medicine, University of Pennsylvania, Philadelphia, Pennsylvania

Address for correspondence: Dr. Robert G. Kalb, Children's Hospital of Philadelphia, Abramson Research Center 814, 3615 Civic Center Blvd., Philadelphia, PA 19104. kalb@email.chop.edu

The neuronal dendritic tree is a key determinant of how neurons receive, compute, and transmit information. During early postnatal life, synaptic activity promotes dendrite elaboration. Spinal motor neurons utilize GluA1-containing AMPA (2-amino-3-(3-hydroxy-5-methyl-isoxazol-4-yl) propanoic acid) receptors (AMPA-R) to control this process. This form of developmental dendrite growth can occur independently of N-methyl-D-aspartate receptors (NMDA-R). This review focuses on the mechanism by which the GluA1 subunit of AMPA-R transforms synaptic activity into dendrite growth, and describes the essential role of the GluA1 binding partner SAP97 (synapse-associated protein of 97 kDa molecular weight) in this process. This work defines a new mechanism of activity-dependent development, which might be harnessed to stimulate the recovery of function following insult to the central nervous system.

Keywords: dendrite; activity dependent; AMPA; GluA1; growth; development

Introduction

Over the past several decades, it has become increasingly clear that nervous system development can be roughly divided into two periods: (1) a genetically driven, activity-independent phase that sets a rough wiring diagram, and (2) an activity-dependent phase that refines connectivity under the supervision of active synapses. During embryonic and very early postnatal life, spontaneous synaptic activity drives circuit maturation. Subsequently, environmentally evoked synaptic activity (experience-dependent development) plays the prominent role in shaping nervous system maturation. Together, these processes lead to the precisely patterned connectivity among neurons that underlies purposeful behavior.

Activity-dependent development

Activity-dependent development during pre- and postnatal life is an important mechanism for the specification of synaptic phenotype and connectivity. Here, we will review some of the key experiments undertaken in the visual system that highlight the cell biological processes underlying activity-dependent development. These introductory remarks provide the context for thinking about the role of activity-dependent processes in motor system development.

In the mammalian visual system, retinal ganglion cells (RGC) project to the lateral geniculate nucleus (LGN) of the thalamus. Thalamocortical (TC) connections from the LGN project to the visual cortex. In mature animals, right and left eye afferents are segregated from each other in both the LGN and visual cortex.[1,2] For example, in adult cats and primates, TC afferents projecting to layer four of the visual cortex are organized into discrete right and left eye patches, referred to as ocular dominance columns (OCD).[3] This circuitry arrangement subserves high acuity vision.[4]

In their classic work, Hubel and Weisel showed that this pattern of visual cortex innervation is not present at birth, as right and left eye afferents demonstrate extensive overlap.[5] Most remarkably, they show that the segregation of right and left eye

doi: 10.1111/nyas.12053

afferents is driven by visual experience.[5] Monocular deprivation of one eye (by suturing the eyelid shut) during a discrete period in early postnatal life leads to a dramatic shift in TC innervation of the visual cortex.[5] The afferents from the nondeprived eye innervate a larger territory of the visual cortex, while those from the deprived eye innervate a smaller territory.[5] This experience-dependent shift in ocular dominance leaves a permanent imprint on visual system organization.[5]

How does environmentally evoked synaptic activity lead to changes in synaptic strength and connectivity? Substantial evidence supports the view that synapses will undergo strengthening and stabilization when the activity of pre- and postsynaptic elements is coincident. This model of synaptic plasticity was originally suggested in theoretical work by Daniel Hebb and has been most rigorously tested in long-term potentiation (LTP) paradigms.[6] Many forms of LTP depend upon the activation of N-methyl-D-aspartate receptors (NMDA-Rs). NMDA-Rs are believed to be the coincidence detectors responsible for detecting the simultaneous activity of pre- and postsynaptic elements.[7] The ionic mechanism underlying this process has been linked to the voltage-dependent block of NMDA-Rs by Mg^{2+}.[7] Patterned afferent input, sufficient to remove the Mg^{2+} block, allows NMDA-Rs to conduct Ca^{2+} influx, which inactivates several protein kinases, including Ca^{2+}/calmodulin-dependent protein kinase II (CamKII).[7] CamKII is necessary for the maintenance of LTP and phosphorylation of 2-amino-3-(3-hydroxy-5-methyl-isoxazol-4-yl) propanoic acid receptor (AMPA-R) subunits to increase their conductance.[8] These and other calcium-activated processes drive the observed synaptic plasticity.[8,9]

In the visual system, activity-dependent processes drive large-scale alterations in the architecture of axons and dendrites. How do activity-dependent changes in synapses control the growth and distribution of axons and dendrites? Vaughn proposed the synaptotropic hypothesis: dendritic branches are formed near active synapses and synapse stabilization consequently stabilizes dendrites.[10] Beautiful *in vivo* work from the Haas lab implicates beta-neurexin (NRX) and neuroligin-1 (NLG1) in this process, in which NMDA-R–dependent synapse maturation was required for persistent NRX-NLG1 function in dendritogenesis.[11] Blocking synaptogenesis thereby blocks dendrite outgrowth stabiliza-

tion. The linkage of synaptic plasticity to neurite architecture is a fundamental principle in developmental neuroscience and provides an explanation for earlier observations that the size and complexity of the dendrite tree controls the qualitative and quantitative nature of the afferent input.[12] In studies of rabbit ciliary ganglia, the number of ganglion cell primary dendrites is highly correlated with the number of innervating axons.[13] This is not true of some of the cells in the neonate, where the initial set of inputs is confined to the cell body, allowing only one axon to survive.[13] It is hypothesized that the complexity of some cells allows for a higher number of afferents to innervate the ganglion cell.[13] Thus, a competition-based model of synapse formation holds for dendrite growth.

While many studies implicate AMPA-Rs in the control of dendrite growth, a consistent picture has yet to emerge. Blocking AMPAergic transmission in retinotectal neurons decreases synapse stabilization, and subsequent dendrite growth and stabilzation.[14] Conversely, Casticas *et al.* showed that enhanced conductance of Ca^{2+}-permeable AMPA-Rs inhibited neurite outgrowth in dissociated chick retinal neurons.[15] Outside of the visual system, blockade of AMPA-Rs in chick motoneurons has also been seen to increase dendritic outgrowth in chick motoneurons, but only at certain time points in embryonic development.[16] It is unclear what role NMDA-Rs played in these processes because blocking AMPA-Rs will prevent NMDA-R activation.

In the central nervous system (CNS), it is understood that activation of AMPA-Rs, sufficient to relieve the voltage-dependent magnesium block of NMDA-Rs, drives activity-dependent plasticity, synaptic stabilization, and patterned innervation.[17] Less understood is the extent to which NMDA-R–independent mechanisms can drive activity-dependent developmental processes. Below we describe work in the spinal cord showing how AMPA-Rs assembled with the GluA1 subunit can promote activity-dependent development by an NMDA-R–independent process.

GluA1 promotes activity-dependent dendrite growth

Neonatal motor neurons express very high levels of GluA1 (both mRNA and protein). The properties of GluA1 can be modified by alternative splicing and editing at the glutamine/arginine

Q/R site. The GluA1 expressed during this developmental period contains the flip alternatively spliced exon and is unedited in the Q/R site.[18,19] Previous work has shown that neonatal motor neurons express Ca^{2+}-permeable AMPA-Rs (as one would expect if they were enriched with GluA1(Q)).[19] Taken together with the electrophysiological data, the work suggests that many AMPA-Rs are assembled with GluA1 homomers at this point in the development of motor neurons.

The unusually high level of GluA1 expression by neonatal motor neurons raises the possibility that AMPA-Rs assembled with GluA1 play a special role in activity-dependent motor system development. To examine this notion, we began by asking whether manipulation of GluA1 influenced spinal neuron dendritic architecture. Several approaches were taken. First, we found that knockdown of GluA1 expression inhibits dendrite growth. Conversely, overexpression of GluA1 in spinal neurons *in vitro* stimulates dendritic growth; this growth effect was blocked by the AMPA-R antagonist, CNQX (6-cyano-7-nitroquinoxaline-2, 3-dione).[20] Second, we compared the effects of two types of overexpressed GluA1 in motor neurons *in vivo*. We used a version that robustly passes current (GluA1(Q)) and compared that with a version that passes very little current (GluA1(R)). Only overexpressed GluA1(Q) stimulated dendritic branching.[21] These results suggest that the activity of AMPA-Rs assembled with GluA1 is a crucial step for dendrite growth and this effect is Ca^{2+} dependent. Subsequent *in vitro* work indicates that the degree of calcium permeability of AMPA-Rs assembled with GluA1 controls the dendritic growth process.[21]

One interpretation of the above results is that overexpression of GluA1 enhances neuronal depolarization, thereby promoting NMDA-R–mediated events. We think this is not true for a number of reasons. First, our *in vivo* observations were made in juvenile rodents at a time when motor neurons do not express NMDA-R.[22] It is possible that prior *in situ* hybridization and immunohistological studies were insufficiently sensitive to detect NMDA-Rs in juvenile motor neurons. To address this possibility, we expressed GluA1(Q) in juvenile animals and simultaneously treated them with the NMDA-R antagonist, MK-801.[23] We know that MK-801 was administered in an effective dose because LTP could not be evoked in these animals.[23] Nonetheless, MK-

801 did not block the pro-dendrite growth actions of overexpressed GluA1(Q).[23] Second, we undertook *in vitro* pharmacological studies. Administration of MK-801 did not block the dendrite growth–promoting actions of GluA1(Q); in contrast, administration of the L-type calcium channel blocker nifedipine did block the GluA1(Q) effect.[24] Taken together, these results suggest that GluA1 is sufficient to promote dendrite growth in an NMDA-R–independent manner.

The work described previously primarily focuses on the effects of GluA1 on dendrite architecture *in vitro*. What about *in vivo*? To address this question, studies of the GluA1$^{-/-}$ mouse have been informative.[20] Analysis of the dendritic tree reveals that motor neurons from GluA1$^{-/-}$ animals are smaller and less branched at P10 and P23 (see Fig. 1).[20] This suggests that GluA1$^{-/-}$ motor neurons develop over a different trajectory than wild-type motor neurons. How does this decrease in the size and complexity of the motor neuron dendrite tree affect motor circuitry and behavior of the animal? To study the innervation of motor neurons within the segmental spinal cord, a recombinant pseudorabies virus engineered to express green fluorescence protein (PRV-GFP) was used. PRV-GFP labeling experiments revealed a distinct pattern of interneuronal connectivity in the spinal cord of GluA1$^{-/-}$ mice in comparison to wild-type mice. The greatest difference between genotypes was found in the number of contralaterally located interneurons, especially in Rexed's lamina VIII.[20] The stunted dendrite tree and change in interneuron connectivity correlated with a locomotor defect in the GluA1$^{-/-}$ animals. In comparison to wild-type counterparts, GluA1$^{-/-}$ mice showed poorer performance in grip strength, treadmill, and rotarod at P23 and adulthood.[20] This suggests that GluA1 is important not only for dendrite growth, but also for patterning segmental spinal cord circuitry and motor behavior. Furthermore, changes in development during the postnatal period lead to deficits throughout adulthood.[20]

By what molecular mechanism does the activity of AMPA-Rs assembled with GluA1 control the morphology of motor neuron dendritic architecture? A series of experiments have indicated that the multidomain scaffolding protein, synapse-associated protein of 97 kDa molecular weight (SAP97), interacts with GluA1 and plays a key role in this process.[25] The C-terminal seven amino acids of GluA1

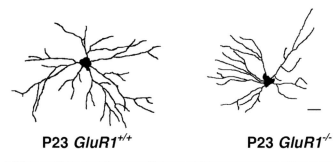

Cell Parameter	$GluR1^{+/+}$ (n=26)	$GluR1^{-/-}$ (n=26)	p
Soma area (μm^2)	1497 ± 120	1495 ± 106	0.83
1^0 dendrites	5.6 ± 0.2	5.3 ± 0.2	0.24
Branches, #	25 ± 1.6	16 ± 1.8	0.00006
Σ length (μm)	4013 ± 1280	2752 ±997	0.004
Average dendrites	627 ± 256	519 ± 188	0.33
Longest dendrite (μm)	576 ± 43	560 ± 37	0.71

Figure 1. Deletion of GluA1 from mice results in motor neurons with a smaller and simplified dendritic arbor (from Ref. 20). Spinal motor neurons of GluA1$^{-/-}$ mice compared to WT mice at P23. Mice at P23 have motor neurons with significantly fewer branches (the average WT dendritic tree has 25 ± 1.6 branches, compared to that of GluA1$^{-/-}$, which has only 16 ± 1.8 branches ($P < 0.0001$)) and a smaller total tree size (the average dendritic tree of WT has a total length of 4013 ± 1280 μM, compared to 2750 ± 997 μM in the GluA1$^{-/-}$ mice ($P < 0.05$)). This effect is unique to the dendritic arbor, in that motor neuron soma size remains unchanged.

physically interact with the second PDZ domain of SAP97. AMPA-R physiology and synaptic plasticity is entirely normal when the physical interaction between GluA1 and SAP97 is disrupted.[26] This was demonstrated using mice in which the wild-type allele of GluA1 was replaced by a version of GluA1 lacking the C-terminal seven amino acids (GluA1Δ7 mice).[26] Contrary to expectations, GluA1 also traffics normally to the cell surface in these mice, suggesting GluA1 chaperones SAP97 to synapses—not vice versa. This was demonstrated further using biochemical and imaging methodologies.[25] However, the normal elaboration of motor neuron dendrites requires SAP97, which was demonstrated *in vitro*: knockdown of SAP97 decreases the total size of the dendritic tree and prevents the pro-dendrite growth effect of GluA1 overexpression.[25] This was further confirmed *in vivo*: a smaller dendritic tree was found in the GluA1Δ7 mice (where SAP97 does not traffic to the cell surface) as well as mice with a conditional deletion of SAP97 from motor neurons.[25] Finally, co-overexpression of GluA1 and SAP97 *in vitro* has

a synergistic pro-dendrite growth effect, which depends solely on their colocalization at the plasma membrane, not on their physical association.[25] This was demonstrated using a membrane-targeted version of SAP97 in conjunction with GluA1Δ7 or a membrane-targeted version of SAP97 with mutations in its PDZ2 domain that make it incapable of binding GluA1.[25]

Together, these experiments suggest that the endogenous GluA1–SAP97 complex is the necessary platform upon which GluA1-containing AMPA-R activity is translated into signals that stimulate dendritic growth and branching in the spinal cord.

Identifying downstream machinery and mechanisms

How do GluA1 and SAP97 work together to promote dendrite growth? One hypothesis is that GluA1 and SAP97 help form a complex during GluA1-containing AMPA-R activity to activate downstream effectors capable of stimulating dendritic growth. At this time, the molecular mechanism by which

SAP97 and GluA1 promote dendritic growth and branching is unknown. We do, however, have evidence that the PDZ3 domain of SAP97 is crucial for these pro-growth effects of SAP97 and GluA1 (unpublished observations). Overexpression of SAP97 with a mutation in the PDZ3 domain no longer increases dendrite growth. Thus, it is logical to ask whether proteins that bind to the PDZ3 domain of SAP97 are part of the machinery to translate activity into growth. This is one current avenue for research, and key downstream players in the GluA1-mediated process are beginning to be identified.

Conclusions

Much research in the field of activity-dependent development has concentrated on the NMDA-subtype of glutamate receptors, although excellent work indicates that NMDA-R–independent forms of activity-dependent development exist. The mechanism by which AMPA-R assembled with GluA1 promote dendrite growth and branching in the spinal cord is one key example of a form of plasticity that is NMDA-R independent. In addition, one challenge in identifying the functions of many MAGUK proteins, such as PSD-95 or SAP97, has been their redundant functions in relation to synaptic plasticity.[27,28] Remaining members of the membrane-associated guanylate kinases (MAGUK) family can compensate for loss of one MAGUK, but other MAGUK family members in the SAP97 conditional knockout do not compensate for the pro-growth effects of SAP97 on the dendrite tree. This suggests that this property of SAP97 is unique, and raises the possibility that proteins in the postsynaptic density involved in this process are unique as well.

Finally, understanding how dendrite growth is promoted during early postnatal life can provide insight into how one may intervene later in life to promote recovery following CNS insult. Part of the evidence supporting this idea lies in the fact that, following spinal cord injury, repetitive activation of specific neuronal pathways can result in a significant improvement in motor behavior via the increase in AMPA-ergic transmission, resulting in the stabilization of dendrites and synapses.[29,30] If we can better understand the mechanism underlying this phenomenon, we might be able to restore the features unique to early postnatal life, allowing for dendrite growth in patients following injury.

Acknowledgments

We wish to acknowledge the excellent work of Lei Zhang and Weiguo Zhou, which makes up a large portion of the work described here. This work was supported by the National Institutes of Health Grants R21NS060754 and R01NS052325 (to R.G.K.).

Conflicts of interest

The authors declare no conflicts of interest.

References

1. Katz, L.C. & C.J. Shatz. 1996. Synaptic activity and the construction of cortical circuits. *Science* **274:** 1133–1138.
2. Mooney, R., A.A. Penn, R. Gallego & C.J. Shatz. 1996. Thalamic relay of spontaneous retinal activity prior to vision. *Neuron* **17:** 863–874.
3. Shatz, C.J. 1996. Emergence of order in visual system development. *Proc. Natl. Acad. Sci. USA* **93:** 602–608.
4. Butts, D.A., P.O. Kanold & C.J. Shatz. 2007. A burst-based "Hebbian" learning rule at retinogeniculate synapses links retinal waves to activity-dependent refinement. *PLoS Biol.* **5:** e61.
5. Hubel, D.H., T.N. Wiesel & S. LeVay. 1977. Plasticity of ocular dominance columns in monkey triate cortex. *Phil. Trans. R. Sec. Lond.* **278:** 377–409.
6. Hebb, D.O. 1949. *The Organization of Behavior*. John Wiley & Sons, Inc. New York.
7. Seeburg, P.H., N. Burnashev, G. Kohr, *et al.* 1995. The NMDA receptor channel: molecular design of a coincidence detector. *Recent Progr. Hormone Res.* **50:** 19–34.
8. Shaywitz, A.J. & M.E. Greenberg. 1999. CREB: a stimulus-induced transcription factor activated by a diverse array of extracellular signals. *Ann. Rev. Biochem.* **68:** 821–861.
9. Luthi, A., M.A. Wikstrom, M.J. Palmer, *et al.* 2004. Bidirectional modulation of AMPA receptor unitary conductance by synaptic activity. *BMC Neurosci.* **5:** 44.
10. Vaughn, J.E. 1989. Fine structure of synaptogenesis in the vertebrate central nervous system. *Synapse* **3:** 255–285.
11. Chen, S.X., P.K. Tari, K. She & K. Haas. 2010. Neurexin-neuroligin cell adhesion complexes contribute to synaptotropic dendritogenesis via growth stabilization mechanisms in vivo. *Neuron* **67:** 967–983.
12. Oswald, S. 1989. *Principles of Cellular, Molecular, and Devlopmental Neuroscience*. Springer-Verlag. New York.
13. Hume, R.I. & D. Purves. 1981. Geometry of neuonatal neurones and the regulation of synapse of elimination. *Nature* **293:** 469–471.
14. Haas, K., J. Li & H.T. Cline. 2006. AMPA receptors regulate experience-dependent dendritic arbor growth in the intact brain. *Proc. Natl. Acad. Sci.* **103:** 12127–12131.
15. Casticas, M., S. Allcorn & P. Mobbs. 2001. Early activation of Ca(2+)-permeable AMPA receptors reduces neurite outgrowth in embryonic chick retinal neurons. *J. Neurobiol.* **49:** 200–211.
16. Ni, X. & M. Martin-Caraballo. 2010. Differential effect of glutamate receptor blockade on dendritic outgrowth in

chiken lumbar motoneurons. *Neuropharmacology* **58:** 593–604.

17. Peng, Y.R., S. He, H. Marie, *et al*. 2009. Coordinated changes in dendritic arborization and synaptic strength during neural circuit development. *Neuron* **61:** 71–84.

18. Jakowec, M.W., L. Yen & R.G. Kalb. 1995. In situ hybridization analysis of AMPA receptor subunit gene expression in the developing rat spinal cord. *Neuroscience* **67:** 909–920.

19. Jakowec, M.W., A.J. Fox, L.J. Martin & R.G. Kalb. 1995. Quantitative and qualitative changes in AMPA receptor expression during spinal cord development. *Neuroscience* **67:** 893–907.

20. Zhang, L., J. Schessl, M. Werner, *et al*. 2008. Role of GluR1 in activity-dependent motor system development. *J. Neurosci.* **28:** 9953–9968.

21. Jeong, G.B., M. Werner, V.R. Gazula, *et al*. 2006. Bidirectional control of motor neuron dendrite remodeling by the calcium permeability of AMPA receptors. *Mol. Cell Neurosci.* **32:** 299–314.

22. Stegenga, S.L. & R.G. Kalb. 2001. Developmental regulation of N-methyl-D-aspartate- and kainate-type glutamate receptor expression in the rat spinal cord. *Neuroscience* **105:** 499–507.

23. Inglis, F.M., R. Crockett, S. Korada, *et al*. 2002. The AMPA receptor subunit GluR1 regulates dendritic architecture of motor neurons. *J. Neurosci.* **22:** 8042–8051.

24. Kalb, R.G., L. Zhang & W. Zhou. 2013. The molecular basis of experience-dependent motor system development. *Progr. Motor Contr.* **782:** 23–38.

25. Zhou, W., L. Zhang, X. Guoxiang, *et al*. 2008. GluR1 controls dendrite growth through its binding partner, SAP97. *J. Neurosci.* **28:** 10220–10233.

26. Kim, C.H., K. Takamiya, R.S. Petralia, *et al*. 2005. Persistent hippocampal CA1 LTP in mice lacking the C-terminal PDZ ligand of GluR1. *Nat. Neurosci.* **8:** 985–987.

27. Howard, M.A., G.M. Elias, L.A. Elias, *et al*. 2010. The role of SAP97 in synaptic glutamate receptor dynamics. *Proc. Natl. Acad. Sci.* **107:** 3805–3810.

28. Schluter, O.M., W. Xu & R.C. Malenka. 2006. Alternative N-terminal domains of PSD-95 and SAP97 govern activity-dependent regulation of synaptic AMPA receptor function. *Neuron* **51:** 99–111.

29. Spooren, A.I., Y.J. Janssen-Potten, G.J. Snoek, *et al*. 2008. Rehabilitation outcome of upper extremity skilled performance in persons with cervical spinal cord injuries. *J. Rehabil. Med.* **40:** 637–644.

30. Engesser-Cesar, C., A.J. Anderson, D.M. Basso, *et al*. 2005. Voluntary wheel running improves recovery from a moderate spinal cord injury. *J. Neurotrauma* **22:** 157–171.

Ann. N.Y. Acad. Sci. ISSN 0077-8923

ANNALS OF THE NEW YORK ACADEMY OF SCIENCES

Issue: *Neurons, Circuitry, and Plasticity in the Spinal Cord and Brainstem*

Optical imaging of the spontaneous depolarization wave in the mouse embryo: origins and pharmacological nature

Yoko Momose-Sato[1] and Katsushige Sato[2]

[1]Department of Health and Nutrition, Kanto Gakuin University, College of Human and Environmental Studies, Kanazawa-ku, Japan. [2]Department of Health and Nutrition Sciences, Komazawa Women's University, Faculty of Human Health, Inagi-shi, Tokyo, Japan

Address for correspondence: Katsushige Sato, Department of Health and Nutrition Sciences, Komazawa Women's University Faculty of Human Health, 238 Sakahama, Inagi-shi, Tokyo 206-8511, Japan. katsu-satoh@komajo.ac.jp

Spontaneous embryonic movements, called embryonic motility, are produced by correlated spontaneous activity in the cranial and spinal nerves, which is driven by brainstem and spinal networks. Using optical imaging with a voltage-sensitive dye, we revealed previously in the chick and rat embryos that this correlated activity is a widely propagating wave of neural depolarization, which we termed the *depolarization wave*. One important consideration is whether a depolarization wave with similar characteristics occurs in other species, especially in different mammals. Here, we provide evidence for the existence of the depolarization wave in the mouse embryo by summarizing spatiotemporal characteristics and pharmacological natures of the widely propagating wave activity. The findings show that a synchronized wave with common characteristics is expressed in different species, suggesting its fundamental roles in neural development.

Keywords: optical recording; voltage-sensitive dye; depolarization wave; spontaneous activity; development

Introduction

During development, embryos show spontaneous movement well before sensory pathways are functionally organized.[1–4] This behavior, called *embryonic motility*, is produced by correlated spontaneous activity in the cranial and spinal nerves, which is driven by brainstem and spinal networks.[5–7] Using a voltage-sensitive dye imaging technique, we revealed previously that this correlated activity is a widely propagating wave of neural depolarization, which we referred to as the *depolarization wave*.[8–15] In the chick and rat embryos, we observed that activity arising in one location of the central nervous system (CNS) spread over a wide region, including the spinal cord, hindbrain, cerebellum, midbrain, and forebrain.

Correlated activity in the primitive nervous system is considered to play a fundamental role in neural development.[6,16–21] Previous studies have provided evidence that correlated activity in the embryonic spinal cord, which exhibits primordial patterns similar to those of the electrical discharges associated with the depolarization wave, is important for axonal path finding,[22,23] regulation of synaptic strength,[24] and proper development of locomotor function.[25] Despite these demonstrations, the developmental role of the activity, especially the significance of its large-scale propagation, is still unclarified.

One important consideration is whether a depolarization wave with similar characteristics is observed in different species. The depolarization wave was described first in the chick embryo[8–10,13–15,26,27] and then in the rat embryo.[11,12] In this review, we focus on the developing mouse CNS and demonstrate that the depolarization wave, with similar spatiotemporal characteristics to those in the chick and rat embryos, is present in the mouse embryo. We also illustrate the origin of the depolarization wave and show that the wave initiator moves from the rostral spinal cord to the caudal cord as development proceeds. We further review the pharmacological nature of the mouse depolarization wave and describe

doi: 10.1111/j.1749-6632.2012.06806.x

Ann. N.Y. Acad. Sci. 1279 (2013) 60–70 © 2013 New York Academy of Sciences.

two types of change in pharmacological characteristics that occur during development. One is that the depolarization wave is strongly dependent on nicotinic acetylcholine receptors early in development, but is dominated by glutamate later on. The second is that GABA (γ-aminobutyric acid), which acts as an excitatory mediator of the depolarization wave during the early phase, becomes an inhibitory modulator in the later stages. Furthermore, we discuss that the second switching is the possible mechanism underlying the loss of synchronization over the network, resulting in a disappearance of the depolarization wave and a replacement with mature rhythmogenerators. Results described here have been published recently in our research papers.[28,29] Details of the optical recording with voltage-sensitive dyes have been described in several reviews[30–38] and will not be discussed in this review.

Widely propagating spontaneous depolarization wave in the mouse embryo

Figure 1 shows examples of multiple-site optical recordings of spontaneous activity obtained from the brainstem–whole spinal cord (Fig. 1A) and whole brain–whole spinal cord (Fig. 1B) preparations dissected from E12 and E13 embryos, respectively. The distribution of spontaneous optical signals was broad, covering the mid-brain, diencephalon, and part of the cerebrum in addition to the hindbrain and spinal cord. In some preparations, signals were also detected in the cerebellum, although they were very small in amplitude. These results show that the wide region of the CNS, extending from the spinal cord to the forebrain, is spontaneously and synchronously active during E12–E13. At E11, similar spontaneous activity was detected from the brainstem and spinal cord, although the signals were smaller than in E12–E13 preparations. Taken together, it is reasonable to conclude that the large-scale correlated activity, which was previously termed the depolarization wave, is also expressed in the mouse embryo.

In past studies, primordial activity has often been analyzed using isolated hindbrains and/or spinal cords, which has caused some confusion, as though the activities in these structures are different, independent phenomena. In truth, synchronized activity propagates over a wide area, including the spinal cord, hindbrain, mid-brain, cerebellum, and cerebrum (Fig. 1), and thus these regions are function-

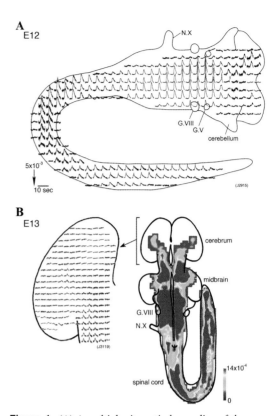

Figure 1. (A) A multiple-site optical recording of the spontaneous activity detected from an E12 brainstem–whole spinal cord preparation stained with a voltage-sensitive merocyanine-rhodanine dye, NK2761. The preparation was set to the recording chamber with a horseshoe-like shape to monitor the whole of the brainstem and spinal cord. The optical signals were detected with a photodiode array–based optical recording system constructed in our laboratory. Incident light was collimated using an interference filter with a transmission maximum at 699 ± 13 nm. The objective (×2 or ×4) and photographic eyepiece (×2.5) lenses projected a real image of the preparation onto a 34 × 34 element silicon photodiode matrix array mounted on a microscope. Changes in the transmitted light intensity through the preparation were detected with the photodiode array and recorded with a 1,020-site optical recording. The optical signals are arranged according to the positions of the photodiode array elements, and signals from outside of the preparation are omitted for clarity. The direction of the arrow on the lower left indicates the increase in transmitted light intensity (a decrease in dye absorption), and the length of the arrow represents the stated value of the fractional change. The upward direction of the signal corresponds to membrane depolarization. In this and other figures, the recordings were obtained in single sweeps. (B) Color-coded representation of maximum signal amplitudes (right) and the waveform of the optical signals in the cerebrum (left). The spontaneous activity was detected from an E13 whole brain–whole spinal cord preparation. G.V, trigeminal ganglion; G.VIII, vestibulo-cochlear ganglion; N.X, vagus nerve. Modified from Ref. 28.

ally correlated. This large-scale, nonspecific propagation across anatomical boundaries suggests that the depolarization wave may not serve as a simple regulator of specific neuronal circuit formation, but plays more global roles in the development of the CNS.

The propagation velocity of the widely spreading spontaneous wave was 10–20 mm/s,[28] which was the same range of the propagation velocity of the chick and rat depolarization waves.[12] This propagation rate is too slow to be explained by axonal conduction along the unmyelinated fiber at the corresponding stage (0.1–0.2 m/s).[39] Potential mechanisms underlying the spread of the activity include sequential synaptic activation of adjacent regions coupled by short range synaptic connections[5,40] and the coordination of chemical transmitters with gap junctions as well as electrical interactions between neighboring neurons.[41,42] In a recent study in the mouse spinal cord, it has been reported that a nonsynaptic release of glycine from radial cells influences the excitability of spinal neurons, thereby modulating the propagation of the spontaneous activity.[43] Such a nonsynaptic release of transmitters might also be one of the mechanisms underlying the large-scale propagation of the primordial activity.

Origin of the depolarization wave

Where does the depolarization wave originate? In other words, which part of the CNS acts as a conductor of the large-scale correlated wave spreading over the brain and spinal cord? Figure 2A shows examples of optical images acquired during two independent episodes of wave activity in an E12 preparation. In case 1, the optical signals appeared first in the cervical region (green circle in the inset), and propagated bidirectionally to the brainstem and caudal cord. Whereas, in case 2 the activity was initiated in the lumbosacral cord (red ellipse in the inset) and propagated rostrally to the brainstem. Regardless of the initiation sites, the final pattern of activation (the third and fifth rows of cases 1 and 2, respectively) was mostly the same between the images. These results show that a wave of similar spatial distribution is initiated in multiple locations at the E12 developmental stage.

Figure 2B shows developmental profiles of the origin of the depolarization wave in E11 to E13 preparations. In the E11 preparation, the origins were located in the rostral cord, and no wave orig-

inating in the caudal cord and hindbrain was detected. At E12, a mixed pattern of wave initiation was observed, with the activity appearing either in the cervical cord or in the lumbosacral cord. At E13, waves originating in the rostral cord disappeared, and every wave was initiated in the lumbar cord.

The developmental shift in the origin of the correlated activity from the rostral to caudal spinal cord has also been demonstrated for the chick and rat depolarization waves[12,13] and for the spinal nerve discharges in the rat[44,45] and mouse.[46] Possible mechanisms underlying the shift in the origin include a decrease and increase in neural excitability in the rostral and caudal spinal cords, respectively, assuming that the depolarization wave is paced by the region with the highest excitability. In the chick embryo, it has been reported that in the early developmental stage (E4–E5), electrical stimulation was effective in inducing the evoked depolarization wave only when it was applied to the upper cervical cord and lower medulla near the obex, while at E7–E8, regional differences in neural excitability were less prominent.[13] This developmental profile suggests that there is a gradient of neural responsiveness along the rostrocaudal axis of the spinal cord, and that this gradient changes dynamically with development.

In general, there are two hypotheses regarding the mechanism underlying the genesis of the spontaneous activity. One is the *pacemaker hypothesis*, according to which the rhythm arises in pacemaker neurons; the other is the *network hypothesis*, according to which the activity is generated by neural network interaction. The widely propagating depolarization wave appears not to depend on a single population of oscillatory neurons since the activity originates at various levels of the spinal cord even in the same preparation (Fig. 2). Concerning the mechanism by which a network generates rhythmic activity in the absence of a pacemaker, O'Donovan *et al.* proposed a model based on work on the embryonic chick spinal cord.[47–49] They hypothesize that transient depolarization and firing in motoneurons, originating from random fluctuations of interneuronal synaptic activity, activate R-interneurons, which then trigger the recruitment of the rest of the spinal interneuronal network. In this model, activity-dependent depression occurs after an episode of activity: the membrane is hyperpolarized, and the evoked and spontaneous

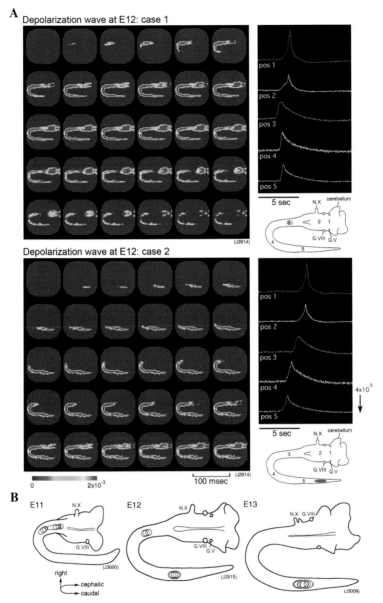

Figure 2. (A) Pseudocolor images of the spontaneous depolarization wave in an E12 brainstem–whole spinal cord preparation. Images in the upper and lower panels were acquired during two independent episodes in the same preparation. The signals on the right were detected from five positions indicated in the lower-right insets. Note that in the upper panel (case 1), the activity was initiated in the upper cervical cord (green circle in the inset), while in the lower panel (case 2), the wave originated in the lumbosacral cord (red oval in the inset). (B) Origin of the spontaneous depolarization wave at E11–E13. The origins of the spontaneous depolarization wave are indicated with solid circles for typical preparations at each developmental stage. In each preparation, one circle corresponds to one spontaneous wave, and variations in the circles' locations show variations in the origin of the depolarization wave. As development proceeded, the origin shifted to the caudal spinal cord. Modified from Ref. 28.

synaptic potentials fall to a minimum. One mechanism generating the postepisode depression is chloride's extrusion through the $GABA_A$ receptor-gated chloride channel.[50,51] During the interepisode interval, the intracellular chloride concentration is restored, and network excitability recovers gradually. When the network excitability reaches a threshold, neurons will be explosively recruited throughout the network, and an episode will recur with a time constant of minutes. One important characteristic of this model is that activity can be initiated in any part of the preparation where the excitability is high enough to sustain regenerative recruitment of the entire network, and no exact, highly precise circuit needs to exist to produce the activity pattern.[16,47] This model of activity-dependent depression also seems applicable to the mouse spinal cord, in which the network undergoes refractoriness following an episode of activity.[41]

Developmental changes in pharmacological natures of the depolarization wave

In the chick and rat embryos, it has been shown that the depolarization wave is mediated by multiple neurotransmitters and possibly gap junctions.[9,11,12,15] Furthermore, the depolarization wave in the chick is strongly dependent on nicotinic acetylcholine receptors at the early stage, while it

is dominated by glutamate at the later stage.[15] As summarized below, pharmacological characteristics of the mouse depolarization wave exhibit many similarities with those in the chick and rat embryos.

Figure 3 shows examples of electrical discharges in association with the depolarization wave at E11 (Fig. 3A) and E14 (Fig. 3B). When d-tubocurarine, a general blocker of nicotinic acetylcholine receptors, was applied, the spontaneous activity at E11 was slowed in frequency, and slightly decreased in amplitude (Fig. 3A). However, at E14, neither the activity restricted to the hindbrain (asterisks) nor that synchronized between the hindbrain and spinal cords (arrowheads) was affected by d-tubocurarine (Fig. 3B). Figure 3C summarizes the effects of d-tubocurarine on the frequency of the spontaneous activity at each developmental stage. The inhibitory action of d-tubocurarine was age dependent, and the activity at early stages was more sensitive to d-tubocurarine. Application of nicotinic acetylcholine receptor subunit blockers revealed that the

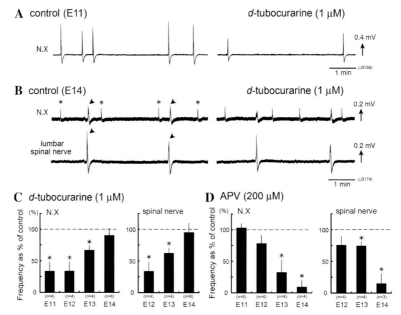

Figure 3. Effects of cholinergic and glutamatergic blockers on the spontaneous activity. (A and B) The effects of d-tubocurarine (1 μM) on the spontaneous activity at E11 (A) and E14 (B). Electrical discharges were detected from the vagus nerve (N.X) (A) or simultaneously recorded from the vagus and lumbar spinal nerves (B). The spontaneous activity was inhibited by d-tubocurarine at E11 (A), but not at E14 (B). Asterisks and arrowheads in B indicate the activity restricted to the vagus nerve and that synchronized between the vagus and spinal nerves, respectively. (C and D) Histograms showing the effects of d-tubocurarine (1 μM) (C) and APV (200 μM) (D) on the frequency of the spontaneous activity. Data in C and D were obtained from the vagus nerve (left) and lumbar spinal nerve (right) at E11–E14. The effects of blockers are presented as the percent changes in the activity frequency relative to the control (mean ± SEM). "n" indicates the number of preparations. For each preparation, four to six spontaneous episodes were sampled, and interepisode intervals were measured. When the activity was blocked for 10–15 minutes after the application of the drug, the frequency change was evaluated as 0%. *The effect of the blocker was significant ($P < 0.05$). Modified from Ref. 29.

activity at E11 is more dominantly mediated by the non-α7 type receptor than the α7 subtype.[29]

Contradictory to the results obtained with *d*-tubocurarine, APV (DL-2-amino-5-phosphono-valeric acid), a blocker of NMDA (*N*-methyl-D-aspartate)-type glutamate receptors, inhibited the spontaneous activity more prominently at later stages (Fig. 3D). Similar results were obtained for CNQX (6-cyano-7-nitroquinoxaline-2,3-dione), a blocker of non-NMDA-type glutamate receptors, suggesting that both the NMDA and non-NMDA subtypes contribute to the depolarization wave at the later stage, but not in the early phase.[29]

GABA is also responsible for the depolarization wave in the mouse embryo. At E11–E13, the application of GABA$_A$ receptor antagonists, such as bicuculline and picrotoxin, inhibited the spontaneous activity, while at E14, the activity was augmented in amplitude.[29] Compared with the role of GABA, any contribution of glycine to the depolarization wave is minor. Strychnine, a glycine receptor antagonist, slightly reduced the frequency of the spontaneous activity at E12–E13 only at a higher concentration (1 μM), and enhancement of the signal amplitude at E14 was not observed.[29]

In many structures of the developing brain, GABA and glycine depolarize the cell membrane, as the Cl$^-$ reversal potential (E$_{Cl}$) is more positive than the resting potential, due to the high concentration of intracellular Cl$^-$ (Ref. 52). The developmental change in GABA/glycinergic responses from depolarizing to hyperpolarizing seems to be regulated by a shift in the expression of chloride transporters, and the role of KCC2, a Cl$^-$ extruder coupled with K$^+$, is especially highlighted.[52–54] In the mouse embryo, application of DIOA (*R*-(+)-[(2-*n*-butyl-6,7-dichloro-2-cyclopenyl-2,3-dihydro-1-oxo-1H-inden-5-yl) oxy] acetic acid), a KCC2 blocker, increased the frequency of the spontaneous activity at E14, although the effect was significant only for the spinal nerve activity.[29] The result suggests that KCC2 is functional at E14, which might underlie the switching of GABA's actions on the depolarization wave from excitatory to inhibitory.

The pharmacological nature of the depolarization wave summarized previously is consistent with those of the spontaneous activity observed in the isolated spinal cord, although the role of glycine was less obvious in the study of the depolarization wave. In the mouse spinal cord, it has been reported

that spontaneous electrical discharges display two distinct periods in which different transmitter systems are responsible for generating the activity.[25,55] During the early phase, from E12.5 to E14.5, spontaneous activity is dependent on acetylcholine, glycine, and to a lesser extent, GABA$_A$ receptors. At these stages, glycine and GABA act as excitatory mediators that promote spontaneous bursts. From E15.5 onward, the acetylcholine-mediated drive is replaced by a glutamatergic drive, and glycine and GABA suppress, rather than promote, the spontaneous activity.[25,55] The differences in the contribution of glycine might be due to the differences in the strychnine concentration used in the experiment. In the isolated spinal cord, strychnine at 3–5 μM was applied,[25,41,43] while in the study of the depolarization wave, the concentration was 1 μM or less to avoid the side effects of strychnine on GABA$_A$ receptors.[56] In previous studies in chick and rat embryos, the pharmacological nature of the depolarization wave was examined using a higher concentration of strychnine (10 μM).[9,11,12,15] Since strychnine has nonspecific actions at high concentrations,[56,57] any positive conclusion concerning the role of glycine in the depolarization wave will require careful confirmation in future experiments.

A notable finding obtained in the study of the mouse depolarization wave is that the pharmacological switching seems to occur earlier in the hindbrain than in the spinal cord. In the hindbrain, the glutamate dependence of the depolarization wave was observed from E12, and vagal discharges were augmented by GABAergic blockers even at E13 in some experiments.[29] These results suggest that there is a rostrocaudal gradient of developmental processes regarding the switching of the pharmacological characteristics.

Concerning the mechanisms by which the pharmacological characteristics of the spontaneous activity changes during development, interactions of different transmitter systems have been suggested. In the retina of embryonic mice lacking the β2-subunit of nicotinic acetylcholine receptors, the absence of cholinergic waves induces a precaucious transition to glutamate-mediated waves, showing that the expression of the later glutamatergic wave is under the control of early cholinergic activity.[58] In a study using the chick embryo *in ovo*, Liu *et al.*[59] showed that the chronic blockade of nicotinic acetylcholine receptors prevented the conversion of GABA/glycinergic responses from excitation to inhibition. In another study using choline acetyltransferase (ChAT)-deficient mouse spinal cords, Myers et al.[25] reported that cholinergic signaling determines the timing of glutamatergic activity and also the transition of glycinegic responses from excitation to inhibition. Expression of the glutamatergic drive appeared to precede the switching of glycine function, and thus it has been speculated that the cholinergic activity dictates the glutamatergic function, which in turn leads to the maturation of inhibitory signaling.[25] In the developing retinotectal circuit, it has been reported that depolarizing GABAergic transmission is necessary for the formation of glutamatergic synapses, which subsequently regulate the development of inhibitory GABAergic inputs.[60]

With regard to the depolarization wave in the mouse embryo, the expression of glutamate dependence slightly precedes the switching of GABAergic responses, with the former appearing in some brainstems at E12 and brainstem–whole spinal cords at E13, while the latter is observed from E13 in the brainstem and E14 in the brainstem–whole spinal cord. This developmental sequence appears to be consistent with the scheme of cholinergic–glutamatergic–GABA/glycinergic interactions suggested in other studies. However, in the chick embryo, the GABAergic response is still to promote the spontaneous activity even at E10–E11,[61] while cholinergic to glutamatergic transition occurs at around E6.[15] It is possible that the pharmacological mechanisms differ between the species.

Disappearance of the large-scale synchronization at E14

The widely propagating correlated depolarization wave was observed during a specific time window, E11–E13.[28] At E14, the correlation of the vagus and spinal nerve discharges declined, and synchronized activity was absent in a solution containing a lower $[K^+]_o$ (3.0 mM). In E14 preparations, optical responses were usually observed in restricted regions of the medulla and lumbosacral cord (Fig. 4A), suggesting that the initially synchronized network segregates into more specialized subnetworks at E14.

In the mouse hindbrain, the embryonic parafacial population becomes active at E14.5 as the first respiratory oscillator, and the pre-Bötzinger complex emerges subsequently at E15.5.[62] Thus, the optical response identified in the E14 medulla seems to correspond to the respiratory activity originating in the parafacial respiratory group. In the isolated mouse spinal cord, the spinal central pattern generator of locomotion is in the process of development from E15.[63] Considering these, we reasonably conclude that E14–E15 is the stage when the segregated respiratory and locomotor networks appear in the brainstem and spinal cord, and that this age may be a critical stage at which mature circuits differentiate to substitute for the larger primordial neuronal assemblies mediating the depolarization wave.

One possible mechanism underlying the substitution of the primordial network with the mature circuits is that the action of GABA changes from excitatory to inhibitory at E14, and this switching is responsible for the decrease in the network excitability of the brainstem and spinal cord, resulting in the disappearance of the large-scale synchronization. It has also been suggested that descending pathways from the brainstem, especially monoaminergic projections, may have some influences.[7,64] Consistent with the first hypothesis, when bicuculline or picrotoxin was applied in the E14 brainstem–whole spinal cord preparation, large optical signals appeared spontaneously in the rostral cord and propagated bidirectionally to the hindbrain and caudal cord (Fig. 4B). The waveform, spatial distribution, and propagation velocity of the optical signal were similar to those of the depolarization wave identified in the E12–E13 preparations in the control solution (Figs. 1 and 2A). These results suggest that the neural network at E14 has the ability to produce the depolarization wave, but is inhibited by GABA. The blockade of GABA receptors disinhibits the neural network, resulting in the generation of the large-scale depolarization wave.

Figure 4C illustrates a typical example of a map showing the origin of the depolarization wave that appeared in the presence of the blockers at E14. In this map, the depolarization wave was initiated in the lower cervical to thoracic cord (Fig. 4C), not the region in which the local spontaneous activity was generated in the control solution (Fig. 4A). This observation suggests that the shift in the origin of the depolarization wave from the rostral cord at E11 to the lumbosacral region from E13 (Fig. 2B) is not

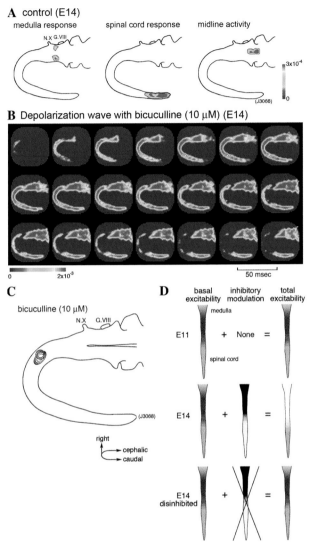

Figure 4. (A) Spatial distribution of the spontaneous activity detected from an E14 preparation in the control solution. The activity was localized to a restricted region of the medulla (left), caudal spinal cord (middle), or midline of the brainstem (right). The maximum amplitude of the optical signal is presented with color. (B) In the presence of bicuculline (10 μM), spontaneous activity spread over the brainstem and spinal cord, which was similar to the spatiotemporal pattern of the depolarization wave observed in the control solution at the earlier stages. Propagation patterns of the spontaneous wave are presented with color images. The frame interval was 50 ms. (C) Origin of the depolarization wave that appeared in the presence of bicuculline (10 μM) at E14. One circle corresponds to one spontaneous wave, and variations in circles' locations show variations in the origin of the depolarization wave. (D) Schematic hypothesis for the generation of the depolarization wave. The left and right panels show the ability of neurons/neural networks to generate the spontaneous activity without and with inhibitory modulation, respectively. The dark area is more responsible for generating the spontaneous activity. In the middle panel, the strength of inhibitory modulation is presented with gray scaling. The top row shows the E11 stage, in which the spontaneous depolarization wave is initiated in the rostral cord, and no inhibitory modulation is expressed. The second row shows the E14 stage, in which the inhibitory modulation of GABA suppresses the generation and propagation of the depolarization wave. The bottom row shows the E14 preparation with application of the GABAergic blocker, in which the inhibitory modulation is deprived, resulting in the appearance of the depolarization wave that originates in the rostral spinal cord. Modified from Ref. 29.

due to the change in the rostrocaudal gradient of the basal excitability (the excitability of neural networks without inhibition, Fig. 4D, left panels), but is the result of heterogeneous expression of inhibitory modulation (Fig. 4D, middle panels), which is possibly stronger in the rostral segments.

Future perspectives

In this review, we have provided evidence for the existence of the depolarization wave in the mouse embryo by summarizing spatiotemporal characteristics and the pharmacological nature of the widely propagating wave activity. Recent advances in our understanding of the widely propagating synchronized activity in the brainstem and spinal cord are due to progress using multiple techniques, including electrophysiology, molecular biology, and optical imaging. Although much consensus has been obtained concerning the global features of the activity, there are still several unanswered questions that should be addressed in future investigations. Perhaps the most important issue is the functional significance of the activity. In the embryonic brainstem and spinal cord, several investigations have provided evidence that the primordial spontaneous activity is indispensable for proper development of the neural network and expression of behavior. Despite these demonstrations, the functional significance of the activity, especially the necessity of its large-scale correlation, is still unresolved. Primordial activity having characteristics similar to those described here has been observed in several other structures of the developing nervous system, and investigations from multiple aspects would improve our understanding of the activity and its role in the development of the nervous system.

Acknowledgments

We are grateful to Kohtaro Kamino for his advice and help during this work. We also thank Tomoharu Nakamori for his collaboration in the experiments. This study was supported by grants from the Ministry of Education–Science–Culture of Japan and research funds from the Human Frontier Science Program and Opto-Medical Institute Inc.

Conflicts of interest

The authors declare no conflicts of interest.

References

1. Hamburger, V. & M. Balaban. 1963. Observations and experiments on spontaneous rhythmical behavior in the chick embryo. *Dev. Biol.* **7:** 533–545.
2. Narayanan, C.H., M.W. Fox & V. Hamburger. 1971. Prenatal development of spontaneous and evoked activity in the rat (*Rattus norvegicus albinus*). *Behaviour* **40:** 100–134.
3. Suzue, T. 1992. Physiological activities of late-gestation rat fetuses in vitro. *Neurosci. Res.* **14:** 145–157.
4. Suzue, T. & Y. Shinoda. 1999. Highly reproducible spatiotemporal patterns of mammalian embryonic movements at the developmental stage of the earliest spontaneous motility. *Eur. J. Neurosci.* **11:** 2697–2710.
5. Fortin, G., F. Kato, A. Lumsden & J. Champagnat. 1995. Rhythm generation in the segmented hindbrain of chick embryos. *J. Physiol.* **486:** 735–744.
6. O'Donovan, M.J. 1999. The origin of spontaneous activity in developing networks of the vertebrate nervous system. *Curr. Opin. Neurobiol.* **9:** 94–104.
7. Marder, E. & K.J. Rehm. 2005. Development of central pattern generating circuits. *Curr. Opin. Neurobiol.* **15:** 86–93.
8. Momose-Sato, Y., K. Sato, H. Mochida, *et al.* 2001. Spreading depolarization waves triggered by vagal stimulation in the embryonic chick brain: optical evidence for intercellular communication in the developing central nervous system. *Neuroscience* **102:** 245–262.
9. Momose-Sato, Y., N. Miyakawa, H. Mochida, *et al.* 2003. Optical analysis of depolarization waves in the embryonic brain: a dual network of gap junctions and chemical synapses. *J. Neurophysiol.* **89:** 600–614.
10. Momose-Sato, Y., H. Mochida, S. Sasaki & K. Sato. 2003. Depolarization waves in the embryonic CNS triggered by multiple sensory inputs and spontaneous activity: optical imaging with a voltage-sensitive dye. *Neuroscience* **116:** 407–423.
11. Momose-Sato, Y., Y. Honda, H. Sasaki & K. Sato. 2005. Optical imaging of large-scale correlated wave activity in the developing rat CNS. *J. Neurophysiol.* **94:** 1606–1622.
12. Momose-Sato, Y., K. Sato & M. Kinoshita. 2007. Spontaneous depolarization waves of multiple origins in the embryonic rat CNS. *Eur. J. Neurosci.* **25:** 929–944.
13. Momose-Sato, Y., H. Mochida & M. Kinoshita. 2009. Origin of the earliest correlated neuronal activity in the chick embryo revealed by optical imaging with voltage-sensitive dyes. *Eur. J. Neurosci.* **29:** 1–13.
14. Mochida, H., K. Sato, Y. Arai, *et al.* 2001. Optical imaging of spreading depolarization waves triggered by spinal nerve stimulation in the chick embryo: possible mechanisms for large-scale coactivation of the CNS. *Eur. J. Neurosci.* **14:** 809–820.
15. Mochida, H., K. Sato & Y. Momose-Sato. 2009. Switching of the transmitters that mediate hindbrain correlated activity in the chick embryo. *Eur. J. Neurosci.* **29:** 14–30.
16. Feller, M.B. 1999. Spontaneous correlated activity in developing neural circuits. *Neuron* **22:** 653–656.
17. Roerig, B. & M.B. Feller. 2000. Neurotransmitters and gap junctions in developing neural circuits. *Brain Res. Rev.* **32:** 86–114.

18. Ben-Ari, Y. 2001. Developing networks play a similar melody. *Trends Neurosci.* **24:** 353–359.

19. Chatonnet, F., M. Thoby-Brisson, V. Abadie, *et al.* 2002. Early development of respiratory rhythm generation in mouse and chick. *Respir. Physiol. Neurobiol.* **131:** 5–13.

20. Moody, W.J. & M.M. Bosma. 2005. Ion channel development, spontaneous activity, and activity-dependent development in nerve and muscle cells. *Physiol. Rev.* **85:** 883–941.

21. Blankenship, A.G. & M.B. Feller. 2010. Mechanisms underlying spontaneous patterned activity in developing neural circuits. *Nat. Rev. Neurosci.* **11:** 18–29.

22. Hanson, M.G. & L.T. Landmesser. 2004. Normal patterns of spontaneous activity are required for correct motor axon guidance and the expression of specific guidance molecules. *Neuron* **43:** 687–701.

23. Hanson, M.G., L.D. Milner & L.T. Landmesser. 2008. Spontaneous rhythmic activity in early chick spinal cord influences distinct motor axon pathfinding decisions. *Brain Res. Rev.* **57:** 77–85.

24. Gonzalez-Islas, C. & P. Wenner. 2006. Spontaneous network activity in the embryonic spinal cord regulates AMPAergic and GABAergic synaptic strength. *Neuron* **49:** 563–575.

25. Myers, C.P., J.W. Lewcock, M.G. Hanson, *et al.* 2005. Cholinergic input is required during embryonic development to mediate proper assembly of spinal locomotor circuits. *Neuron* **46:** 37–49.

26. Komuro H., Y. Momose-Sato, T. Sakai, *et al.* 1993. Optical monitoring of early appearance of spontaneous membrane potential changes in the embryonic chick medulla oblongata using a voltage-sensitive dye. *Neuroscience* **52:** 55–62.

27. O'Donovan, M., S. Ho & W. Yee. 1994. Calcium imaging of rhythmic network activity in the developing spinal cord of the chick embryo. *J. Neurosci.* **14:** 6354–6369.

28. Momose-Sato, Y., T. Nakamori & K. Sato. 2012. Spontaneous depolarization wave in the mouse embryo: origin and large-scale propagation over the CNS identified with voltage-sensitive dye imaging. *Eur. J. Neurosci.* **35:** 1230–1241.

29. Momose-Sato, Y., T. Nakamori & K. Sato. 2012. Pharmacological mechanisms underlying the switching from the large-scale depolarization wave to segregated activity in the mouse CNS. *Eur. J. Neurosci.* **35:** 1242–1252.

30. Cohen, L.B. & B.M. Salzberg. 1978. Optical measurement of membrane potential. *Rev. Physiol. Biochem. Pharmacol.* **83:** 35–88.

31. Salzberg, B.M. 1983. Optical recording of electrical activity in neurons using molecular probes. In *Current Methods in Cellular Neurobiology. Electrophysiological Techniques.* Vol. 3. J.L. Barker & J.F. McKelvy, Eds.: 139–187. Wiley. New York.

32. Grinvald, A. 1985. Real-time optical mapping of neuronal activity: from single growth cones to the intact mammalian brain. *Annu. Rev. Neurosci.* **8:** 263–305.

33. Grinvald, A., R.D. Frostig, E. Lieke & R. Hildesheim. 1988. Optical imaging of neuronal activity. *Physiol. Rev.* **68:** 1285–1366.

34. Wu, J.-Y. & L.B. Cohen. 1993. Fast multisite optical measurement of membrane potential. In *Fluorescent and Luminescent Probes for Biological Activity.* W.T. Mason, Ed.: 389–404. Academic Press. Boston.

35. Grinvald, A. & R. Hildesheim. 2004. VSDI: a new era in functional imaging of cortical dynamics. *Nature Rev. Neurosci.* **5:** 874–885.

36. Baker, B.J., E.K. Kosmidis, D. Vucinic, *et al.* 2005. Imaging brain activity with voltage- and calcium-sensitive dyes. *Cell Mol. Neurobiol.* **25:** 245–282.

37. Glover, J.C., K. Sato & Y. Momose-Sato. 2008. Using voltage-sensitive dye recording to image the functional development of neuronal circuits in vertebrate embryos. *Dev. Neurobiol.* **68:** 804–816.

38. Momose-Sato, Y., K. Sato & K. Kamino. 2010. Monitoring population membrane potential signals during functional development of neuronal circuits in vertebrate embryos. In *Membrane Potential Imaging in the Nervous System: Methods and Applications.* M. Canepari & D. Zecevic, Eds.: 83–96. Springer-Verlag. New York.

39. Sakai, T., H. Komuro, Y. Katoh, *et al.* 1991. Optical determination of impulse conduction velocity during development of embryonic chick cervical vagus nerve bundles. *J. Physiol.* **439:** 361–381.

40. O'Donovan, M.J., A. Bonnot, G.Z. Mentis, *et al.* 2008. Imaging the spatiotemporal organization of neural activity in the developing spinal cord. *Dev. Neurobiol.* **68:** 788–803.

41. Hanson, M.G. & L.T. Landmesser. 2003. Characterization of the circuits that generate spontaneous episodes of activity in the early embryonic mouse spinal cord. *J. Neurosci.* **23:** 587–600.

42. Ren, J., Y. Momose-Sato, K. Sato & J.J. Greer. 2006. Rhythmic neuronal discharge in the medulla and spinal cord of fetal rats in the absence of synaptic transmission. *J. Neurophysiol.* **95:** 527–534.

43. Scain, A.-L., H. Le Corronc, A.-E. Allain, *et al.* 2010. Glycine release from radial cells modulates the spontaneous activity and its propagation during early spinal cord development. *J. Neurosci.* **30:** 390–403.

44. Nakayama, K., H. Nishimaru, M. Iizuka, *et al.* 1999. Rostrocaudal progression in the development of periodic spontaneous activity in fetal rat spinal motor circuits in vitro. *J. Neurophysiol.* **81:** 2592–2595.

45. Ren, J. & J.J. Greer. 2003. Ontogeny of rhythmic motor patterns generated in the embryonic rat spinal cord. *J. Neurophysiol.* **89:** 1187–1195.

46. Yvert, B., P. Branchereau & P. Meyrand. 2004. Multiple spontaneous rhythmic activity patterns generated by the embryonic mouse spinal cord occur within a specific developmental time window. *J. Neurophysiol.* **91:** 2101–2109.

47. O'Donovan, M.J., N. Chub & P. Wenner. 1998. Mechanisms of spontaneous activity in developing spinal networks. *J. Neurobiol.* **37:** 131–145.

48. Tabak, J., J. Rinzel & M.J. O'Donovan. 2001. The role of activity-dependent network depression in the expression and self-regulation of spontaneous activity in the developing spinal cord. *J. Neurosci.* **21:** 8966–8978.

49. Wenner, P. & M.J. O'Donovan. 2001. Mechanisms that initiate spontaneous network activity in the developing chick spinal cord. *J. Neurophysiol.* **86:** 1481–1498.

50. Chub, N. & M.J. O'Donovan. 2001. Post-episode depression of GABAergic transmission in spinal neurons of the chick embryo. *J. Neurophysiol.* **85:** 2166–2176.

51. Chub, N., G.Z. Mentis & M.J. O'Donovan. 2006. Chloride-sensitive MEQ fluorescence in chick embryo motoneurons following manipulations of chloride and during spontaneous network activity. *J. Neurophysiol.* **95:** 323–330.

52. Ben-Ari, Y., J.-L. Gaiarsa, R. Tyzio & R. Khazipov. 2007. GABA: a pioneer transmitter that excites immature neurons and generates primitive oscillations. *Physiol. Rev.* **87:** 1215–1284.

53. Fiumelli, H. & M.A. Woodin. 2007. Role of activity-dependent regulation of neuronal chloride homeostasis in development. *Curr. Opin. Neurobiol.* **17:** 81–86.

54. Stil, A., C. Jean-Xavier, S. Liabeuf, *et al.* 2011. Contribution of the potassium-chloride co-transporter KCC2 to the modulation of lumbar spinal networks in mice. *Eur. J. Neurosci.* **33:** 1212–1222.

55. Ladle, D.R., E. Pecho-Vrieseling & S. Arber. 2007. Assembly of motor circuits in the spinal cord: driven to function by genetic and experience-dependent mechanisms. *Neuron* **56:** 270–283.

56. Jonas, P., J. Bischofberger & J. Sandkühler. 1998. Corelease of two fast neurotransmitters at a central synapse. *Science* **281:** 419–424.

57. Hinckley, C.A., B. Seebach & L. Ziskind-Conhaim. 2005. Distinct roles of glycinergic and GABAergic inhibition in coordinating locomotor-like rhythms in the mouse spinal cord. *Neuroscience* **131:** 745–758.

58. Bansal, A., J.H. Singer, B.J. Hwang, *et al.* 2000. Mice lacking specific nicotinic acetylcholine receptor subunits exhibit dramatically altered spontaneous activity patterns and reveal a limited role for retinal waves in forming ON and OFF circuits in the inner retina. *J. Neurosci.* **20:** 7672–7681.

59. Liu, Z., R.A. Neff & D.K. Berg. 2006. Sequential interplay of nicotinic and GABAergic signaling guides neuronal development. *Science* **314:** 1610–1613.

60. Akerman, C.J. & H.T. Cline. 2006. Depolarizing GABAergic conductances regulate the balance of excitation to inhibition in the developing retinotectal circuit *in vivo. J. Neurosci.* **26:** 5117–5130.

61. Chub, N. & M.J. O'Donovan. 1998. Blockade and recovery of spontaneous rhythmic activity after application of neurotransmitter antagonists to spinal networks of the chick embryo. *J. Neurosci.* **18:** 294–306.

62. Thoby-Brisson, M., M. Karlén, N. Wu, *et al.* 2009. Genetic identification of an embryonic parafacial oscillator coupling to the preBötzinger complex. *Nat. Neurosci.* **12:** 1028–1036.

63. Branchereau, P., D. Morin, A. Bonnot, *et al.* 2000. Development of lumbar rhythmic networks: from embryonic to neonate locomotor-like patterns in the mouse. *Brain Res. Bull.* **53:** 711–718.

64. Vinay, L., F. Brocard, F. Clarac, *et al.* 2002. Development of posture and locomotion: an interplay of endogenously generated activities and neurotrophic actions by descending pathways. *Brain Res. Rev.* **40:** 118–129.

Ann. N.Y. Acad. Sci. ISSN 0077-8923

Imaging spinal neuron ensembles active during locomotion with genetically encoded calcium indicators

Christopher A. Hinckley and Samuel L. Pfaff

Howard Hughes Medical Institute and Gene Expression Laboratory, The Salk Institute for Biological Studies, La Jolla, California

Address for correspondence: Samuel L. Pfaff, Ph.D., Howard Hughes Medical Institute, Gene Expression Laboratory, Salk Institute, 10010 North Torrey Pines Rd., La Jolla, CA 92037. pfaff@salk.edu

Advances in molecular-genetic tools for labeling neuronal subtypes, and the emerging development of robust genetic probes for neural activity, are likely to revolutionize our understanding of the functional organization of neural circuits. In principle, these tools should be able to detect activity at cellular resolution for large ensembles of identified neuron types as they participate in specific behaviors. This report describes the use of genetically encoded calcium indicators (GECIs), combined with two-photon microscopy, to characterize V1 interneurons, known to be critical for setting the duration of the step cycle. All V1 interneurons arise from a common precursor population and express engrailed-1 (*En1*). Our data show that although neighboring interneurons that arise from the same developmental lineage and share many features, such as projection patterns and neurotransmitter profiles, they are not irrevocably committed to having the same pattern of activity.

Keywords: Engrailed-1; GCaMP3; two-photon microscopy

Introduction

In the nervous system, the interconnected activity of neurons that make up a circuit is transformed into a coherent output that drives behavior. A complete understanding of information processing in any mammalian neural circuit has proved difficult because it requires the integration of many disparate perspectives. Approaches used to piece together circuits include tracing techniques to define synaptic connectivity, electrophysiological recordings to investigate the properties of specific synapses, and biophysical experiments to examine how neurons compute their outputs. Nevertheless, there remains a large gap in our understanding of the relationship between activity at the cellular and circuit levels. Filling this gap requires recording the activity of many neurons simultaneously so that circuit level activity can be related to a behavioral output. Two emerging techniques that are being applied in this way are multielectrode array recordings and neuronal activity imaging.

Extracellular recordings from electrode arrays have the advantage of high temporal resolution, although it is difficult to extract detailed information about the identities of the recorded neurons. Likewise, optical imaging with nonspecific voltage or calcium dyes can be difficult to correlate with cell types and suffers further from weak signals and/or slow temporal responses of the dye. The gap in relating neuronal function to neuronal subtypes has long been acknowledged and is being addressed using molecular genetics to link gene expression patterns with cell types. In genetically modifiable systems, such as the zebrafish and mouse, it is now possible to use genetically encoded reporters expressed in neuron types to anchor the relationship between neuronal identity, connectivity, biophysical properties, and function. This approach has reached a new level of potential with the emergence of sensitive types of genetically encoded calcium indicators (GECIs), such as the GCaMPs, which simplify the task of establishing the relationship between neuronal subtype identity and cellular activity within the context of a defined circuit.

The locomotor circuitry that makes up the central pattern generator within the lumbar spinal cord is a well-suited system for understanding the

doi: 10.1111/nyas.12092

cellular basis of circuit function. Because spinal motor circuits directly link the central nervous system (CNS) to behavior, the significance of motor activity is relatively easy to interpret. An intensive characterization of spinal cord development, gene expression profiling, and the development of Cre lines for labeling and toxins for ablating specific cell types in the spinal cord, has led to the identification of neurons that are required for locomotor function, and the creation of genetic tools that can be leveraged to study the cellular constituents of locomotor circuitry.[1]

Two-photon imaging with GECIs

Optical detection of neuronal activity at a cellular resolution has been achieved with a variety of indicators for either voltage or calcium.[2] Of the available indicators, calcium-binding molecules generate the largest changes in fluorescent signals. When paired with the deep tissue imaging potential of fast scanning two-photon (2P) microscopes, calcium imaging offers a powerful compromise between spatial and temporal resolution, and sensitivity. Cellular resolution imaging studies of mammalian spinal locomotor circuits have used 2P calcium imaging to monitor the activity of neuron subtypes using genetically encoded fluorescent reporters, which allow labeling of molecularly defined neurons in combination with nonspecific bolus loading of calcium dyes. This approach has been used to characterize the activity of Hb9-expressing interneurons, Chx10-expressing V2a excitatory interneurons, and lamina V/VI gamma aminobutyric acid (GABA)-ergic interneurons during fictive locomotion.[3–6]

While these early experiments have mainly been used to complement electrophysiological observations and to explore technical aspects of imaging and data analysis, their observations highlight the promise of imaging locomotor network activity. One common observation emerging from imaging studies is that even very small regions of the spinal cord seem to contain neurons oscillating in a variety of phases. For example, neighboring Chx10-expressing V2a interneurons, as well as unidentified lamina VII/VIII interneurons, can be rhythmically active but in distinct phases relative to motor output.[5,7] A major advantage of imaging is that oscillations in neighboring neurons can be readily compared. Studies that imaged groups of Hb9-expressing interneurons during successive bouts of

fictive locomotion found that individual Hb9 interneurons are active only in a subset of locomotor cycles.[3] Although Hb9 interneuron oscillations were sparse at the level of single and small groups of cells, it remains to be determined how Hb9 interneuron activity at a population level relates to motor output.

Despite success using 2P microscopy to image single cells deep within the spinal cord, this approach also has drawbacks that limit the detailed characterization of locomotor circuitry. Key among these are the difficulty in labeling all of the relevant neurons using nonspecific bolus loading of dyes, and the trade-offs between field of view and frame rate inherent in laser scanning microscopy. For these reasons, we have explored the use of fast scanning 2P imaging in combination with GECIs.

A new generation of promising GECIs has become available during the last few years. Genetic sensors of neuronal activity have long been sought as a tool by neuroscientists, but have required multiple iterations of development to produce GECIs that detect calcium at physiological levels without being toxic, while producing fluorescent signals that are easily separated from noise. The early sensors were based on fusing the calcium-binding protein, calmodulin, between two different fluorescent proteins whose fluorescence resonance energy transfer efficiency changed when calcium was bound.[8,9] New, more sensitive, forms of GECIs have been created by fusing calmodulin to circular permutations of fluorescent proteins, such as green fluorescent protein (GFP), to create a series of probes with different sensitivities and spectral qualities. Structure-guided evolution of these probes continues to produce variants with improved kinetics and fluorescence changes on calcium binding.[10] For example, the widely adopted GECI, GCaMP3, reliably reports even small groups of two to three spikes *in vitro* and *in vivo*.[11] GCaMP3-based experiments have allowed examination of neural activity over long periods,[12] and large areas[13] of the vertebrate brain. The ongoing development of GCaMP sensors recently led to the description of a family of improved GCaMP5 sensors.[10] As the development of GECI molecules continues, neuroscientists will have access to indicators with spectral, affinity-related, and kinetic properties tailored to a specific experiment. A variety of vehicles may be employed to deliver GECIs into neurons, including viruses, transgenic mice, and recombinase-dependent systems, which allow flexible integration of GECI expression across the gamut of experimental designs. Moreover, GECIs may be readily utilized in animals carrying other mutations that alter locomotor activity or to study wild-type locomotion.

A primary advantage of GECIs is that they allow specific, reproducible, population-wide labeling with the reporter. In circuit studies, the genetic markers used to identify a neuron of interest may or may not have any direct functional bearing on the cell; nevertheless, they serve as a proxy for studying the labeled neuron. For example, engrailed-1 (*En1*) has been used extensively as a tool to label and study spinal interneurons arising from the V1 cell class.[14,15] While *En1* is dispensable for many properties of V1 interneurons,[16] it has been invaluable for characterizing the development and connectivity of the V1 population. The advent of GECIs with *En1* gene drivers now provides a means to study neuronal activity of the V1 population during locomotion. For example, mice expressing Cre recombinase under the control of the *En1* promoter may be crossed with a Cre-dependent GCaMP3 mouse available to the research community[17] (ROSA:Lox-STOP-Lox-GCaMP3) to allow GCaMP3 expression to be maintained specifically in V1 interneurons from embryonic development into adulthood (abbreviated En1:GCaMP3 mice). Viral-based GECI expression strategies may also be used to image-specific interneuron subsets. However, these approaches necessitate the invasive injection of a virus into the animal several days before imaging, complicating early postnatal experiments. Furthermore, much like the bolus loading of cell permeable dyes, viral-driven expression of GECIs will lead to variable indicator expression in a small area of the CNS surrounding the injection site. For these reasons, we have focused on a genetic GECI expression system to stably and reproducibly express the indicator in specific cells throughout the CNS.

Genetic labeling strategies can take advantage of the wide variety of Cre lines driving expression in specific sets of neurons. One important starting point is Cre drivers for the major ventral interneuron domains: V0 Dbx1:Cre, V1 En1:Cre, V2a Chx10:Cre, VMN Hb9:Cre, Isl1:Cre, and V3 Sim1:Cre. Genetic strategies may also take advantage of expression patterns defined by many other genes, including those of neurotransmitters,

axon guidance and cell adhesion molecules, as well as transcription factors not directly related to the developmental classes of spinal neurons.[1,18] A variety of exploratory online tools such as the Gensat project (www.gensat.org/cre.jsp), Allen Institute gene expression atlas (mouse.brain-map.org/), and a comprehensive list of Cre lines in published research (www.informatics.jax.org/recombinase.shtml) are available to facilitate the development of mouse genetic expression strategies for cell types of interest.

From photons to physiology

The use of imaging to monitor neuronal activity in large numbers of identified cells offers unprecedented opportunities, but these types of experiments generate large multidimensional data sets that introduce new technical challenges. Analysis of calcium imaging data generally begins with choosing regions of interest (ROIs). Manually drawing ROIs can introduce biases into the sampling and becomes tedious with large cell populations. A solution is to design algorithms that automatically identify ROIs based on neuron morphology.[19] We have found methods based on grouping pixels with similar intensity fluctuations over time to be particularly powerful for extracting interesting signals from our data sets.[20–23]

Cell type–specific GECI labeling facilitates automated morphological approaches by restricting labeling to particular neural classes, increasing the contrast between neurons of interest and the surrounding neuropil. In cases where neurons are recruited sparsely within a population, ROI detection by neuronal morphology may be advantageous. By segmenting images independent of time series data, morphology-based approaches will similarly detect silent and active neurons. In contrast, segmentation methods based on time series information generally rely on the presence of non-noise signals recorded on particular pixels. For example, silent neurons may be grouped with background noise. Correlated activity between pixels or procedures, such as independent component analysis, can group pixels with related activity patterns. Because time series–based methods are designed to detect similar patterns of activity from pixels, individual neurons with similar activity profiles are often grouped, requiring a final morphology-based approach to break cell groups into individual somata.[17] While it is intuitive to segment imaging data into individual neurons, it is also important to characterize groups of neurons with similar activity patterns. Our preliminary results indicate that a combination of principal component analysis and k-means clustering may be used to detect groups of spinal motoneurons with related activity, bypassing the need to segment images into single cells.[23]

Once imaging data are segmented into specific ROIs, further analyses can be performed comparing the detected regions. To date, spinal cord imaging has focused on identifying rhythmically active neurons coupled to particular phases of motor output. Methods based on continuous wavelet transforms[24] and coherence analysis[7] can be used to automate detection of common frequencies and to identify oscillation phase relationships between single neuron calcium signals and electrically recorded motoneuron firing.

Because calcium signals are a reflection of neuronal spiking,[11,25] their waveforms can be used to infer important information about the firing patterns of neurons. While waveform analysis of GECI signals may be complicated by slow kinetics, nonlinear responses, and the difficulty of comparing signal amplitudes between neurons, a variety of information may be extracted from these neural activity-derived signals. Because the slow decay kinetics of sensors results in summing of fluorescence when a train of spikes occurs, the relative position of a calcium oscillation peak is related to the timing of the final spike in a burst of firing in the imaged neuron. Likewise, despite the difficulty comparing signal amplitudes between neurons, it remains possible to extract novel information about a single neuron's circuit integration by correlating its signal amplitude over time with other neurons (Fig. 1D).

Rhythmic activity in the V1 interneuron population

The cardinal progenitor cell domains, pV0–pV3, generate multiple ventral neuron subtypes, including V0, V1, V2a, V2b, and V3 interneurons. We have examined the ensemble activity of En1[+] V1 interneurons, which, although they arise from a common progenitor population, represent a mixture of incompletely defined interneuron subtypes that include at least two cell groups first identified in classic studies using electrophysiology: Renshaw cells and 1A inhibitory interneurons,[26,27] both of which provide monosynaptic inhibitory inputs onto motor neurons. Renshaw cells receive collateral inputs from motor neurons, and dampen activity in homonymous and synergistic motor neurons. Likewise, 1A inhibitory interneurons dampen motor activity, but target antagonistic motor pools as part of a simple reflex circuit that includes 1A sensory afferents. The precise contribution of Renshaw cells and 1A inhibitory interneurons to locomotor network activity and behavior remains to be determined, but it is clear that the burst duration of the step cycle is elongated when the function of the entire V1 cell population is disrupted, or when cholinergic feedback from motor neurons to Renshaw cells is prevented.[14,28] Thus, the activity of V1 cells influences the rate of fictive stepping.

To investigate the activity of En1-expressing interneurons, the expression of GCaMP3 was targeted to the V1 lineage using the En1:GCaMP3 mouse described previously. We have found that the expression of GCaMP3 driven from the ROSA locus did not cause obvious fictive locomotor phenotypes in animals where recombination was activated in V1 interneurons (data not shown). Likewise, we have generated CMV:GCaMP3 transgenic mice that express high levels of GCaMP3 throughout all tissues in embryos and adults, and did not observe fictive locomotor phenotypes as a byproduct of widespread expression of this calcium binding sensor (K. Hilde, C.A. Hinckley & S.L. Pfaff, unpublished observations).

We have noticed fewer labeled neurons with the ROSA:Lox-STOP-Lox-GCaMP3 allele compared to similar ROSA-targeted conditional reporter strains (e.g., ROSA:Lox-STOP-Lox-tdTom) when crossed to a variety of cell-specific Cre lines labeling spinal neurons (data not shown). A between-animal comparison of GCaMP3 and tdTomato recombination efficiency with a robust Cre driver suggested that 50–75% of the neurons observed with tdTomato would clearly express GCaMP3. Whether these differences reflect low expression of GCaMP3 in a subset of neurons or subtle differences in the recombination efficiencies of the ROSA-targeted transgenes remains unknown. Consistent with previous studies,[14,29] we have not observed any clear patterns of bias or a lack of specificity in En1:Cre-driven GCaMP3 recombination, suggesting that Cre-dependent expression of GCaMP3 is a powerful tool to examine the activity of defined neuron populations with cellular

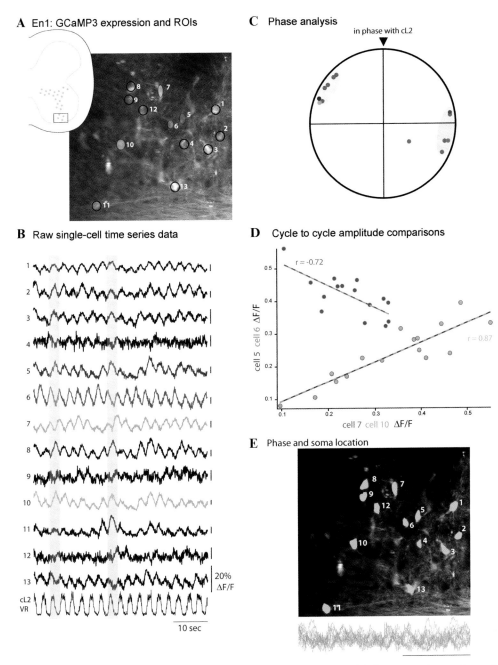

A En1: GCaMP3 expression and ROIs

B Raw single-cell time series data

10 sec

C Phase analysis

in phase with cL2

D Cycle to cycle amplitude comparisons

r = -0.72

r = 0.87

cell 5 cell 6 ΔF/F

cell 7 cell 10 ΔF/F

E Phase and soma location

Figure 1. Rhythmic activity and heterogeneity among neighboring V1 interneurons. (A) Fluorescence image of upper lumbar, ventral horn En1:GCaMP3 neurons imaged through the ventral white matter in a P1 mouse. Thirteen neurons were circled as ROIs in this 300×300 μm field of view. Lateral is up. Inset illustrates En1:GCaMP3 expression in upper lumbar spinal cord. Box indicates approximate location of imaged neurons within the ventral horn as ~50 μm deep within the ventral gray matter. (B) Time series traces of fluorescence intensity fluctuations from ROIs defined in A. Cells 5, 6, 7, and 10 are highlighted in panels C and D. During neurochemically induced fictive locomotion, 12/13 interneurons in this field of view showed rhythmic activity significantly phase locked to electrically recorded ventral root output (bottom trace). Oscillation amplitudes varied between 10–60% $\Delta F/F$. Gray shading highlights two arbitrary cycles of activity to clarify phase relationships between the imaged neurons. (C) The phase of V1 interneuron bursts identified in 1B cluster into two groups. Phase relationships relative to the ventral root were calculated from smoothed imaging traces. Each group contains similar numbers of cells and are ~180° out of phase. In this example, 7/13

resolution. Even though baseline GCaMP3 fluorescence is dim relative to GFP, we have generally been able to identify individual GCaMP3-expressing En1^{+} neurons in the quiescent spinal cord. Although it was necessary to use Ti:sapphire laser powers several fold higher than those for GFP to image GCaMP3 fluorescence, we routinely imaged individual fields of view for 100 s without significant bleaching or photo damage.

We investigated the activity heterogeneity of upper lumbar (L2) V1 interneurons during neurochemically induced fictive locomotion. The activity of individual En1:GCaMP3 neurons in the Renshaw cell area[16] (Fig. 1A, inset) were imaged through the ventral surface of isolated spinal cord at postnatal days 0–2 using 2P microscopy. Following pharmacological induction of fictive locomotor activity with N-methyl-D,L-aspartate (NMA) and serotonin GCaMP3 fluorescence oscillations in V1 interneurons routinely exceeded 30% $\Delta F/F$, similar to other reports using the conditional CGaMP3 reporter mouse[17] (Fig. 1A and B). These observations suggest that GCaMP3 has the required sensitivity and kinetics to resolve locomotor oscillations with single-cell resolution within an en bloc tissue preparation.

Preliminary analyses have extensively characterized the activity heterogeneity among 13 GCaMP3^{+} V1 interneurons present in a single 300×300 μm optical section, ~50 μm deep in the ventral gray matter, scanned at 13–24 frames/s. Similar types of heterogeneity were evident in a set of experiments ($n = 4$; see below) where ROIs were manually drawn in image J, while subsequent phase and amplitude measurements were performed on 0.1–1.0 Hz bandpass-filtered traces by custom-written scripts in IGOR Pro (WaveMetrics, OR). Changes in rhythmic fluorescence intensity, phase locked

to ventral root bursting, were detected ($P < 0.01$ for $r > 0.6$, Rayleighs test) in 12 out of 13 cells (92%) in the presence of locomotor-evoking drugs (Fig. 1B, supporting video V1). While the amplitude of oscillations varied among these neighboring V1 interneurons, even V1 interneurons with low signal-to-noise oscillations (e.g., cells 4 and 9) retained tight coupling to electrically recorded ventral root activity ($r = 0.87$ and 0.89, respectively).

Next, we examined the phase relationship of the rhythmic V1 interneuron oscillations and found two dominant patterns. Seven cells were clustered around a similar bursting phase to cells 5 and 10 (Fig. 1C, blue shading), whereas six cells displayed bursting patterns that were ~180° out of phase compared to this grouping (Fig. 1C, orange shading), similar to a bi-phasic distribution of bursts observed with single-cell recordings of Renshaw cell activity during fictive locomotion.[30] This preliminary example suggests that GECI-based imaging can recapitulate phase measurements from whole cell recordings;[30] however, further analysis will be necessary to determine whether all fields of view contain similar heterogeneity. In these studies, the cells were distributed in nearly equal numbers in the two phase groups. One interpretation is that the Renshaw cells, and more generally, the V1 interneuron population studied here, are linked to the opposing flexor-extensor activity and therefore fall into two phase groups.

Although this interpretation is a strong possibility, there are several observations that suggest that this bi-phasic pattern of activity may be associated with a different layer of CPG regulation. First, the finding that the number of L2 V1 interneurons is evenly distributed between the two phases seems inconsistent with the flexor-dominated L2 motor output. Second, genetic ablation of V1

neurons are approximately in phase with the cyan neuron (blue shading), whereas the remaining six neurons are out of phase with the cyan neuron (orange shading). (D) Pair-wise cell comparisons of relative burst amplitudes. Trough-to-peak amplitudes were calculated for each locomotor cycle in each cell and plotted against one another; linear regressions determined whether correlated amplitude fluctuations were present between neurons. A high amplitude oscillation in neuron 10 (cyan) predicts a high amplitude oscillation in neuron five (blue, $R = 0.87$). In contrast, high amplitude oscillations in neuron seven (orange) predict a low amplitude oscillation in neuron six (red, $R = -0.72$). (E) Neighboring V1 interneurons are not spatially segregated by phase. ROIs were color coded according to the phase groups identified in C. Neighboring V1 interneurons were often active in opposite phases (e.g., cells 5 and 6), such that no clear spatial segregation was noticed. Superimposed, expanded traces from all 13 neurons in the field of view are shown below the image. Traces are color coded according to their phase group from C, highlighting the two alternating cell groups in the field of view. Imaging trace amplitudes scaled arbitrarily; scale bar 10 seconds. Representative example from $n = 4$ experiments.

interneuron function altered the bursting rate of the CPG and failed to cause defects in flexor-extensor coordination.[14] It will be intriguing to monitor the pattern of V1 interneuron activity in mutants that lack flexor-extensor control in order to determine whether V1 activity can be dissociated from the circuitry that controls flexion and extension.

Waveform analysis: circuit integration from GECI imaging

It is possible that additional heterogeneity could be identified among V1 interneurons by extending our analysis beyond the characterization of phase relationships. For example, the examination of GECI waveforms might help identify neurons similarly integrated into locomotor circuits because neurons with common sources of synaptic drive will likely have highly correlated activity. To investigate this possibility, the cycle-to-cycle fluctuations in the GCaMP3 signal amplitude were analyzed, which were thought to vary depending on the amount of synaptically driven spiking during a cycle. Pair-wise comparisons were performed between this set of 13 interneurons and examples were found of cells whose bursts were strongly correlated in amplitude, either positively or negatively (Fig. 1D). Interestingly, cells with in-phase activity could have positive or negative correlations. Future experiments will determine the precise relationship between spike numbers and GCaMP3 signals. The most parsimonious explanation for these findings is that spike numbers numbers/rates may be modulated independently of oscillation phase. Because the V1 interneurons sharing the same burst phase could be further divided into groups that were positively, negatively, and nonsignificantly correlated with burst amplitudes, it is clear that the activity patterns of V1 cells are more heterogeneous than would be suggested by phase comparisons alone.

Finally, this 300×300 μm field of view was examined for spatial organization among V1 interneurons that were active during the two main phases. In contrast to the organization of motoneurons in functionally related homonymous motor pools, initial analyses suggest that V1 interneurons active in different phases were intermingled (Fig. 1E). Amplitude correlations may provide another measure of network-driven activity that could be used to examine whether neighboring neurons have similar activity (Fig. 1D). For example, neurons with highly correlated cycle-to-cycle amplitude fluctuations may spatially segregate from uncorrelated neurons.

Because only small fields of view (300×300 μm) have been examined, the large-scale organization among V1 interneurons remains unknown. Motoneurons, the ultimate targets of locomotor networks, are highly spatially organized, segregating into stereotyped columns and pools that spread rostrocaudally over multiple spinal segments.[31,32] It would be reasonable to predict that any spatial organization among spinal interneurons would reflect this rostrocaudal columnar pattern. Therefore, reconstructing neural activity over multisegment regions of the spinal cord is a likely prerequisite to uncover any spatial organization in the network.

Conclusion

The preliminary work presented here is intended to illustrate our approach to analyzing cell type–specific imaging data. A surprising degree of heterogeneity has been detected in the activity of V1 interneurons by comparing the phase of rhythmic oscillations and correlations of amplitude between neighboring interneurons. In the same way that differences in oscillation phase can be attributed to differences in synaptic drive, we hypothesize that differences in amplitude modulation and imaging signal waveforms are likely to reflect distinct patterns of network integration. If this interpretation is correct, these novel measures provide an activity-based readout of the heterogeneity within a cell population. The observation of many distinct functionally related groups within a defined neural class would suggest that the spinal locomotor network is built up from numerous semi-independent modules.

Our preliminary experiments indicate that neurons from a common genetically defined lineage are specified, such that even neighboring neurons can differentially integrate into locomotor networks. Our observation that similar numbers of neurons fall into two distinct phase groups may suggest a tightly regulated developmental process producing equal numbers of functionally defined V1 subtypes in a specific area, similar to the balanced production of V2a and V2b subtypes from a common progenitor domain.[33] This result is perhaps most surprising because, based on their ventral location, many of the V1 neurons sampled are likely Renshaw cells, which are generally believed to be synaptically driven by nearby motoneurons. Given that the fictive locomotor output from L2 motoneurons is dominated by a single phase of motoneuron activity, local motoneuron input seems unable to drive two equal antiphase populations of V1 Renshaw cells. Alternatively, this heterogeneity could reflect spatial intermixing between distinct subtypes of V1 interneurons. Our experiments point to the promise of GECI-based imaging to uncover functional heterogeneity within molecularly defined cell populations. Future experiments will systematically examine many aspects of V1 interneuron activity over larger areas of the lumbar spinal cord, potentially providing clues about how these neurons contribute to locomotor networks.

Acknowledgments

We thank Ariel Levine for helpful comments. Conditional GCaMP3 mice were provided by Hongkui Zeng, the Allen Institute for Brain Science; Engrailed-1 Cre mice were available from Jackson Labs, line 007916. C.A.H. is supported by a NINDS NRSA fellowship, and S.L.P. is Benjamin H. Lewis Chair in Neuroscience and Investigator with the Howard Hughes Medical Institute.

Conflicts of interest

The authors declare no conflicts of interest.

Supporting Information

Additional supporting information may be found in the online version of this article.

Video S1: Raw and processed images from the cells analyzed in Figure 1 replayed at $5\times$ speed. Top left: Raw images from data acquisition. Top right: Raw images filtered with the 3D-hybrid median filter in image J. This filter uses adjacent averaging to reduce spatial and temporal noise. Bottom left: Pixel by pixel 1 Hz low pass filter on raw images. This image approximates the preprocessing used to extract phase and amplitude information. Bottom right: same 1 Hz low pass filtration as bottom left with each pixel's minimum intensity value set to zero by subtraction.

References

1. Alaynick, W.A., T.M. Jessell & S.L. Pfaff. 2011. SnapShot: spinal cord development. *Cell* **146:** 178–178.e1.

2. Knöpfel, T. 2012. Genetically encoded optical indicators for the analysis of neuronal circuits. *Nat. Rev.* **13:** 687–700.

3. Kwan, A.C., S.B. Dietz, W.W. Webb & R.M. Harris-Warrick. 2009. Activity of Hb9 interneurons during fictive locomotion in mouse spinal cord. *J. Neurosci.* **29:** 11601–11613.

4. Wilson, J.M., D.A. Dombeck, M. Díaz-Ríos, *et al.* 2007. Two-photon calcium imaging of network activity in XFP-expressing neurons in the mouse. *J. Neurophysiol.* **97:** 3118–3125.

5. Zhong, G. *et al.* 2010. Electrophysiological characterization of V2a interneurons and their locomotor-related activity in the neonatal mouse spinal cord. *J. Neurosci.* **30:** 170–182.

6. Wilson, J.M., E. Blagovechtchenski & R.M. Brownstone. 2010. Genetically defined inhibitory neurons in the mouse spinal cord dorsal horn: a possible source of rhythmic inhibition of motoneurons during fictive locomotion. *J. Neurosci.* **30:** 1137–1148.

7. Kwan, A.C., S.B. Dietz, G. Zhong, *et al.* 2010. Spatiotemporal dynamics of rhythmic spinal interneurons measured with two-photon calcium imaging and coherence analysis. *J. Neurophysiol.* **104:** 3323–3333.

8. Romoser, V.A., P.M. Hinkle & A. Persechini. 1997. Detection in living cells of Ca^{2+}-dependent changes in the fluorescence emission of an indicator composed of two green fluorescent protein variants linked by a calmodulin-binding sequence: a new class of fluorescent indicators. *J. Biol. Chem.* **272:** 13270–13274.

9. Miyawaki, A. *et al.* 1997. Fluorescent indicators for Ca^{2+} based on green fluorescent proteins and calmodulin. *Nature* **388:** 882–887.

10. Akerboom, J. *et al.* 2012. Optimization of a GCaMP calcium indicator for neural activity imaging. *J. Neurosci.* **32:** 13819–13840.

11. Tian, L. *et al.* 2009. Imaging neural activity in worms, flies and mice with improved GCaMP calcium indicators. *Nature Methods* **6:** 875–881.

12. Huber, D. *et al.* 2012. Multiple dynamic representations in the motor cortex during sensorimotor learning. *Nature* **484:** 473–478.

13. Ahrens, M.B. *et al.* 2012. Brain-wide neuronal dynamics during motor adaptation in zebrafish. *Nature* **485:** 471–477.

14. Gosgnach, S. *et al.* 2006. V1 spinal neurons regulate the speed of vertebrate locomotor outputs. *Nature* **440:** 215–219.

15. Benito-Gonzalez, A. & F.J. Alvarez 2012. Renshaw cells and Ia inhibitory interneurons are generated at different times from p1 progenitors and differentiate shortly after exiting the cell cycle. *J. Neurosci.* **32:** 1156–1170.

16. Sapir, T. *et al.* 2004. Pax6 and engrailed 1 regulate two distinct aspects of renshaw cell development. *J. Neurosci.* **24:** 1255–1264.

17. Zariwala, H.A. *et al.* 2012. A Cre-dependent GCaMP3 reporter mouse for neuronal imaging in vivo. *J. Neurosci.* **32:** 3131–3141.

18. Garcia-Campmany, L., F.J. Stam & M. Goulding. 2010. From circuits to behaviour: motor networks in vertebrates. *Curr. Opin. Neurobiol.* **20:** 116–125.

19. Ohki, K., S. Chung, Y.H. Ch'ng, *et al.* 2005. Functional imaging with cellular resolution reveals precise microarchitecture in visual cortex. *Nature* **433:** 597–603.

20. Ozden, I., H.M. Lee, M.R. Sullivan & S.S.-H. Wang. 2008. Identification and clustering of event patterns from in vivo multiphoton optical recordings of neuronal ensembles. *J. Neurophysiol.* **100:** 495–503.

21. Miri, A., K. Daie, R.D. Burdine, *et al.* 2011. Regression-based identification of behavior-encoding neurons during large-scale optical imaging of neural activity at cellular resolution. *J. Neurophysiol.* **105:** 964–980.

22. Mukamel, E.A., A. Nimmerjahn & M.J. Schnitzer. 2009. Automated analysis of cellular signals from large-scale calcium imaging data. *Neuron* **63:** 747–760.

23. Hinckley, C., S. Driscoll & S. Pfaff. 2012. Large scale imaging and reconstruction of spinal motoneuron activity. *Soc. Neurosci. Abstract* **788.23.**

24. Gallarda, B.W., T.O. Sharpee, S.L. Pfaff & W.A. Alaynick. 2010. Defining rhythmic locomotor burst patterns using a continuous wavelet transform. *Ann. N.Y. Acad. Sci.* **1198:** 133–139.

25. Smetters, D., A. Majewska & R. Yuste. 1999. Detecting action potentials in neuronal populations with calcium imaging. *Methods* **18:** 215–221.

26. Renshaw, B. 1941. Influence of discharge of motoneurons upon excitation of neighboring. *J. Neurophysiol.* **4:** 167–183.

27. Eccles, J.C., P. Fatt & K. Koketsu. 1954. Cholinergic and inhibitory synapses in a pathway from motor-axon collaterals to motoneurones. *J. Physiol.* **126:** 524–562.

28. Myers, C.P. *et al.* 2005. Cholinergic input is required during embryonic development to mediate proper assembly of spinal locomotor circuits. *Neuron* **46:** 37–49.

29. Zariwala, H.A. *et al.* 2012. A Cre-dependent GCaMP3 reporter mouse for neuronal imaging in vivo. *J. Neurosci.* **32:** 3131–3141.

30. Nishimaru, H., C.E. Restrepo & O. Kiehn. 2006. Activity of Renshaw cells during locomotor-like rhythmic activity in the isolated spinal cord of neonatal mice. *J. Neurosci.* **26:** 5320–5328.

31. Romanes, G.J. 1951. The motor cell columns of the lumbo-sacral spinal cord of the cat. *J. Comp. Neurol.* **94:** 313–363.

32. McHanwell, S. & T.J. Biscoe. 1981. The localization of motoneurons supplying the hindlimb muscles of the mouse. *Philos. Trans. R. Soc. Lond. Ser. B* **293:** 477–508.

33. Peng, C.-Y. *et al.* 2007. Notch and MAML signaling drives Scl-dependent interneuron diversity in the spinal cord. *Neuron* **53:** 813–827.

Ann. N.Y. Acad. Sci. ISSN 0077-8923

ANNALS OF THE NEW YORK ACADEMY OF SCIENCES

Issue: *Neurons, Circuitry, and Plasticity in the Spinal Cord and Brainstem*

Glutamatergic reticulospinal neurons in the mouse: developmental origins, axon projections, and functional connectivity

Marie-Claude Perreault[1] and Joel C. Glover[2]

[1]Department of Physiology, Emory University, Atlanta. [2]Department of Physiology, University of Oslo, Oslo, Norway

Address for correspondence: Marie-Claude Perreault, Emory University School of Medicine, Whitehead Biomedical Research Building, Department of Physiology, 615 Michael Street, Atlanta, GA 30322. m-c.perreault@emory.edu

Subcortical descending glutamatergic neurons, such as reticulospinal (RS) neurons, play decisive roles in the initiation and control of many motor behaviors in mammals. However, little is known about the mechanisms used by RS neurons to control spinal motor networks because most of the neuronal elements involved have not been identified and characterized. In this review, we compare, in the embryonic mouse, the timing of developmental events that lead to the formation of synaptic connections between RS and spinal cord neurons. We then summarize our recent research in the postnatal mouse on the organization of synaptic connections between RS neurons and lumbar axial motoneurons (MNs), hindlimb MNs, and commissural interneurons. Finally, we give a brief account of some of the most recent studies on the intrinsic capabilities for plasticity of the mammalian RS system. The present review should give an updated insight into how functional specificity in RS motor networks emerges.

Keywords: voluntary movement; descending motor control; reticular formation; spinal cord; motoneurons; commissural interneurons

Introduction

Motor commands from the brain to spinal motoneurons (MNs) are implemented primarily by glutamatergic neurons in the brainstem that project to the spinal cord. The largest source of glutamatergic descending input to the spinal cord is the reticulospinal (RS) system, which has long been known to play a decisive role in the initiation and control of locomotion,[1,2] and the regulation of autonomic functions.[3] During the last few years, the RS system has received renewed attention because of recent work in cats and monkeys that also implicates it in the control of reaching,[4,5] grasping,[6] and fine finger movements.[7]

Despite the fact that individual RS axons have extensive terminal distributions[8,9] that presumably control numerous and spatially widespread spinal neurons, the identification of the spinal targets of the RS system has traditionally relied on microelectrode recording that samples one or only a few neurons at

a time.[10–13] This is a challenging and low throughput approach, and more innovative and productive technologies would permit a more rapid characterization of the complex interactions between RS neurons and the large number of spinal neurons in mammals.

Over the last two decades, technical advances have made the recording of extended neuronal networks in the spinal cord of mammals possible. Functional multineuron calcium imaging (fMCI) has been used successfully to monitor the activity of hundreds of individual neurons simultaneously in the isolated neonatal mouse spinal cord.[14,15] We have employed fMCI in different brainstem–spinal cord preparations of the neonatal mouse and have begun to explore the connections between brainstem descending glutamatergic neurons and various populations of spinal neurons. In this review, after a short recapitulation of the ontogeny of the RS system in the embryonic mouse, we discuss our recent data on the organization of medullary RS neuron

doi: 10.1111/nyas.12054

Ann. N.Y. Acad. Sci. 1279 (2013) 80–89 © 2013 New York Academy of Sciences.

connections to lumbar MNs and commissural interneurons (CINs) in the neonatal mouse. We also provide a survey of the most recent work in mammals on RS plasticity and its contribution to functional motor recovery after brain or spinal cord injury.

The RS system in the developing mouse

As in lower vertebrates (e.g., amphibians,[16] fish,[17] and birds[18]), RS neurons in the mouse are among the very first neurons generated in the entire central nervous system. Murine RS neurons are generated starting around embryonic (E) day 8,[19,20] before spinal MNs are born.[21,22] Following the birth of RS neurons, a series of prenatal developmental events takes place that contribute to the formation of functional synaptic connections between RS and spinal cord neurons. In the mouse, these events occur between E11 and E17. Early during this period (through about E13), the rhombomeric organization that subdivides the brainstem along its anteroposterior axis is clearly seen. This rhombomeric patterning is under the control of developmental regulatory genes such as the Hox genes and their regulatory agents.[23] Careful mapping using retrograde tracing[24] has shown that specific RS neuron groups lie in defined rhombomeric territories r1–r8 (Fig. 1A). However, because rhombomeres are expressed only ephemerally, a direct correlation of rhombomere-related embryonic RS neuron clusters to adult RS neuron populations requires fate mapping of rhombomeric territories. Although we have not yet completed a genetically based fate mapping of RS neuron groups (as we and others have done for vestibular projection neurons,[25] Fig. 1B), the general topographical distribution of RS neurons as discrete clusters at distinct anteroposterior and mediolateral (dorsoventral) locations observed during embryonic development[24] (Fig. 1A) appears to be largely maintained in the neonatal (unpublished observations, M.C. Perreault and J.C. Glover) and adult mouse.[27] The largest coherent group of RS neurons is located in the medulla in the nucleus reticularis gigantocellularis, with an estimated 10,000 RS neurons on the ipsilateral side versus 4,000 on the contralateral side (retrogradely labeled from ipsilateral spinal cord), and the second largest group is found in the pons in the nucleus pontis oralis and caudalis, with about 3,000 RS neurons on each side.[28]

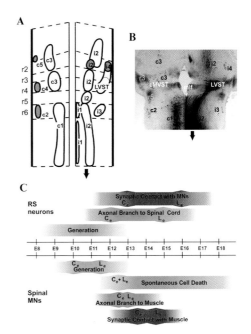

Figure 1. RS neurons in the mouse embryo are organized into coherent groups occupying specific domains related to hindbrain rhombomeres. (A) Schematic redrawn and modified from Ref. 24, illustrating the RS groups whose axons project down one side of the spinal cord (arrow), identified in whole mount preparations of the E16.5 mouse embryo brainstem. Each group is labeled in sequence from caudal to rostral locations and according to whether it projects contralaterally (c1, c2, etc.) or ipsilaterally (i1, i2, etc.). Groups that are colored gray exhibit lateral displacements during embryonic development. Rhombomeres (r2–r6) are indicated to the left (r1 not shown). Some groups correspond to well-defined neuron populations; for example, i1 represents the raphespinal neurons and i2, c1, and c3 contain the glutamatergic medullary and pontine RS neurons. (B) A single 50 μm section from a transgenic E16.5 mouse embryo brainstem in which RS and vestibulospinal groups have been labeled retrogradely with biotin dextran amine (brown) from one side of the spinal cord (arrow). Neuronal somata and their descending axons are visualized by the biotin labeling, and the RS groups shown in (A) that are visible in this section are indicated. The two vestibulospinal groups LVST (lateral vestibulospinal tract group) and cMVST (contralateral medial vestibulospinal tract group) are also indicated. The caudal part of the i2 RS group is largely obscured by the descending LVST axons. The blue staining indicates the expression of beta-galactosidase driven by an r3/r5-specific enhancer element. Modified from Ref. 25. (C) Schematic showing the timeline of major events in the development of RS neurons and their synaptic connections (above the timeline) and MNs and their synaptic connections (below the timeline) in the embryonic mouse spinal cord. These events, whose approximate durations are illustrated by the colored bands, typically do not begin simultaneously in the different segments of the spinal cord but progress from upper cervical to sacrocaudal segments. Approximate onsets in the cervical and lumbar enlargements (Ce and Le, respectively) are indicated by the enlargements of the bands.

The clustering of RS neurons into identifiable subpopulations with defined anteroposterior and dorsoventral domains suggests a differentiation based on molecular determinants. Identification of such determinants has barely begun, but initial studies show that about 30% of RS neurons fall into one of three subpopulations that transiently express the LIM (Lin-11, Isl-1, and Mec-3) homeobox proteins LHX1/5 (one ipsilateral and one contralateral subpopulation) or LHX3/4 and Chox10 (one ipsilateral subpopulation).[29] Given the number of RS neuron clusters identified by retrograde tracing alone,[24] a more exhaustive survey of transcription factor expression should reveal many more combinatorial subsets that can be linked to specific clusters.

Soon after they are born, RS neurons start extending axons toward the spinal cord. Although many details remain to be described regarding the precise trajectories that different groups of RS axons take within the brainstem, the axons of several RS groups join the medial longitudinal fasciculus (MLF) either ipsilaterally or contralaterally,[24] while the axons of others descend laterally (unpublished observations, M.C. Perreault and J.C. Glover). The first RS axons reach the spinal cord by E11[24] (Fig. 1C). The precise time at which RS axons make synaptic contacts with cervical spinal neurons has not yet been determined in the mouse, but Sholomenko and O'Donovan[30] have shown that activation of RS neurons in the chicken embryo can trigger rhythmic activity in the caudal spinal cord at E6, or three days after RS axons enter the cervical spinal cord.[31] This corresponds to about stage E14 in the mouse. Interestingly, the time window between E11 and E17, during which RS neurons likely establish first synaptic contacts with spinal neurons, begins earlier and overlaps with the period of spontaneous cell death in spinal MNs[32,33] (Fig. 1C), a process associated with the numerical matching of MNs and muscle fibers. The influence of RS inputs on MN survival during this developmental period remains to be investigated in the mouse, but removal of descending inputs in the chicken embryo substantially enhances motoneuronal cell death at least in the lateral motor column (LMC) of the lumbar cord.[34,35] Finally, as is typical for many neurons, RS neurons are likely to be capable of releasing neurotransmitters before making synaptic contacts. The vesicular glutamate transporter Vglut2, which may already be expressed

in RS neurons at E12.5,[36] is clearly present in these neurons in the newborn mouse.[37,38]

Thus, consistent with a prime role for RS neurons in the control of movement, the timeline along which RS neurons and their axonal projections develop in the mouse embryo indicates that RS synaptic connections to MNs, and probably other spinal neurons, are very likely to be functional already at birth.

Specificity of RS connections in the newborn mouse

We have employed fMCI to investigate the functionality and specificity of synaptic connections between RS neurons of the medullary reticular formation (MRF) and three populations of spinal cord neurons in the lumbar (L) segments: MNs of the medial motor column (MMC, Fig. 2), MNs of the lateral motor column (LMC, Fig. 2), and descending commissural interneurons (dCINs, Fig. 3). Microstimulation of the MRF, where medullary RS neurons reside, leads to the recruitment of MNs in the MMC and the LMC in L2 (Fig. 2B), and L5 segments in the newborn mouse.[39] Single pulses are often sufficient to evoke widespread activation of lumbar MNs, but two pulses are generally required to generate response magnitudes sufficiently above the background signal level (>2 SD) to permit unequivocal detection. Simultaneous recordings of Ca^{2+} responses in spinal MN somata, and electrical activity in ventral roots, indicate that somatic Ca^{2+} transients in MNs relate directly to the presence of action potentials in their axons.[40–42]

Increasing the number of stimuli further increases the magnitudes of the evoked responses in lumbar MNs and reveals differences in the shapes of their waveforms in the LMC (compare the responses to train stimulation of MNs 1, 3, and 4, and of MNs 2, 5, and 6 in Fig. 2B). The reason for the differences in response waveforms in the LMC remains to be determined, but our working hypothesis is that it is linked to the recruitment of different proportions of excitatory and inhibitory premotor interneurons that specifically target distinct task-related MN groups in the LMC, which are likely to be more diverse than those in the MMC (flexor, extensor, and bifunctional hindlimb MNs versus trunk MNs). The increase in response magnitudes with sequential increase in number of stimulation pulses in

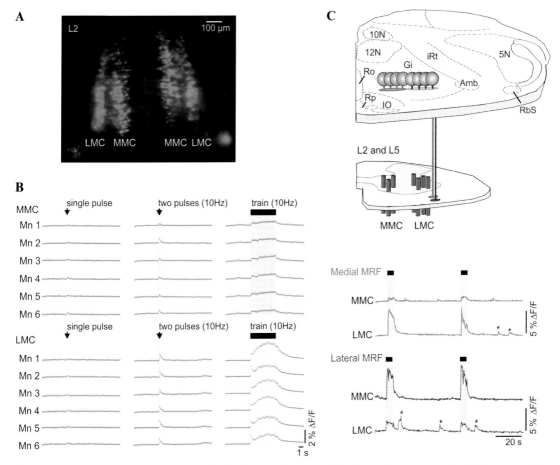

Figure 2. RS neurons and lumbar MNs are already functionally connected in the newborn mouse and exhibit a connectivity pattern that would permit differential control over trunk and hindlimb muscles. Synaptically induced calcium transients were optically recorded in individual MNs of the ipsilateral and contralateral MMC and LMC following focal electrical stimulation of the medullary reticular formation in an isolated brainstem–spinal cord preparation. (A) Lumbar MNs were retrogradely loaded with the fluorescent calcium indicator, calcium green-1 (CG-1) conjugated dextran amine. In the newborn mouse, individual MNs are easily distinguishable through the ventral white matter. Modified from Ref. 42. (B) Stimulation with a single pulse induces widespread activation of the lumbar MNs. Increasing the number of pulses to 50 increases the magnitudes of calcium responses in the recruited MNs in both motor columns and reveals differences in the shape of the response waveforms in the LMC. The differences in shape of responses might be linked to the fact that the LMC has a greater diversity of task-related groups compared to the MMC (flexor, extensor, and bifunctional hindlimb MNs versus trunk MNs). (C) Stimulation of the ventral MRF at different mediolateral locations reveals a functional organization wherein distinct lateral and medial populations of RS neurons predominantly activate MMC or LMC MNs of the lumbar segments. Modified from Ref. 38.

both motor columns is compatible with a progressive recruitment of premotor interneurons.

Supporting a topographic specificity in RS connections to spinal cord neurons, stimulation of the ventral MRF at different mediolateral locations in the newborn mouse reveals a functional organization wherein distinct lateral and medial populations of RS neurons predominantly activate MMC or LMC MNs, respectively, in lumbar segments.[39,43]

This reciprocally organized connectivity pattern has been observed in both ipsilateral and contralateral L2 and L5 segments (Fig. 2C), and we are currently investigating whether a similar mediolateral organization exists for RS control over thoracic axial and forelimb musculature. Reports of excitatory connections from RS neurons to limb MNs, and also more recently to lower trunk MNs,[44] have been published in adult mammals, but it remains to be determined

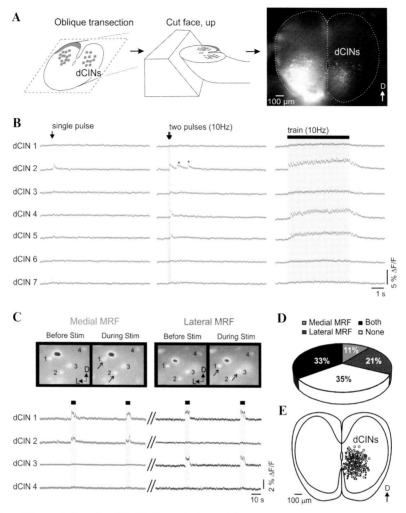

Figure 3. Differential control of trunk and hindlimb muscles by medullary RS neurons may be mediated, in part, by identifiable dCIN populations. (A) An obliquely cut, face-up brainstem–spinal cord preparation allows visualization of interneurons located deep in the gray matter, including dCINs. (B) Stimulation of the MRF at one specific locus selectively recruits dCINs. The fact that the number of recruited dCINs increased little when the number of stimuli increased from 1 to 50 suggests a high degree of specificity in the connections between the medullary RS neurons and dCINs. (C, D) A majority of dCINs in the upper lumbar cord responds to stimulation of the medial MRF, lateral MRF, or both. (E) The three responsive dCIN populations have largely overlapping spatial distributions. Panels C, D, and E are modified from Ref. 45.

whether the mediolateral organization within the MRF that we report in newborn mammals is also present in adult mammals.

The responses evoked by stimulation of the medullary RS neurons in the lumbar MNs of the neonatal mouse are strongly reduced following restriction to monosynaptic transmission by bath application of mephenesin to the spinal cord.[45] This supports the idea that RS-evoked activation of MNs is mediated to a large extent via interneurons. A similar suggestion was made earlier by Floeter and

Lev-Tov[46] with regard to activation of lumbar MNs by stimulation of the pontine RS neurons whose axons run in the MLF. Of all the populations of interneurons physiologically identified thus far in the mammalian spinal cord, very few have been classified as excitatory.[45,47,48] However, several of the transcription factor–defined populations of interneurons have a glutamatergic phenotype,[49] and it is likely a matter of time before their identity as excitatory interneurons is confirmed physiologically. Evidence in the neonatal rodent indicates that some of the interneurons within the dCIN population in the lumbar cord can excite contralateral MNs both locally[50] and in more caudal segments.[51] Thus, the question arises as to whether dCINs are involved in mediating the RS-evoked excitatory responses in lumbar MNs of the neonatal mouse. Reduction of RS-evoked responses in both MMC and LMC MNs upon lesion of the midline[45] is compatible with such a possibility, and supports a role for excitatory CINs in the descending control not only of hindlimb MNs, but also of trunk MNs.

In a modified brainstem–spinal cord preparation[43] (Fig. 3A), we directly investigated the effects of MRF stimulation in more than 500 individual dCINs in the L2 segment.[45] In contrast to lumbar MNs, stimulation of the MRF at one specific locus did not lead to a widespread activation of the dCINs. In the experiment shown in Figure 3B, single-pulse stimulation to the lateral MRF recruited one out of seven dCINs. The number of recruited dCINs increased to three with two pulses, but no additional recruitment or shift from one set of recruited dCINs to another was observed when the number of stimuli increased to either 25 or 50. Such selective recruitment argues for a high degree of specificity in the connections between the medullary RS neurons and dCINs.

Consistent with this idea, we find that responsive dCINs can be divided into three populations according to their inputs from the MRF: dCINs that only respond to stimulation of the medial MRF (dCIN2 in Fig. 3C), dCINs that only respond to stimulation of the lateral MRF (dCIN3 in Fig. 3C), and lastly, dCINs that respond to stimulation of the medial and of the lateral MRF (dCIN1 in Fig. 3C). While many dCINs responded to stimulation of the medial and/or lateral MRF, we found a substantial proportion of dCINs that did not respond to stimulation of either MRF region (Fig. 3D). It is worth noting that nearly all of these unresponsive dCINs were activated by stimulation of sensory afferents in the L2 dorsal root, demonstrating that their lack of response to RS stimulation was not due to damage but was part of a selective RS innervation pattern.

The three populations of responsive dCINs had overlapping spatial distributions and could not be differentiated on the basis of their location in the ventral horn (Fig. 3E). Thus, at least in the L2 segment, the patterns of RS–dCINs connections cannot be readily predicted by the actual locations of the dCINs in the transverse plane of that segment. In fact, we have some evidence that the functional identity of the presumed motoneuron targets of the dCINs (i.e., flexor, extensor, or trunk MNs) may be a better predictor of the connections between brainstem descending glutamatergic neurons and dCINs. In that context, it will be important to assess the longitudinal extent of the terminal collaterals of the glutamatergic dCIN axons, as well as their arborization patterns in the different projection segments, similar to what has been done previously in the adult cat.[52] Such information would be a significant step toward determining whether individual dCINs innervate the same types of neurons within each of their target segments.

In summary, the response patterns evoked in dCINs suggest that they could mediate the full complement of RS-evoked responses in lumbar MNs and support a role for dCINs in the control of both lumbar axial muscles and hindlimb muscles. An important next step will be to determine whether excitatory dCINs establish synaptic connections with MMC MNs as well as with LMC MNs. To assess this, it might be advantageous to use techniques with higher throughput than dual whole-cell recording, such as glutamate uncaging by laser scanning photostimulation.[53]

Plasticity and role in recovery of motor function

Spinal cord injuries (SCIs) are rarely complete, often leaving a substantial number of descending (and ascending) axons intact. The descending axons that remain intact after incomplete SCI have the capacity to sprout new collaterals both above and below the injury level.[54,55] Compared to the corticospinal system,[56] however, little attention has been devoted to the study of brainstem-derived descending systems and their capacity for plasticity after SCI, at least in mammals (see Refs. 57 and 58 for reviews of work performed in nonmammalian vertebrates). While the strong focus on the corticospinal tract can be attributed to its significance in recovery of voluntary motor control, it is noteworthy that brainstem-derived descending axons are the largest contingent of all descending axons ($> 60\%$ in the mouse[28]) and mediate much of the mammalian motor repertoire, including movements that require a substantial component of cortical control, the latter through corticobulbospinal connections.[59,60]

Mammalian RS axons have been shown to exhibit SCI-induced anatomical plasticity,[61–63] and evidence indicates that they may have substantially greater regenerative capacity than corticospinal axons.[64] Interestingly, CINs are the only spinal interneurons that have been shown to regenerate spontaneously after SCI in mammals.[65] Perhaps more importantly, the regenerating axons of CINs are able to conduct action potentials and evoke monosynaptic potentials in MNs.[66] It will be important in the future to determine whether RS neurons can maintain existing synaptic connections, as well as form new synaptic connections with CINs after SCI. Demonstrating that regenerating or sprouting RS axons are capable of making functional synaptic connections with their normal or new synaptic targets after SCI will be an important step in evaluating the success of combinatorial treatment strategies aimed at promoting the regeneration of descending axons.[63]

The intrinsic capabilities of the RS system during both normal motor control and recovery of function have largely been underestimated. This is well illustrated by the recent work of Baker and collaborators in nonhuman primates, which demonstrates the involvement of the pontine RS system in the control of hand and finger movements,[7,67–69] long considered to be the prerogative of the corticospinal system. Their work also illustrates that each descending system that contributes to recovery of function appears to do so within the limitations of its own specific range of normally possible connections, but not necessarily in a consistent manner, sometimes producing detrimental motor effects.[70] Although little is known about the normal constraints under which the different brainstem-derived descending systems operate, this is a crucial issue as the balance between detrimental and beneficial motor outcomes will ultimately determine the success of recovery.

Concluding remarks

Given their early development, multiple sources of inputs, and distributed connections, RS neurons probably participate in diverse motor behaviors depending on functional context. Functional mapping with high-throughput approaches, such as fMCI, goes beyond static connectivity maps, and should help determine the underlying functional architecture of the RS motor circuits controlling these different motor behaviors. Despite earlier reports describing the organization of RS neurons into discrete clusters in both birds and mammals[24,31] and the existence of a topographical organization in the reticular formation,[71–73] the fact that the reticular formation does not exhibit easily discernible laminar organization of the type seen at the level of the spinal cord or the cortex has led to a general misconception that it is a poorly organized structure. In this review, we describe RS synaptic connections to spinal MNs and INs in the mouse that not only function at birth, but also exhibit a substantial degree of topographic specificity. The level of specificity we observe would allow for both independent and integrated control of trunk and limb muscles in different functional contexts.

A better characterization of the nature of the synaptic connections between RS neurons and spinal cord neurons should not only help determine how much the mammalian brain relies on the RS system during normal motor behavior, but also provide useful information about the role of RS neurons in the context of recovery of function after spinal cord or brain injury. Success in this endeavor may require a general recognition that despite the inherent complexity of the RS system (various medullary and pontine components with extensive axonal arborization patterns), each of its components is likely to exhibit specific intrinsic functional attributes. A substantial challenge for the future will be to elucidate how these different functional components contribute individually and interactively to integrated motor regulation.

Acknowledgments

We are grateful to many past and present members of the Laboratory for Neural Development and Optical Recording at the University of Oslo for their essential contributions to the ideas and work summarized here. This work has been supported by grants from the European Union (J.C.G.), the Marie Curie Training Program (J.C.G.), Human Frontiers Science Program (J.C.G), the Christopher and Dana Reeve Foundation (M-C.P. and J.C.G.), the Norwegian Research Council (M-C.P. and J.C.G.), the University of Oslo Medical Faculty (J.C.G.), and the International Foundation for Research in Paraplegia (M-C.P.).

Conflicts of interest

The authors declare no conflicts of interest.

References

1. Armstrong, D.M. 1988. The supraspinal control of mammalian locomotion. *J. Physiol.* **405:** 1–37.

2. Jordan, L.M. 1998. Initiation of locomotion in mammals. *Ann. N.Y. Acad. Sci.* **860:** 83–93.

3. Guyenet, P.G. 2006. The sympathetic control of blood pressure. *Nat. Rev.* **7:** 335–346.

4. Schepens, B., P. Stapley & T. Drew. 2008. Neurons in the pontomedullary reticular formation signal posture and movement both as an integrated behavior and independently. *J. Neurophysiol.* **100:** 2235–2253.

5. Davidson, A.G. & J.A. Buford. 2006. Bilateral actions of the reticulospinal tract on arm and shoulder muscles In the monkey: stimulus triggered averaging. *Exp. Brain Res.* **173:** 25–39.

6. Pettersson, L.-G. *et al.* 2007. Skilled digit movements in feline and primate—recovery after selective spinal cord lesions. *Acta Physiol.* **189:** 141–154.

7. Soteropoulos, D., E.R. Williams & S.N. Baker. 2012. Cells in the monkey ponto-medullary reticular formation modulate their activity with slow finger movements. *J. Physiol.* **590:** 4011– 4027.

8. Matsuyama, K. *et al.* 1993. Termination mode and branching patterns of reticuloreticular and reticulospinal fibers of the nucleus reticularis pontis oralis in the cat: an anterograde PHA-L tracing study. *Neurosci. Res.* **17:** 9–21.

9. Matsuyama, K. *et al.* 1999. Morphology of single pontine reticulospinal axons in the lumbar enlargement of the cat: a study using the anterograde tracer PHA-L. *J. Comp. Neurol.* **410:** 413–430.

10. Davies, H.E. & S.A. Edgley. 1994. Inputs to group II-activated midlumbar interneurones from descending motor pathways in the cat. *J. Physiol.* **479:** 463–473.

11. Jankowska, E. *et al.* 2005. Functional differentiation and organization of feline midlumbar commissural interneurones. *J. Physiol.* **565:** 645–658.

12. Hammar, I., K. Stecina & E. Jankowska. 2007. Differential modulation by monoamine membrane receptor agonists of reticulospinal input to lamina VIII feline spinal commissural interneurons. *Eur. J. Neurosci.* **26:** 1205–1212.

13. Jankowska, E. & K. Stecina. 2007. Uncrossed actions of feline corticospinal tract neurones on lumbar interneurones evoked via ipsilaterally descending pathways. *J. Physiol.* **580:** 133–147.

14. O'Donovan, M.J. *et al.* 2005. Calcium imaging of network function in the developing spinal cord. *Cell Calcium* **37:** 443–450.

15. Wilson, J.M. *et al.* 2007. Two-photon calcium imaging of network activity in XFP-expressing neurons in the mouse. *J. Neurophysiol.* **97:** 3118–3125.

16. Vargas-Lizardi, P. & K.M. Lyser. 1974. Time of origin of Mauthner's neuron in Xenopus laevis embryos. *Dev. Biol.* **38:** 220–228.

17. Mendelson, B. 1986. Development of reticulospinal neurons of the zebrafish. I: time of origin. *J. Comp. Neurol.* **251:** 160–171.

18. Sechrist, J. & M. Bronner-Fraser. 1991. Birth and differentiation of reticular neurons in the chick hindbrain: ontogeny of the first neuronal population. *Neuron* **7:** 947–963.

19. Pierce, E.T. 1973. Time of origin of neurons in the brain stem of the mouse. *Prog. Brain Res.* **40:** 53–65.

20. McConnell, J.A. 1981. Identification of early neurons in the brainstem and spinal cord. II: an autoradiographic study in the mouse. *J. Comp. Neurol.* **200:** 273–288.

21. Nornes, H.O. & M. Carry. 1978. Neurogenesis in spinal cord of mouse: an autoradiographic analysis. *Brain Res.* **159:** 1–16.

22. Benito-Gonzalez, A. & F.J. Alvarez. 2012. Renshaw cells and Ia Inhibitory interneurons are generated at different times from p1 progenitors and differentiate shortly after exiting the cell cycle. *J. Neurosci.* **32:** 1156–1170.

23. Glover, J.C., J.S. Renaud & F.M. Rijli. 2006. Retinoic acid and hindbrain patterning. *J. Neurobiol* **66:** 705–725.

24. Auclair, F., R. Marchand & J.C. Glover. 1999. Regional patterning of reticulospinal and vestibulospinal neurons in the hindbrain of mouse and rat embryos. *J. Comp. Neurol.* **411:** 288–300.

25. Pasqualetti, M. *et al.* 2007. Fate-mapping the mammalian hindbrain: segmental origins of vestibular projection neurons assessed using rhombomere-specific Hoxa2 enhancer elements in the mouse embryo. *J. Neurosci.* **27:** 9670–9681.

26. Chen, Y. *et al.* 2012. Hoxb1 controls anteroposterior identity of vestibular projection neurons. *PLoS One* **7:** e34762.

27. VanderHorst, V.G. & B. Ulfhake. 2006. The organization of the brainstem and spinal cord of the mouse: relationships between monoaminergic, cholinergic, and spinal projection systems. *J. Chem. Neuroanat.* **31:** 2–36.

28. Liang, H., G. Paxinos & C. Watson. 2011. Projections from the brain to the spinal cord in the mouse. *Brain Struct. Funct.* **215:** 159–186.

29. Cepeda-Nieto, A.C., S.L. Pfaff & A. Varela-Echavarria. 2005. Homeodomain transcription factors in the development of subsets of hindbrain reticulospinal neurons. *Mol. Cell Neurosci.* **28:** 30–41.

30. Sholomenko, G.N. & M.J. Odonovan. 1995. Development and characterization of pathways descending to the spinal-cord in the embryonic chick. *J. Neurophysiol.* **73:** 1223–1233.

31. Glover, J.C. & G. Petursdottir. 1988. Pathway specificity of reticulospinal and vestibulospinal projections in the 11-day chicken embryo. *J. Comp. Neurol.* **270:** 25–38, 60–21.

32. Lance-Jones, C. 1982. Motoneuron cell death in the developing lumbar spinal cord of the mouse. *Brain Res.* **256:** 473–479.

33. Yamamoto, Y. & E. Christopher. 1999. Patterns of programmed cell death in populations of developing spinal motoneurons in chicken, mouse, and rat. *Dev. Biol.* **214:** 60–71.

34. Okado, N. & R.W. Oppenheim. 1984. Cell death of motoneurons in the chick embryo spinal cord. IX. The loss of motoneurons following removal of afferent inputs. *J. Neurosci.* **4:** 1639–1652.

35. Qin-Wei, Y. *et al.* 1994. Cell death of spinal motoneurons in the chick embryo following deafferentation: Rescue effects of tissue extracts, soluble proteins, and neurotrophic agents. *J. Neurosci.* **14:** 7629–7640.

36. Borgius, L. *et al.* 2010. A transgenic mouse line for molecular genetic analysis of excitatory glutamatergic neurons. *Mol. Cell Neurosci.* **45:** 245–257.

37. Hagglund, M. *et al.* 2010. Activation of groups of excitatory neurons in the mammalian spinal cord or hindbrain evokes locomotion. *Nat. Neurosci.* **13:** 246–252.

38. Martin, E.M. *et al.* 2011. Molecular and neuroanatomical characterization of single neurons in the mouse medullary gigantocellular reticular nucleus. *J. Comp. Neurol.* **519:** 2574–2593.

39. Szokol, K., J.C. Glover & M.-C. Perreault. 2008. Differential origin of reticulospinal drive to motoneurons innervating trunk and hindlimb muscles in the mouse revealed by optical recording. *J. Physiol.* **586:** 5259–5276.

40. O'Donovan, M.J. *et al.* 1993. Real-time imaging of neurons retrogradely and anterogradely labelled with calcium-sensitive dyes. *J. Neurosci. Methods* **46:** 91–106.

41. Lev-Tov, A. & M.J. O'Donovan. 1995. Calcium imaging of motoneuron activity in the en-bloc spinal cord preparation of the neonatal rat. *J. Neurophysiol.* **74:** 1324–1334.

42. Kasumacic, N., J.C. Glover & M.-C. Perreault. 2010. Segmental patterns of vestibular-mediated synaptic inputs to axial and limb motoneurons in the neonatal mouse assessed by optical recording. *J. Physiol.* **588:** 4905–4925.

43. Szokol, K. & M.-C. Perreault. 2009. Imaging synaptically mediated responses produced by brainstem inputs onto identified spinal neurons in the neonatal mouse. *J. Neurosci. Methods* **180:** 1–8.

44. Galea, M.P. *et al.* 2010. Bilateral postsynaptic actions of pyramidal tract and reticulospinal neurons on feline erector spinae motoneurons. *J. Neurosci.* **30:** 858–869.

45. Szokol, K., J.C. Glover & M.-C. Perreault. 2011. Organization of functional synaptic connections between medullary reticulospinal neurons and lumbar descending commissural interneurons in the neonatal mouse. *J. Neurosci.* **31:** 4731–4742.

46. Floeter, M.K. & A. Lev-Tov. 1993. Excitation of lumbar motoneurons by the medial longitudinal fasciculus in the in vitro brain stem spinal cord preparation of the neonatal rat. *J. Neurophysiol.* **70:** 2241–2250.

47. Jankowska, E. 1992. Interneuronal relay in spinal pathways from proprioceptors. *Prog. Neurobiol.* **38:** 335–378.

48. Jankowska, E. & S.A. Edgley. 2010. Functional subdivision of feline spinal interneurons in reflex pathways from group Ib and II muscle afferents; an update. *Eur. J. Neurosci.* **32:** 881–893.

49. Alaynick, W.A., T.M. Jessell & S.L. Pfaff. 2011. SnapShot: spinal cord development. *Cell* **146:** 178.

50. Quinlan, K.A. & O. Kiehn. 2007. Segmental, synaptic actions of commissural interneurons in the mouse spinal cord. *J. Neurosci.* **27:** 6521–6530.

51. Butt, S.J.B. & O. Kiehn. 2003. Functional identification of interneurons responsible for left-right coordination of hindlimbs in mammals. *Neuron* **38:** 953–963.

52. Matsuyama, K. *et al.* 2004. Lumbar commissural interneurons with reticulospinal inputs in the cat: morphology and discharge patterns during fictive locomotion. *J. Comp. Neurol.* **474:** 546–561.

53. Perreault, M.-C. 2012. Focal release of caged glutamate onto neurons in the ventral spinal cord. In *Cellular and Network Functions in the Spinal Cord Conference.* L. Ziskind-Conhaim, Ed.: 32. Madison, Wisconsin.

54. Tuszynski, M.H. & O. Steward. 2012. Concepts and methods for the study of axonal regeneration in the CNS. *Neuron* **74:** 777–791.

55. van den Brand, R. *et al.* 2012. Restoring voluntary control of locomotion after paralyzing spinal cord injury. *Science* **336:** 1182–1185.

56. Oudega, M. & M.A. Perez. 2012. Corticospinal reorganization after spinal cord injury. *J. Physiol.* **590:** 3647–3663.

57. Steeves, J.D. *et al.* 1994. Permissive and restrictive periods for brainstem-spinal regeneration in the chick. *Prog. Brain Res.* **103:** 243–262.

58. ten Donkelaar, H. 2000. Development and regenerative capacity of descending supraspinal pathways in tetrapods: a comparative approach. *Adv. Anat. Embryol. Cell Biol.* **154:** 1–145.

59. Alstermark, B. & J. Ogawa. 2004. In vivo recordings of bulbospinal excitation in adult mouse forelimb motoneurons. *J. Neurophysiol.* **92:** 1958–1962.

60. Fisher, K.M., B. Zaaimi & S.N. Baker. 2012. Reticular formation responses to magnetic brain stimulation of primary motor cortex. *J. Physiol.* **590:** 4045–4060.

61. Jin, Y. *et al.* 2002. Transplants of fibroblasts genetically modified to express BDNF promote axonal regeneration from supraspinal neurons following chronic spinal cord injury. *Exp. Neurol.* **177:** 265–275.

62. Ballermann, M. & K. Fouad. 2006. Spontaneous locomotor recovery in spinal cord injured rats is accompanied by anatomical plasticity of reticulospinal fibers. *Eur. J. Neurosci.* **23:** 1988–1996.

63. Lu, P. *et al.* 2012. Motor axonal regeneration after partial and complete spinal cord transection. *J. Neurosci.* **32:** 8208–8218.

64. Blesch, A. & M.H. Tuszynski. 2008. Spinal cord injury: plasticity, regeneration and the challenge of translational drug development. *Trends Neurosc.* **32:** 41–47.

65. Fenrich, K.K. *et al.* 2007. Axonal regeneration and development of de novo axons from distal dendrites of adult feline commissural interneurons after a proximal axotomy. *J. Comp. Neurol.* **502:** 1079–1097.

66. Fenrich, K.K. & P.K. Rose. 2009. Spinal interneuron axons spontaneously regenerate after spinal cord injury in the adult feline. *J. Neurosci.* **29:** 12145–12158.

67. Soteropoulos, D.S., S.A. Edgley & S.N. Baker. 2011. Lack of evidence for direct corticospinal contributions to control of the ipsilateral forelimb in monkey. *J. Neurosci.* **31:** 11208–11219.

68. Baker, S.N. 2011. The primate reticulospinal tract, hand function and functional recovery. *J. Physiol.* **589:** 5603–5612.

69. Riddle, C.N., S.A. Edgley & S.N. Baker. 2009. Direct and indirect connections with upper limb motoneurons from the primate reticulospinal tract. *J. Neurosci.* **29:** 4993–4999.

70. Zaaimi, B. *et al*. 2012. Changes in descending motor path-way connectivity after corticospinal tract lesion in macaque monkey. *Brain* **135:** 2277–2289.

71. Peterson, B.W. 1979. Reticulospinal projections to spinal motor nuclei. *Annu. Rev. Physiol.* **41:** 127–140.

72. Drew, T. & S. Rossignol. 1990. Functional organization within the medullary reticular formation of intact unanes-thetized cat. I: movements evoked by microstimulation. *J. Neurophysiol.* **64:** 767–781.

73. Perreault, M.-C., T. Drew & S. Rossignol. 1993. Activity of medullary reticulospinal neurons during fictive locomo-tion. *J. Neurphysiol.* **69:** 2232–2247.

Ann. N.Y. Acad. Sci. ISSN 0077-8923

ANNALS OF THE NEW YORK ACADEMY OF SCIENCES
Issue: *Neurons, Circuitry, and Plasticity in the Spinal Cord and Brainstem*

Pre- and postsynaptic inhibitory control in the spinal cord dorsal horn

Rita Bardoni,[1] Tomonori Takazawa,[2] Chi-Kun Tong,[3] Papiya Choudhury,[3] Gregory Scherrer,[5] and Amy B. MacDermott[3,4]

[1]Department of Biomedical, Metabolic and Neural Science, University of Modena and Reggio Emilia, Modena, Italy. [2]Department of Anesthesiology, Gunma University Graduate School of Medicine, Gunma, Japan. [3]Department of Physiology and Cellular Biophysics, [4]Department of Neuroscience, Columbia University, New York, New York. [5]Department of Anesthesia, Stanford Institute for Neuro-Innovation and Translational Neurosciences, Stanford University, Palo Alto, California

Address for correspondence: Rita Bardoni, Department of Biomedical, Metabolic and Neural Sciences, University of Modena and Reggio Emilia, Via Campi, 287, 41125, Modena, Italy. bardoni@unimo.it

Sensory information transmitted to the spinal cord dorsal horn is modulated by a complex network of excitatory and inhibitory interneurons. The two main inhibitory transmitters, GABA and glycine, control the flow of sensory information mainly by regulating the excitability of dorsal horn neurons. A presynaptic action of GABA has also been proposed as an important modulatory mechanism of transmitter release from sensory primary afferent terminals. By inhibiting the release of glutamate from primary afferent terminals, activation of presynaptic GABA receptors could play an important role in nociceptive and tactile sensory coding, while changes in their expression or function could be involved in pathological pain conditions, such as allodynia.

Keywords: dorsal horn; pain; GABA; glycine; inhibition

Introduction

The superficial and deep laminae of the spinal cord dorsal horn are under strong inhibitory control, importantly exerted by two neurotransmitters, gamma-aminobutyric acid (GABA) and glycine. These transmitters, released by both local interneurons and inhibitory descending fibers, bind to their cognate anion permeable receptors, GABA$_A$ and glycine, respectively. GABA additionally binds to its G protein–coupled receptor, the GABA$_B$ receptor. Activation of GABA and glycine receptors depresses neuronal excitation through hyperpolarization of the postsynaptic membrane and/or activation of a shunting conductance. Indeed, application of bicuculline or strychnine (blockers of GABA$_A$ and glycine receptors, respectively) to a spinal cord slice preparation shows the effect of ambient GABA and glycine on excitability of dorsal horn neurons (Fig. 1). As we will describe subsequently, GABA can also directly decrease glutamate release from primary afferent fibers (PAFs).[1] Behavioral studies in rodents have shown that intrathecal administration of bicuculline or strychnine induces nocifensive responses and lowers nociceptive threshold in rats, while injection of GABA or glycine is antinociceptive under most circumstances.[2–4] Furthermore, enhancing GABA$_A$ receptor function by spinal application of GABA or a positive allosteric modulator, such as midazolam, depresses noxious stimulus-evoked activity in spinal cord neurons.[5,6] Loss of synaptic inhibition is widely accepted as an important factor contributing to the generation and maintenance of chronic pain. Therefore, we will consider recent advances in our understanding of inhibition in the spinal cord dorsal horn.

Region-specific inhibition by GABA and glycine in the spinal cord dorsal horn

The relative contributions of GABA and glycine to the control of the flow of information in the dorsal horn varies among different laminae and somatosensory modalities. Immunohistochemical studies and, more recently, transgenic mice in which enhanced green fluorescent protein (eGFP)

doi: 10.1111/nyas.12056

Ann. N.Y. Acad. Sci. 1279 (2013) 90–96 © 2013 New York Academy of Sciences.

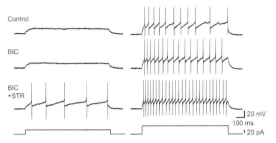

Figure 1. Blockade of GABA$_A$ and glycine receptors with bicuculline and strychnine enhances excitability of inhibitory neurons in mouse dorsal horn. Shown are examples of action potentials induced by current injection under three conditions: (1) control; NBQX [10 μM] and AP5 [50 μM]; (2) BIC; NBQX, AP5, and bicuculline [10 μM]; and (3) BIC+STR; NBQX, AP5, bicuculline, and strychnine [1 μM]. Resting membrane potential was kept at –65 mV. Left panel shows responses to a small current injection in the three conditions, and the right panel shows responses to a larger current injection. Bottom traces show injected currents. Taken with permission from Ref. 15.

is expressed under the promoter of the gene encoding the enzyme glutamic acid decarboxylase 67 (GAD67), involved in the synthesis of GABA (GAD67-eGFP mice), have shown that GABAergic interneurons are abundant in the spinal cord dorsal horn. This is especially true in superficial laminae,[7–10] where nociceptive fibers terminate. A subpopulation of GABAergic interneurons also express glycine (33%, 43%, and 64% of GABAergic neurons in lamina I, II, and III, respectively, express glycine).[10,11] *In situ* hybridization studies and observations from glycine transporter 2 (GlyT2)-eGFP mice have shown that glycinergic neurons are more abundant in the deeper dorsal layers (laminae III–V)[12,13] that receive tactile sensory inputs as well as some nociceptive inputs.

Analysis of miniature inhibitory postsynaptic currents (mIPSCs) mediated by GABA$_A$ and glycine receptors, reflecting quantal release of these transmitters, has confirmed that the contribution of GABA and glycine to fast synaptic inhibition changes between different laminar regions of the dorsal horn. GABAergic mIPSCs seem to predominate in laminae I and II outer (IIo), while glycine mIPSCs play a dominant role in lamina III.[14] Our group has recently characterized two populations of inhibitory interneurons in GAD67-eGFP mice. One group of neurons predominantly receives strongly bicuculline-sensitive mIPSCs with slower decay kinetics (GABA-dominant) while another class of inhibitory neurons predominantly receives fewer

bicuculline-sensitive mIPSCs with fast decay kinetics (glycine-dominant).[15] Consistent with previous studies,[14,16,17] inhibitory interneurons are mainly GABA-dominant in laminae I–IIo, while glycine-dominant neurons are prevalent at the lamina II–III border, as illustrated in Figure 2.

Extrasynaptic GABA$_A$ and glycine receptors mediate tonic currents in dorsal horn neurons,[18–20] where they play important roles in regulating neuronal excitability. We have shown in mature mice that tonic GABA currents predominate in GABA-dominant neurons, while tonic glycine currents are critical in regulating the inhibitory tone of glycine-dominant neurons at the lamina II–III border.[15] This border area receives inputs from low-threshold mechanosensitive afferents and is critically involved in the generation of dynamic mechanical allodynia.[21] Excitatory interneurons expressing the γ isoform of protein kinase C (PKCγ), located at the ventral border of inner lamina II (lamina IIi), and in lamina III, are essential contributors to mechanical allodynia.[22] These PKCγ-expressing interneurons receive projections from low-threshold mechanoreceptors[23] and are normally inhibited by glycinergic interneurons.[24] Removal of glycine inhibition allows activation of a polysynaptic excitatory pathway triggered by low-threshold mechanical input, leading to the excitation of nociceptive-specific projection neurons in the superficial dorsal horn.[21] This is similar to the polysynaptic excitatory pathway between low-threshold Aβ fibers and lamina I projection neurons observed in the presence of bicuculline and strychnine in another study.[25] Further experiments will be required to assess the contribution of synaptic and tonic glycine currents in controlling the excitability of inhibitory and excitatory interneurons, both in control animals and in animal models of chronic pain.

Presynaptic modulation of primary afferent terminals

GABA$_A$ receptors and primary afferent depolarization

GABA has long been known to be one of the inhibitory transmitters mediating presynaptic inhibition of excitatory transmission in the spinal cord, acting through both ionotropic (GABA$_A$) and G protein–coupled receptors (GABA$_B$). In 1957, Frank and Fuortes[26] first proposed the concept of

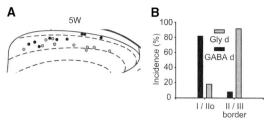

Figure 2. GAD67-eGFP[+] neurons have regionally distinct properties of synaptic inhibitory input. Part A shows soma locations of glycine-dominant (Gly-d, gray circles) and GABA-dominant (GABA-d, black circles) neurons recorded from postnatal day 29–32 (5W) mice. Right and upper sides of the schematic diagram show lateral and dorsal edges of the dorsal horn, respectively. Laminae I, II, and III are separated by dotted lines. (B) Gly-d neurons are the major population at the lamina II/III border ($n = 12$), while GABA-d neurons (black bars) are the major population in lamina I and IIo ($n = 11$). Bars indicate the incidence of neurons located in laminae I/IIo, and at the laminae II/III border. Taken with permission from Ref. 15.

presynaptic inhibition, based on the observation that muscle afferent volleys depressed the size of the monosynaptic excitatory postsynaptic potential of spinal motoneurons produced by other muscle afferents without any changes in the membrane potential or excitability of those motoneurons. This form of presynaptic inhibition is caused by depolarization of the primary afferent terminals (PAD) and is strongly depressed by GABA$_A$ receptor antagonists, such as bicuculline, suggesting that GABAergic interneurons can be involved in this mechanism through a polysynaptic circuit.[27–29]

Histological evidence supporting this hypothesis has been provided by several studies. GABAergic interneurons form axo-axonic synapses in both the ventral and dorsal horns, not only on group Ia, Ib, and II muscle PAFs, but also on cutaneous afferents, particularly those of large diameter.[30] As shown in Figure 3, the stimulation of sensory cutaneous afferents could activate a disynaptic circuit in the dorsal horn, producing release of GABA from inhibitory interneurons, causing decrease of glutamate release from the PADs. Electron microscopy studies performed on the superficial dorsal horn from spinal cord preparations of different mammalian species have demonstrated the presence of complex synaptic structures called glomeruli. These are formed by a central terminal of either myelinated or unmyelinated PAFs, and several dendrites and axons.[11,31] Axon terminals presynaptic to central terminals of unmyelinated PAFs are predominantly GABAergic,

while the majority of those contacting terminals of myelinated PAFs are both GABAergic and glycinergic.[32] In lamina III of the rat spinal cord, axons containing GABA, or GABA plus glycine, form a synaptic triadic arrangement, contacting both the terminal of a hair follicle myelinated afferent and a postsynaptic dendrite.[33] Parvalbumin-expressing inhibitory interneurons, which predominate in lamina III, have recently been shown to be involved in axo-axonic synapses with nonnociceptive Aδ down hair afferents belonging to synaptic glomeruli, or with larger myelinated fibers such as hair follicle afferents.[34]

The mechanisms of presynaptic inhibition mediated by PAD have largely been elucidated.[30] Primary sensory neurons exhibit a higher intracellular concentration of chloride than central neurons. This is due to the high expression of the transporter NKCC1, which transports Cl$^-$, Na$^+$, and K$^+$ into the cell, and low expression of KCC2, which transports Cl$^-$ and K$^+$ out of the cell.[35–37] For this reason, the chloride equilibrium potential in dorsal root ganglion neurons (DRGs) is about –30 mV. Thus, the opening of GABA$_A$ receptors causes the efflux of Cl$^-$ and depolarization of the terminals. Inactivation of voltage-dependent sodium and calcium channels, caused by the terminal depolarization[38] and the shunting effect due to opening of GABA$_A$ receptors, impairs the propagation of action potentials along PAFs into the terminals, and decreases the release of glutamate. Suprathreshold depolarizations can potentially produce the opposite effect by eliciting action potentials in spinal PAF terminals, possibly triggering dorsal root reflexes that may contribute to neurogenic inflammation.[39] The upregulation of the GABA system on DRG neurons[40] and/or the increase of NKCC1 transporter activity[41] could contribute to the shift from presynaptic inhibition to dorsal reflexes and pain sensitization.

Studies performed during the last decade have suggested the need for additional mechanisms for PAD. PAF terminals express other synaptic receptors able to depolarize the terminals. Glutamatergic ionotropic AMPA, NMDA, and kainate receptors have been detected at the central terminals of PAFs. Activation of these receptors causes depolarization of PAFs in the rat spinal cord.[42–44] Furthermore, the observation that PAD is not completely blocked by inhibition of synaptic transmission suggests that it could be partly mediated by either spillover of

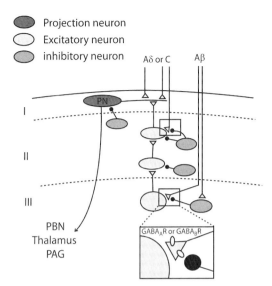

Figure 3. Schematic representation of a proposed polysynaptic excitatory pathway connecting lamina III to lamina I projection neurons. This pathway is normally under powerful inhibitory control mediated by GABA and glycine released from inhibitory neurons. Presynaptic inhibition mediated by GABA$_A$ or GABA$_B$ receptors in the spinal cord dorsal horn is illustrated in detail for the Aβ fiber synaptic terminal onto a lamina III neuron, but also occurs on nociceptor PAF terminals. Stimulation of cutaneous tactile fibers induces GABA release from inhibitory interneurons, causing the activation of GABA$_A$ or GABA$_B$ receptors expressed on primary afferent terminals and the inhibition of glutamate release onto lamina III–IV neurons.

transmitters released from PAFs or a dendroaxonic reciprocal synaptic microcircuit.[45,46] At the glomerular level, several dendrites contacting PAF central terminals contain clear vesicles, so these could be involved in the generation of PAD.[10,47]

The subunit composition of presynaptic GABA$_A$ receptors expressed on PAF terminals has recently been investigated. Central terminals of primary afferents in mouse dorsal horn express four α subunits (α1–3, α5), with a prevalence of α2 or α3 on C-fibers, while myelinated Aδ and Aβ fibers are co-labeled in roughly equal proportion with each subunit.[48] Mice with nociceptors (DRG neurons expressing SNS [or Nav1.8] channels) selectively lacking the GABA$_A$α2 subunits exhibit reduced potentiation of dorsal root potentials and impaired thermal and mechanical antihyperalgesia by diazepam in a model of inflammatory pain,[49] confirming a role of the α2 subunit in regulating sensitization in models of inflammatory and neuropathic pain.[50,51]

Although glycine is present in several axon terminals of GABAergic neurons presynaptic to PAFs

(see previous discussion), strychnine does not block PAD or presynaptic inhibition,[27,52,53] and glycine does not directly depolarize PAFs.[54] Furthermore, glycine receptors have not been detected on PAF terminals,[55] so the putative presynaptic role of glycine on PAF central terminals is still controversial. It is possible that the major site of action of glycine, released by inhibitory interneurons, is on postsynaptic dendrites, belonging to glomeruli or triadic arrangements.

GABA$_B$ receptors

Metabotropic GABA$_B$ receptors are expressed both in DRGs and spinal cord dorsal horn, particularly in the superficial laminae.[56,57] Their activation produces antinociceptive effects: treatment with the GABA$_B$ agonist baclofen induces dose-dependent inhibition of C-fiber and pinch-evoked activity of rat wide dynamic range neurons *in vivo*,[58] and reverses hypersensitivity of these neurons to mechanical stimuli after spinal cord ischemia.[59]

Exogenous activation of GABA$_B$ receptors by baclofen inhibits glutamate and substance P release from PAFs in spinal cord dorsal horn[60] by acting on presynaptic voltage-dependent calcium channels. A postsynaptic effect of baclofen has also been observed in rat superficial and deep dorsal horn, consisting of the generation of an outward current mediated by potassium channels.[57,61,62] GABA$_B$ receptors are also involved in the depression of GABA release from dorsal horn neurons: in lamina I, paired pulse depression of evoked IPSCs is decreased by an antagonist of GABA$_B$ receptors, while it is not affected by GABA$_A$ antagonists.[63] Because GABA$_B$ receptors have a higher sensitivity for GABA than GABA$_A$ receptors do, they could be activated even under conditions of low extracellular concentrations of GABA. The endogenous effect of GABA$_B$ receptors in modulating glutamate release from PAFs in lamina II has recently been investigated:[64] blockade of GABA$_B$ receptors facilitates the evoked action potential–dependent synaptic responses and increases neuronal excitability after dorsal root stimulation.

Function of GABA-mediated presynaptic inhibition in the dorsal horn, and future perspectives

The first synapse in the somatosensory pathway, that is, the synapse between nociceptive or tactile

PAFs and dorsal horn neurons, is modulated by several mechanisms of presynaptic inhibition. Here we have illustrated some aspects of the inhibition mediated by GABA receptors. Both ionotropic GABA$_A$ and metabotropic GABA$_B$ receptors exert an inhibitory action on glutamate release from PAFs, involving different cellular mechanisms. Despite the large number of studies regarding GABA-mediated presynaptic inhibition in dorsal horn, several aspects remain to be elucidated.

Synaptic responses generated by glutamate release from PAFs onto dorsal horn neurons are known to undergo a strong short-term depression (see Ref. 65). We have observed that, in rat lamina III, the level of depression is variable from one postsynaptic cell to another (unpublished observation). This likely reflects (1) different PAF properties and/or (2) different synaptic circuits recruited by dorsal root stimulation. The roles of GABA$_A$ and GABA$_B$ receptors in short-term depression have not been established. We hypothesize that during PAF repetitive stimulation, the activation of presynaptic GABA$_A$ and GABA$_B$ receptors could contribute to shaping the pattern of postsynaptic response amplitudes, both in superficial and deep dorsal horn. Application of bicuculline attenuates paired pulse depression of postsynaptic responses evoked on dorsal horn neurons by low-threshold afferent stimulation *in vivo*.[66] Preliminary results obtained in our laboratories indicate that the GABA$_A$ agonist muscimol is able to modulate glutamate release from Aβ fibers onto rat spinal lamina III neurons and increase the paired pulse ratio of evoked EPSCs. Thus, synaptic depression at central terminals of PAFs is modified by receptor-mediated presynaptic inhibition. This could, in turn, affect the firing pattern of dorsal horn neurons and the sensory coding process. A characterization of the postsynaptic neurons involved in this mechanism (i.e., inhibitory versus excitatory interneurons) would also be critical in understanding the organization and function of presynaptic inhibition in the dorsal horn.

Another important consideration is how presynaptic modulation is affected by the plastic changes that occur during chronic pain. Some studies suggest that peripheral inflammation induces an increase of PAD on PAFs (both nociceptive and non-nociceptive). Application of inflammatory agents on DRGs induces a rapid increase of intracellular Cl$^-$ concentration through the upregulation of the NKCC1 transporter.[67] After persistent peripheral inflammation, GABA-induced depolarization on DRGs increases, partially due to the inhibition of voltage-dependent potassium currents[40] and to enhanced GABA$_A$ receptor function[67] in addition to elevated intracellular chloride. The potentiation of PAD during inflammatory pain could contribute to hyperalgesia by enhancing dorsal root reflexes in nociceptors.[39,68,69]

Modulation of PAD on low-threshold afferents could be involved in the generation of allodynia. If glutamate release from low-threshold mechanoreceptors increased due to augmented excitability of presynaptic terminals, this could help drive the excitatory polysynaptic pathway revealed with use of GABA$_A$ and glycine receptor antagonists, as proposed by Torsney and MacDermott (Fig. 3). However, enhanced PAD could also mediate enhanced presynaptic inhibition. Correspondingly, a recent study has proposed an opposite role for PAD in inflammatory pain.[49] Mice lacking benzodiazepine-sensitive α2-GABA$_A$ receptors in primary nociceptors showed reduced antihyperalgesia in response to intrathecally injected diazepam in an inflammatory pain model. These results suggest that facilitation of GABA$_A$ receptor activation on spinal nociceptor terminals could exert an analgesic action. A more extensive characterization of the dorsal horn synaptic circuits and neuronal types involved in PAD and presynaptic inhibition will be important to clarify these discrepancies and understand the role of presynaptic GABA$_A$ receptors in chronic pain.

Conflicts of interest

The authors declare no conflicts of interest.

References

1. MacDermott, A.B., L.W. Role & S.A. Siegelbaum. 1999. Presynaptic ionotropic receptors and the control of transmitter release. *Annu. Rev. Neurosci.* **22:** 443–485.
2. Beyer, C., L.A. Roberts & B.R. Komisaruk. 1985. Hyperalgesia induced by altered glycinergic activity at the spinal cord. *Life Sci.* **37:** 875–882.
3. Roberts, L.A., C. Beyer & B.R. Komisaruk. 1986. Nociceptive responses to altered GABAergic activity at the spinal cord. *Life Sci.* **39:** 1667–1674.
4. Yaksh, T.L. 1989. Behavioral and autonomic correlates of the tactile evoked allodynia produced by spinal glycine inhibition: effects of modulatory receptor systems and excitatory amino acid antagonists. *Pain* **37:** 111–123.
5. Clavier, N., M.C. Lombard & J.M. Besson. 1992. Benzodiazepines and pain: effects of midazolam on the activities

of nociceptive non-specific dorsal horn neurons in the rat spinal cord. *Pain* **48:** 61–71.

6. Sumida, T. *et al.* 1995. Intravenous midazolam suppresses noxiously evoked activity of spinal wide dynamic range neurons in cats. *Anesth. Analg.* **80:** 58–63.

7. Daniele, C.A. & A.B. MacDermott. 2009. Low-threshold primary afferent drive onto GABAergic interneurons in the superficial dorsal horn of the mouse. *J. Neurosci.* **29:** 686–695.

8. McLaughlin, B.J. *et al.* 1975. Immunocytochemical localization of glutamate decarboxylase in rat spinal cord. *J. Comp. Neurol.* **164:** 305–321.

9. Oliva, A.A., Jr. *et al.* 2000. Novel hippocampal interneuronal subtypes identified using transgenic mice that express green fluorescent protein in GABAergic interneurons. *J. Neurosci.* **20:** 3354–3368.

10. Todd, A.J. & A.C. Sullivan. 1990. Light microscope study of the coexistence of GABA-like and glycine-like immunoreactivities in the spinal cord of the rat. *J. Comp. Neurol.* **296:** 496–505.

11. Todd, A.J. 1990. An electron microscope study of glycine-like immunoreactivity in laminae I-III of the spinal dorsal horn of the rat. *Neuroscience* **39:** 387–394.

12. Hossaini, M., P.J. French & J.C. Holstege. 2007. Distribution of glycinergic neuronal somata in the rat spinal cord. *Brain Res.* **1142:** 61–69.

13. Zeilhofer, H.U. *et al.* 2005. Glycinergic neurons expressing enhanced green fluorescent protein in bacterial artificial chromosome transgenic mice. *J. Comp. Neurol.* **482:** 123–141.

14. Inquimbert, P., J.L. Rodeau & R. Schlichter. 2007. Differential contribution of GABAergic and glycinergic components to inhibitory synaptic transmission in lamina II and laminae III-IV of the young rat spinal cord. *Eur. J. Neurosci.* **26:** 2940–2949.

15. Takazawa, T. & A.B. MacDermott. 2010. Glycinergic and GABAergic tonic inhibition fine tune inhibitory control in regionally distinct subpopulations of dorsal horn neurons. *J. Physiol.* **588:** 2571–2587.

16. Allain, A.E. *et al.* 2006. Expression of the glycinergic system during the course of embryonic development in the mouse spinal cord and its co-localization with GABA immunoreactivity. *J. Comp. Neurol.* **496:** 832–846.

17. Todd, A.J. *et al.* 1996. Colocalization of GABA, glycine, and their receptors at synapses in the rat spinal cord. *J. Neurosci.* **16:** 974–982.

18. Ataka, T. & J.G. Gu. 2006. Relationship between tonic inhibitory currents and phasic inhibitory activity in the spinal cord lamina II region of adult mice. *Mol. Pain* **2:** 36.

19. Mitchell, E.A. *et al.* 2007. GABAA and glycine receptor-mediated transmission in rat lamina II neurones: relevance to the analgesic actions of neuroactive steroids. *J. Physiol.* **583:** 1021–1040.

20. Takahashi, A., T. Mashimo & I. Uchida. 2006. GABAergic tonic inhibition of substantia gelatinosa neurons in mouse spinal cord. *Neuroreport* **17:** 1331–1335.

21. Miraucourt, L.S. *et al.* 2009. Glycine inhibitory dysfunction induces a selectively dynamic, morphine-resistant, and neurokinin 1 receptor- independent mechanical allodynia. *J. Neurosci.* **29:** 2519–2527.

22. Malmberg, A.B. *et al.* 1997. Preserved acute pain and reduced neuropathic pain in mice lacking PKCgamma. *Science* **278:** 279–283.

23. Neumann, S. *et al.* 2008. Innocuous, not noxious, input activates PKCgamma interneurons of the spinal dorsal horn via myelinated afferent fibers. *J. Neurosci.* **28:** 7936–7944.

24. Miraucourt, L.S., R. Dallel & D.L. Voisin. 2007. Glycine inhibitory dysfunction turns touch into pain through PKCgamma interneurons. *PLoS ONE* **2:** e1116.

25. Torsney, C. & A.B. MacDermott. 2006. Disinhibition opens the gate to pathological pain signaling in superficial neurokinin 1 receptor-expressing neurons in rat spinal cord. *J. Neurosci.* **26:** 1833–1843.

26. Frank, K. & M.G.F. Fuortes. 1957. Presynaptic and Postsynaptic Inhibition of Monosynaptic Reflexes. *Fed. Proc.* **16:** 39–40.

27. Eccles, J.C., R. Schmidt & W.D. Willis. 1963. Pharmacological studies on presynaptic inhibition. *J. Physiol.* **168:** 500–530.

28. Feltz, P. & M. Rasminsky. 1974. A model for the mode of action of GABA on primary afferent terminals: depolarizing effects of GABA applied iontophoretically to neurones of mammalian dorsal root ganglia. *Neuropharmacology* **13:** 553–563.

29. Gallagher, J.P., H. Higashi & S. Nishi. 1978. Characterization and ionic basis of GABA-induced depolarizations recorded in vitro from cat primary afferent neurones. *J. Physiol.* **275:** 263–282.

30. Rudomin, P. & R.F. Schmidt. 1999. Presynaptic inhibition in the vertebrate spinal cord revisited. *Exp. Brain Res.* **129:** 1–37.

31. Bernardi, P.S. *et al.* 1995. Synaptic interactions between primary afferent terminals and GABA and nitric oxide-synthesizing neurons in superficial laminae of the rat spinal cord. *J. Neurosci.* **15:** 1363–1371.

32. Todd, A.J. 1996. GABA and glycine in synaptic glomeruli of the rat spinal dorsal horn. *Eur. J. Neurosci.* **8:** 2492–2498.

33. Watson, A.H., D.I. Hughes & A.A. Bazzaz. 2002. Synaptic relationships between hair follicle afferents and neurones expressing GABA and glycine-like immunoreactivity in the spinal cord of the rat. *J. Comp. Neurol.* **452:** 367–380.

34. Hughes, D.I. *et al.* 2012. Morphological, neurochemical and electrophysiological features of parvalbumin-expressing cells: a likely source of axo-axonic inputs in the mouse spinal dorsal horn. *J. Physiol.* **590:** 3927–3951.

35. Alvarez-Leefmans, F.J. *et al.* 1988. Intracellular chloride regulation in amphibian dorsal root ganglion neurones studied with ion-selective microelectrodes. *J. Physiol.* **406:** 225–246.

36. Price, T.J., K.M. Hargreaves & F. Cervero. 2006. Protein expression and mRNA cellular distribution of the NKCC1 cotransporter in the dorsal root and trigeminal ganglia of the rat. *Brain Res.* **1112:** 146–158.

37. Sung, K.W. *et al.* 2000. Abnormal GABAA receptor-mediated currents in dorsal root ganglion neurons isolated from Na-K-2Cl cotransporter null mice. *J. Neurosci.* **20:** 7531–7538.

38. Graham, B. & S. Redman. 1994. A simulation of action potentials in synaptic boutons during presynaptic inhibition. *J. Neurophysiol.* **71:** 538–549.

39. Willis, W.D., Jr. 1999. Dorsal root potentials and dorsal root reflexes: a double-edged sword. *Exp. Brain Res.* **124:** 395–421.

40. Zhu, Y., S.G. Lu & M.S. Gold. 2012. Persistent inflammation increases GABA-induced depolarization of rat cutaneous dorsal root ganglion neurons in vitro. *Neuroscience* **220:** 330–340.

41. Price, T.J., F. Cervero & Y. de Koninck. 2005. Role of cation-chloride-cotransporters (CCC) in pain and hyperalgesia. *Curr. Top Med. Chem.* **5:** 547–555.

42. Bardoni, R. *et al.* 2004. Presynaptic NMDA receptors modulate glutamate release from primary sensory neurons in rat spinal cord dorsal horn. *J. Neurosci.* **24:** 2774–2781.

43. Ikeda, H., T. Kiritoshi & K. Murase. 2008. Effect of excitatory and inhibitory agents and a glial inhibitor on optically-recorded primary-afferent excitation. *Mol. Pain* **4:** 39.

44. Lee, C.J. *et al.* 2002. Functional expression of AMPA receptors on central terminals of rat dorsal root ganglion neurons and presynaptic inhibition of glutamate release. *Neuron* **35:** 135–146.

45. Russo, R.E., R. Delgado-Lezama & J. Hounsgaard. 2000. Dorsal root potential produced by a TTX-insensitive micro-circuitry in the turtle spinal cord. *J. Physiol.* **528**(Pt 1)**:** 115–122.

46. Shreckengost, J. *et al.* 2010. Bicuculline-sensitive primary afferent depolarization remains after greatly restricting synaptic transmission in the mammalian spinal cord. *J. Neurosci.* **30:** 5283–5288.

47. Hiura, A., H. Ishizuka & E.L. Villalobos. 1991. Gabaergic neurons in the mouse superficial dorsal horn with special emphasis on their relation to primary afferent central terminals. *Arch. Histol. Cytol.* **54:** 195–206.

48. Paul, J., H.U. Zeilhofer & J.M. Fritschy. 2012. Selective distribution of GABA(A) receptor subtypes in mouse spinal dorsal horn neurons and primary afferents. *J. Comp. Neurol.* **520:** 3895–3911.

49. Witschi, R. *et al.* 2011. Presynaptic alpha2-GABAA receptors in primary afferent depolarization and spinal pain control. *J. Neurosci.* **31:** 8134–8142.

50. Knabl, J. *et al.* 2008. Reversal of pathological pain through specific spinal GABA(A) receptor subtypes. *Nature* **451:** 330–336.

51. Munro, G. *et al.* 2011. A question of balance—positive versus negative allosteric modulation of GABA(A) receptor subtypes as a driver of analgesic efficacy in rat models of inflammatory and neuropathic pain. *Neuropharmacology* **61:** 121–132.

52. De Groat, W.C., P.M. Lalley & W.R. Saum. 1972. Depolarization of dorsal root ganglia in the cat by GABA and related amino acids: antagonism by picrotoxin and bicuculline. *Brain Res.* **44:** 273–277.

53. Levy, R.A. & E.G. Anderson. 1972. The effect of the GABA antagonists bicuculline and picrotoxin on primary afferent terminal excitability. *Brain Res.* **43:** 171–180.

54. Barker, J.L. & R.A. Nicoll. 1973. The pharmacology and ionic dependency of amino acid responses in the frog spinal cord. *J. Physiol.* **228:** 259–277.

55. Mitchell, K., R.C. Spike & A.J. Todd. 1993. An immunocytochemical study of glycine receptor and GABA in laminae I-III of rat spinal dorsal horn. *J. Neurosci.* **13:** 2371–2381.

56. Towers, S. *et al.* 2000. GABAB receptor protein and mRNA distribution in rat spinal cord and dorsal root ganglia. *Eur. J. Neurosci.* **12:** 3201–3210.

57. Yang, K., D. Wang & Y.Q. Li. 2001. Distribution and depression of the GABA(B) receptor in the spinal dorsal horn of adult rat. *Brain Res. Bull.* **55:** 479–485.

58. Dickenson, A.H., C.M. Brewer & N.A. Hayes. 1985. Effects of topical baclofen on C fibre-evoked neuronal activity in the rat dorsal horn. *Neuroscience* **14:** 557–562.

59. Hao, J.X. *et al.* 1992. Baclofen reverses the hypersensitivity of dorsal horn wide dynamic range neurons to mechanical stimulation after transient spinal cord ischemia; implications for a tonic GABAergic inhibitory control of myelinated fiber input. *J. Neurophysiol.* **68:** 392–396.

60. Malcangio, M. & N.G. Bowery. 1993. Gamma-aminobutyric acidB, but not gamma-aminobutyric acidA receptor activation, inhibits electrically evoked substance P-like immunoreactivity release from the rat spinal cord in vitro. *J. Pharmacol. Exp. Ther.* **266:** 1490–1496.

61. Allerton, C.A., P.R. Boden & R.G. Hill. 1989. Actions of the GABAB agonist, (-)-baclofen, on neurones in deep dorsal horn of the rat spinal cord in vitro. *Br. J. Pharmacol.* **96:** 29–38.

62. Kangrga, I., M.C. Jiang & M. Randic. 1991. Actions of (-)-baclofen on rat dorsal horn neurons. *Brain Res.* **562:** 265–275.

63. Chery, N. & Y. De Koninck. 2000. GABA(B) receptors are the first target of released GABA at lamina I inhibitory synapses in the adult rat spinal cord. *J. Neurophysiol.* **84:** 1006–1011.

64. Yang, K. & H. Ma. 2011. Blockade of GABA(B) receptors facilitates evoked neurotransmitter release at spinal dorsal horn synapse. *Neuroscience* **193:** 411–420.

65. Wan, Y.H. & S.J. Hu. 2003. Short-term depression at primary afferent synapses in rat substantia gelatinosa region. *Neuroreport* **14:** 197–200.

66. De Koninck, Y. & J.L. Henry. 1994. Prolonged GABAA-mediated inhibition following single hair afferent input to single spinal dorsal horn neurones in cats. *J. Physiol.* **476:** 89–100.

67. Funk, K. *et al.* 2008. Modulation of chloride homeostasis by inflammatory mediators in dorsal root ganglion neurons. *Mol. Pain* **4:** 32.

68. Lin, Q., J. Wu & W.D. Willis. 1999. Dorsal root reflexes and cutaneous neurogenic inflammation after intradermal injection of capsaicin in rats. *J. Neurophysiol.* **82:** 2602–2611.

69. Weng, H.R. & P.M. Dougherty. 2005. Response properties of dorsal root reflexes in cutaneous C fibers before and after intradermal capsaicin injection in rats. *Neuroscience* **132:** 823–831.

Ann. N.Y. Acad. Sci. ISSN 0077-8923

ANNALS OF THE NEW YORK ACADEMY OF SCIENCES

Issue: *Neurons, Circuitry, and Plasticity in the Spinal Cord and Brainstem*

Activity-dependent development of tactile and nociceptive spinal cord circuits

Stephanie C. Koch and Maria Fitzgerald

Department of Neuroscience, Physiology and Pharmacology, University College London, London, United Kingdom

Address for correspondence: Professor Maria Fitzgerald, Department of Neuroscience, Physiology & Pharmacology, University College London, Gower Street, London WC1E 6BT, UK. m.fitzgerald@ucl.ac.uk

Developing brain circuits are shaped by postnatal sensory experience, but little is known about this process at the level of the spinal cord. Here we review the mechanisms by which cutaneous sensory input drives the maturation of spinal sensory circuits. Newborn animals are highly sensitive to tactile input and dorsal horn circuits are dominated by low threshold A fiber inputs. We show that this arises from the absence of the functional, targeted glycinergic inhibition of tactile activity that emerges only in the second week of life. Selective block of afferent C fibers in postnatal week 2 delays the maturation of glycinergic inhibition and maintains dorsal horn circuits in a neonatal state. We propose that in the newborn strong tactile A fiber input facilitates activity-dependent synaptic strengthening in the dorsal horn, but that this ends with the arrival of nociceptive C fiber spinal input that drives the maturation of targeted glycinergic inhibition.

Keywords: glycine; inhibition; dorsal horn; neonatal; reflex

Introduction

The developing nervous system depends upon sensory input for tuning and organizing sensory and motor circuits. The newborn mammal is almost continuously exposed to tactile skin stimulation, be it through maternal contact, huddling, or spontaneous twitching,[1] and this tactile input plays an important role in activity-dependent synaptic strengthening and circuit maturation. Maternal licking and grooming during a critical period in early life modifies neural development in newborn rat pups;[2] in human infants, there is evidence that skin contact supports growth and development.[3] As the newborn grows up and begins to explore its environment, the pattern of tactile skin stimulation will change and increasingly include activity from higher threshold mechanoreceptors and nociceptors. Therefore, it is likely that, depending upon the stage of maturation, synaptic connections in different neural circuits, excitatory and inhibitory, will be strengthened or weakened by different patterns and modalities of sensory input.

Here we focus on developing sensory circuits in the rat spinal cord. In the newborn spinal dorsal horn, neural activity is dominated by inputs from low-threshold A fibers, while nociceptive C fiber inputs mature gradually over the first postnatal weeks. We present evidence that while A fiber tactile input is required for synaptic tuning of many aspects of dorsal horn circuitry, the postnatal maturation of glycinergic inhibitory circuits requires C fiber input. We propose that the delay in glycinergic inhibition allows maximal A fiber tactile input during an early period of dorsal horn synaptic strengthening. The onset of functional glycinergic inhibition brings an end to this period and marks the beginning of the rapid sensory gating of tactile and nociceptive circuits required for a mature somatosensory system.

The role of tactile input in the development of spinal sensory circuits

The functional maturation of spinal nociceptive reflexes is modulated by tactile input. Rat pups make a high percentage of errors in removing their tails from a noxious heat stimulus, and maturation of the correct reflex behavior occurs between postnatal day (P) 14 and 21. Blocking low-threshold input to the rat tail over P14–P21 with combined local anesthetic and depilation of the tail prevents the normal

doi: 10.1111/nyas.12033

tuning of nociceptive tail flick responses, and the pups continue to make the errors seen in younger pups.[4] Furthermore, the organization of primary afferent terminals in the postnatal dorsal horn also requires low threshold sensory inputs. In newborn rats, A fibers have more widespread terminal fields than in adults,[5] which extend into lamina IIo and I in the first weeks of life before withdrawing to deeper laminae over the next three weeks, via NMDA-dependent mechanisms.[6] This withdrawal of exuberant terminals is prevented if animals are raised in constant tactile noise over large skin areas by placing pups together with their mother in a vibrating cage.[5] Schouenborg's group has proposed that the spontaneous movements that are ubiquitous in developing mammals, generated intrinsically within motor circuits, are a key source of tactile stimulation that drives the postnatal organization of spinal sensory circuits,[1,7] although dorsal horn lamina I cells in young animals display spontaneous, bursting "pacemaker" type activity that could also provide endogenous drive to developing sensory networks.[8]

Newborn animals are highly sensitive to tactile stimulation

The influence of tactile inputs on the developing central nervous system (CNS) is likely to be facilitated by the notably increased sensitivity of newborns to these inputs compared to adults. Cutaneous reflexes in the newborn rat, kitten, and human are exaggerated in amplitude and duration compared to the adult.[9,10] Cutaneous thresholds are also lower in the neonate, and flexion withdrawal reflexes can be elicited by innocuous stimulation; in humans, thresholds are especially low in preterm infants and increase with age. Repeated tactile skin stimulation results in behavioral sensitization with generalized movements of all limbs, which becomes less pronounced after 29–35 weeks' gestational age in the human and P8 in the rat.[11,12]

This behavioral sensitivity to tactile stimulation is reflected in neonatal dorsal horn cell properties. Many dorsal horn neurons in neonatal rats have large excitatory cutaneous receptive fields, covering a greater percentage of the body surface than in adults, which gradually decrease over the first two postnatal weeks.[13–15] In addition, young dorsal horn neurons have low cutaneous sensory thresholds, and stimulation of their receptive fields can

result in a prolonged afterdischarge of action potentials. Repetitive activation of low-threshold A fibers results in a sensitization of dorsal horn neurons, manifested as a progressive increase in background firing, which disappears by P21.[16] Furthermore, low intensity touch and Aβ stimulation produces a significant Fos response in laminae I and II cells, which is only observed with high-threshold stimulation in adult animals,[17] and NMDA-dependent long-term potentiation in the dorsal horn can be evoked by A fiber stimulation in neonatal rats, but not juveniles or adults.[18] Contralateral inhibitory dorsal horn receptive fields are also relatively larger in young animals, and in contrast to adults, can be evoked by low threshold tactile stimulation,[19] but these fields are not spatially aligned with ipsilateral excitatory fields and so are functionally less effective.

The neonatal dorsal horn is dominated by A fiber afferent inputs

Enhanced tactile sensitivity in the newborn is likely due to greater synaptic input from cutaneous A fiber myelinated afferents. Neurons in the superficial dorsal horn (lamina II) receive a greater degree of low threshold Aβ-mediated input during the neonatal period, either from direct Aβ fiber projections to the region or from low-threshold polysynaptic connections. As discussed previously, neonatal A fiber terminal fields are diffusely distributed in the dorsoventral and rostrocaudal dimensions and undergo synaptic organization in the postnatal period to reach a near-adult pattern by P21.[5,6] In the adult, lamina II is the site of termination of nociceptive C fibers and the more dorsal termination of A fibers in the first postnatal weeks, and the presence of A fiber synaptic contacts in lamina II[20] may explain the slow postnatal maturation of C fiber terminals in this region. A high incidence of Aβ-evoked monosynaptic responses has been observed in patch clamp studies of juveniles, P21–23 lamina II neurons;[21,22] and polysynaptic Aβ fiber synapses onto GABAergic neurons in laminae I–II are more prevalent at P16–18.[23] While the precise time course of these events differ, they all indicate enhanced A fiber input in younger dorsal horn neurons and are consistent with properties of immature dorsal horn cells *in vivo*, where responses recorded from cells in the younger animals are elicited mainly by low-threshold mechanoreceptors.[24]

In contrast to low-threshold A fiber input, high-threshold C fiber synaptic input to the dorsal horn matures slowly over a number of postnatal weeks. Thus, while neonates display strong nociceptive reflexes, these are likely to be mediated by Aδ afferents, as C fiber input into the dorsal horn is weak until the end of the first postnatal week.[24,25] Transient receptor potential cation channel subfamily V member 1 (TRPV1)-positive C fibers begin forming functional synapses before birth, but a marked increase in their synaptic inputs between P5 and P10 coincides with the onset of C fiber–evoked spiking in the second postnatal week.[26] Heat and mustard oil skin stimulation, both mediated by C fibers, evoked little Fos expression in dorsal horn cells at birth,[27,28] and hindpaw capsaicin produces only minimal extracellular signal-regulated kinase (pERK) activation at P3.[29] The Fos and pERK activity evoked by chemical irritants increases markedly in the second week after birth, consistent with the maturation of central C fiber synaptic input at that time.[27–29]

Neonatal tactile sensitivity, A fiber inputs, and immature glycinergic signaling

Neonatal sensitivity to tactile stimulation and enhanced Aβ afferent input to neonatal dorsal horn cells may result from immature inhibitory signaling, in particular from glycinergic networks. In adults, Aβ fibers synapse directly onto glycinergic interneurons[30,31] and glycine receptor antagonists enhance tactile behavioural sensitivity[32,33] and increase A fiber excitation of dorsal horn neurons,[34,35] producing a pattern of activity analogous to that seen in the healthy developing neonate. However, there is little glycinergic activity in newborn lamina II: miniature inhibitory postsynaptic currents are low frequency and mediated by GABA$_A$ receptors,[36] and the glycine transporter GLYT2 in presynaptic glycinergic terminals matures slowly, reaching adult termination patterns in laminae IIi and III by P14.[37,38]

We have recently shown, using *in vivo* recordings of single wide dynamic range neurons, that glycinergic inhibition in the dorsal horn of the spinal cord is not functional until the second week of life. Before that time, strychnine does not lead to an enhancement of tactile responses in the dorsal horn. Indeed, not only is there no glycinergic inhibition, there is glycinergic facilitation of dorsal horn–wide dynamic range neuronal responses to low threshold, dynamic

brushing of the skin. The switch from glycinergic facilitation within the sensory network to glycinergic inhibition occurs around P14.[38] There are a number of possible mechanisms underlying the absence of glycinergic inhibition in the neonatal dorsal horn, but all likely reflect immature synaptic connections rather than an absence of glycinergic receptor signaling, since exogenously applied glycine is inhibitory from birth. Our results point to the ratio of α_2/α_1 subunits as a key indicator of absence of inhibition; the high ratio observed at P3, downregulated by P14, suggests fewer synaptically anchored glycinergic receptors in the neonate, lowering the likelihood of targeted inhibitory signaling.[38–40] β subunits are absent in neonatal α_2 homomeric glycine receptors, unlike the α_1/β heteromers found in the adult, and binding of the anchoring protein gephyrin to the β subunit is necessary for glycine receptor cluster formation, thus only α/β heteromeric glycine receptors will aggregate at inhibitory synapses.[39,41] However, the high α_2/α_1 ratio does not explain glycine facilitation of brush-evoked activity observed in the neonate; we suggest that this arises from a different distribution of glycinergic inhibition across inhibitory and excitatory circuits within the neonatal dorsal horn network. This is supported by a shift in the pattern of brush-evoked Fos activation in lamina III interneurons following strychnine application in P3 versus P14 rats. Blocking glycinergic activity at P3 reveals greater Fos activation in GAD$^+$ inhibitory neurons compared to P14 animals, suggesting greater glycinergic inhibitory interneuronal control in the neonate.[38]

Figure 1 illustrates a proposed mechanism for the developmental change in glycinergic function in dorsal horn circuits. As discussed previously, neonatal dorsal horn cells are characterized by strong A fiber input, large excitatory receptive fields,[42] and inhibitory fields that are not spatially aligned.[19] We propose that against this background of immature excitatory (glutamatergic) and inhibitory (GABAergic) circuitry, glycinergic neurons cannot provide the focused, selective inhibitory control of excitatory and inhibitory networks that is observed in adults,[43] but rather exert a less specific control of predominantly inhibitory networks in neonates. As the animal grows up, glycinergic activity not only increases[36] but also becomes more targeted toward excitatory circuits, which are now appropriately balanced with inhibitory ones.

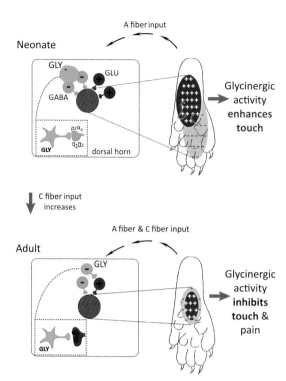

Figure 1. Illustration of the maturation of functional glycinergic inhibition in the dorsal horn of the spinal cord.

What triggers the onset of glycinergic inhibition of tactile activity?

This evidence suggests that the enhanced tactile-evoked responses in neonates are due, at least in part, to weak untargeted glycinergic signalling in the dorsal horn. The timing of onset of glycinergic inhibition in the second week of life coincides with the marked increase of C fiber synaptic input into the dorsal horn, which led us to hypothesize a causal link between the two. This is supported by the finding that chronic selective block of C fibers with the impermeable lidocaine derivative QX-314 over a critical period in the second postnatal week delayed the onset of functional glycinergic inhibition.[38] QX-314 requires the opening of a wide pore ion channel such as TRPV1 to permeate the resting cell membrane.[44] Thus, by injecting a cocktail of QX-314, the local anaesthetic lidocaine, followed by capsaicin (both TRPV1 channel agonists) into the perisciatic space, TRPV1-positive nociceptive C fiber primary afferents can be selectively targeted and silenced.[44] We have adapted this technique to younger animals to produce blockades of peripheral nerve C fibers from P10 to P13, and so delay

the maturation of functional glycinergic inhibition of dorsal horn cells to tactile inputs. This was accompanied (but not necessarily caused) by delayed maturation of subunit expression and a prolonged incidence of glycinergic inhibition of GAD67 interneurons (Fig. 1). In other words, the dorsal horn cell responses in P14 rats whose peripheral C fibers had been blocked were the same as those at P3, with regard to a lack of glycinergic inhibition, a high α_2/α_1 ratio subunit expression, and a wider pattern of inhibitory interneuronal control.[38] Rats recovered after the QX-314 with no C fiber loss, and the same treatment at older ages had no effect on glycinergic transmission.

The changing pattern of activity dependence in the development of tactile and nociceptive spinal cord circuits

Tactile input plays an important role in activity-dependent maturation of the CNS, and it appears that neonatal spinal sensory circuits are designed in such a way as to maximize the chances of tactile afferent input reaching the CNS. The young dorsal horn has exuberant, strong A fiber inputs, and neonatal spinal circuits are highly responsive to tactile stimulation. We propose that this is due, at least in part, to a lack of targeted glycinergic inhibition in the dorsal horn, immature glycinergic subunit expression, likely indicating poorly anchored receptors, and widespread inhibition of inhibitory interneurons.

The situation changes from P10, when C fiber central synaptic input begins to drive activity-dependent targeted glycinergic inhibition toward excitatory inputs, accompanied by a shift in subunit expression. The mature glycinergic inhibition dampens tactile excitability and provides better control of sensory inputs, especially A fiber-evoked activity, which is notably targeted by glycinergic interneurons in adults.[45] It may also contribute to reduced dorsal horn cell receptive field size and alignment of inhibitory and excitatory receptive fields. Earlier reports of disorganized receptive fields and lack of A-evoked inhibition (now known to be glycine mediated) following neonatal C fiber destruction with systemic capsaicin may now be explained as a failure of glycinergic interneurons to mature in these animals.[46,47]

The sensory dorsal horn is shaped by differing patterns of afferent activity. In the immediate neonatal period, low-threshold A fiber input, free from glycinergic control, drives the synaptic organization of afferent terminals and interneuronal connections. In the second postnatal week, C fiber synapses strengthen and begin to drive the maturation of glycinergic inhibition. Thus, the source of activity-dependent development of tactile and nociceptive circuits changes from A fibers in the neonate to C fibers at later postnatal ages.

Conflicts of interest

The authors declare no conflicts of interest.

References

1. Waldenström, A., M. Christensson & J. Schouenborg. 2009. Spontaneous movements: effect of denervation and relation to the adaptation of nociceptive withdrawal reflexes in the rat. *Physiol. Behav.* **98:** 532–536.

2. Kaffman, A. & M.J. Meaney. 2007. Neurodevelopmental sequelae of postnatal maternal care in rodents: clinical and research implications of molecular insights. *J. Child Psychol. Psychiatry* **48:** 224–244.

3. Moore, E.R., G.C. Anderson, N. Bergman & T. Dowswell. 2012. Early skin-to-skin contact for mothers and their healthy newborn infants. *Cochrane Database Syst. Rev.* **5:** CD003519.

4. Waldenström, A., J. Thelin, E. Thimansson, *et al.* 2003. Developmental learning in a pain-related system: evidence for a cross-modality mechanism. *J. Neurosci.* **23:** 7719–7725.

5. Granmo, M., P. Petersson & J. Schouenborg. 2008. Action-based body maps in the spinal cord emerge from a transitory floating organization. *J. Neurosci.* **28:** 5494–5503.

6. Beggs, S., C. Torsney, L.J. Drew & M. Fitzgerald. 2002. The postnatal reorganization of primary afferent input and dorsal horn cell receptive fields in the rat spinal cord is an activity-dependent process. *Eur. J. Neurosci.* **16:** 1249–1258.

7. Petersson, P., A. Waldenström, C. Fåhraeus & J. Schouenborg. 2003. Spontaneous muscle twitches during sleep guide spinal self-organization. *Nature* **424:** 72–75.

8. Li, J. & M.L. Baccei. 2011. Pacemaker neurons within newborn spinal pain circuits. *J. Neurosci.* **31:** 9010–9022.

9. Ekholm, J. 1967. Postnatal changes in cutaneous reflexes and in the discharge pattern of cutaneous and articular sense organs: a morphological and physiological study in the cat. *Acta Physiol. Scand. Suppl.* **297:** 1–130.

10. Fitzgerald, M., A. Shaw & N. MacIntosh. 1988. Postnatal development of the cutaneous flexor reflex: comparative study of preterm infants and newborn rat pups. *Dev. Med. Child Neurol.* **30:** 520–526.

11. Andrews, K. & M. Fitzgerald. 1994. The cutaneous withdrawal reflex in human neonates: sensitization, receptive fields, and the effects of contralateral stimulation. *Pain* **56:** 95–101.

12. Andrews, K. & M. Fitzgerald. 1999. Cutaneous flexion reflex in human neonates: a quantitative study of threshold and stimulus-response characteristics after single and repeated stimuli. *Dev. Med. Child Neurol.* **41:** 696–703.

13. Fitzgerald, M. 1985. The post-natal development of cutaneous afferent fibre input and receptive field organization in the rat dorsal horn. *J. Physiol.* **364:** 1–18.

14. Torsney, C. & M. Fitzgerald. 2002. Age-dependent effects of peripheral inflammation on the electrophysiological properties of neonatal rat dorsal horn neurons. *J. Neurophysiol.* **87:** 1311–1317.

15. Ririe, D.G., L.R. Bremner & M. Fitzgerald. 2008. Comparison of the immediate effects of surgical incision on dorsal horn neuronal receptive field size and responses during postnatal development. *Anesthesiology* **109:** 698–706.

16. Jennings, E. & M. Fitzgerald. 1998. Postnatal changes in responses of rat dorsal horn cells to afferent stimulation: a fibre-induced sensitization. *J. Physiol.* **509**(Pt 3): 859–868.

17. Jennings, E. & M. Fitzgerald. 1996. C-fos can be induced in the neonatal rat spinal cord by both noxious and innocuous peripheral stimulation. *Pain* **68:** 301–306.

18. Wu, J., Q. Hu, D. Huang, *et al.* 2012. Effect of electrical stimulation of sciatic nerve on synaptic plasticity of spinal dorsal horn and spinal c-fos expression in neonatal, juvenile and adult rats. *Brain Res.* **1448:** 11–19.

19. Bremner, L.R. & M. Fitzgerald. 2008. Postnatal tuning of cutaneous inhibitory receptive fields in the rat. *J. Physiol.* **586:** 1529–1537.

20. Coggeshall, R.E., E.A. Jennings & M. Fitzgerald. 1996. Evidence that large myelinated primary afferent fibers make synaptic contacts in lamina II of neonatal rats. *Brain Res. Dev. Brain Res.* **92:** 81–90.

21. Park, J.S., T. Nakatsuka, K. Nagata, *et al.* 1999. Reorganization of the primary afferent termination in the rat spinal dorsal horn during post-natal development. *Brain Res. Dev. Brain Res.* **113:** 29–36.

22. Nakatsuka, T., T. Ataka, E. Kumamoto, *et al.* 2000. Alteration in synaptic inputs through C-afferent fibers to substantia gelatinosa neurons of the rat spinal dorsal horn during postnatal development. *Neuroscience* **99:** 549–556.

23. Daniele, C.A. & A.B. MacDermott. 2009. Low-threshold primary afferent drive onto GABAergic interneurons in the superficial dorsal horn of the mouse. *J. Neurosci.* **29:** 686–695.

24. Fitzgerald, M. & E. Jennings. 1999. The postnatal development of spinal sensory processing. *Proc. Natl. Acad. Sci. USA* **96:** 7719–7722.

25. Fitzgerald, M. 1988. The development of activity evoked by fine diameter cutaneous fibers in the spinal cord of the newborn rat. *Neurosci. Lett.* **86:** 161–166.

26. Baccei, M.L., R. Bardoni & M. Fitzgerald. 2003. Development of nociceptive synaptic inputs to the neonatal rat dorsal horn: glutamate release by capsaicin and menthol. *J. Physiol.* **549:** 231–242.

27. Yi, D.K. & G.A. Barr. 1995. The induction of Fos-like immunoreactivity by noxious thermal, mechanical and chemical stimuli in the lumbar spinal cord of infant rats. *Pain* **60:** 257–265.

28. Williams, S., G. Evan & S.P. Hunt. 1990. Spinal c-fos induction by sensory stimulation in neonatal rats. *Neurosci. Lett.* **109:** 309–314.

29. Walker, S.M., J. Meredith-Middleton, T. Lickiss, *et al.* 2007. Primary and secondary hyperalgesia can be differentiated by postnatal age and ERK activation in the spinal dorsal horn of the rat pup. *Pain* **128:** 157–168.

30. Todd, A.J. 1990. An electron microscope study of glycine-like immunoreactivity in laminae I-III of the spinal dorsal horn of the rat. *Neuroscience* **39:** 387–394.

31. Narikawa, K., H. Furue, E. Kumamoto & M. Yoshimura. 2000. In vivo patch-clamp analysis of IPSCs evoked in rat substantia gelatinosa neurons by cutaneous mechanical stimulation. *J. Neurophysiol.* **84:** 2171–2174.

32. Yaksh, T.L. 1989. Behavioral and autonomic correlates of the tactile evoked allodynia produced by spinal glycine inhibition: effects of modulatory receptor systems and excitatory amino acid antagonists. *Pain* **37:** 111–123.

33. Sherman, S.E. & C.W. Loomis. 1996. Strychnine-sensitive modulation is selective for non-noxious somatosensory input in the spinal cord of the rat. *Pain* **66:** 321–330.

34. Sivilotti, L. & C.J. Woolf. 1994. The contribution of GABAA and glycine receptors to central sensitization: disinhibition and touch-evoked allodynia in the spinal cord. *J. Neurophysiol.* **72:** 169–179.

35. Miraucourt, L.S., X. Moisset, R. Dallel & D.L. Voisin. 2009. Glycine inhibitory dysfunction induces a selectively dynamic, morphine-resistant, and neurokinin 1 receptor-independent mechanical allodynia. *J. Neurosci.* **29:** 2519–2527.

36. Baccei, M.L. & M. Fitzgerald. 2004. Development of GABAergic and glycinergic transmission in the neonatal rat dorsal horn. *J. Neurosci.* **24:** 4749–4757.

37. Spike, R.C., C. Watt, F. Zafra & A.J. Todd. 1997. An ultrastructural study of the glycine transporter GLYT2 and its association with glycine in the superficial laminae of the rat spinal dorsal horn. *Neuroscience* **77:** 543–551.

38. Koch, S.C., K.K. Tochiki, S. Hirschberg & M. Fitzgerald. 2012. C-fiber activity-dependent maturation of glycinergic inhibition in the spinal dorsal horn of the postnatal rat. *Proc. Natl. Acad. Sci. USA* **109:** 12201–12206.

39. Kirsch, M. & B. Synaptic. 1996. Targeting of ionotropic neurotransmitter receptors. *Mol. Cell. Neurosci.* **8:** 93–98.

40. Mangin, J.M. *et al.* 2003. Kinetic properties of the alpha2 homo-oligomeric glycine receptor impairs a proper synaptic functioning. *J. Physiol.* **553:** 369–386.

41. Fritschy, J.-M., R.J. Harvey & G. Schwarz. 2008. Gephyrin: where do we stand, where do we go? *Trends Neurosci.* **31:** 257–264.

42. Fitzgerald, M. 2005. The development of nociceptive circuits. *Nat. Rev. Neurosci.* **6:** 507–520.

43. Labrakakis, C., L.-E. Lorenzo, C. Bories, *et al.* 2009. Inhibitory coupling between inhibitory interneurons in the spinal cord dorsal horn. *Mol. Pain* **5:** 24.

44. Binshtok, A.M., B.P. Bean & C.J. Woolf. 2007. Inhibition of nociceptors by TRPV1-mediated entry of impermeant sodium channel blockers. *Nature* **449:** 607–610.

45. Zeilhofer, H.U., H. Wildner & G.E. Yévenes. 2012. Fast synaptic inhibition in spinal sensory processing and pain control. *Physiol. Rev.* **92:** 193–235.

46. Wall, P.D., M. Fitzgerald, J.C. Nussbaumer, *et al.* 1982. Somatotopic maps are disorganized in adult rodents treated neonatally with capsaicin. *Nature* **295:** 691–693.

47. Wall, P.D. 1982. The effect of peripheral nerve lesions and of neonatal capsaicin in the rat on primary afferent depolarization. *J. Physiol.* **329:** 21–35.

Ann. N.Y. Acad. Sci. ISSN 0077-8923

ANNALS OF THE NEW YORK ACADEMY OF SCIENCES
Issue: *Neurons, Circuitry, and Plasticity in the Spinal Cord and Brainstem*

Force-sensitive afferents recruited during stance encode sensory depression in the contralateral swinging limb during locomotion

Shawn Hochman,[1] Heather Brant Hayes,[2] Iris Speigel,[1] and Young-Hui Chang[3]

[1]Department of Physiology, [2]Department of Rehabilitation Medicine, Emory University School of Medicine, Atlanta, Georgia. [3]School of Applied Physiology, Georgia Institute of Technology, Atlanta, Georgia

Address for correspondence: Shawn Hochman, Ph.D., Department of Physiology, Emory University School of Medicine, 615 Michael St., Atlanta, GA 30322. shawn.hochman@emory.edu

Afferent feedback alters muscle activity during locomotion and must be tightly controlled. As primary afferent depolarization-induced presynaptic inhibition (PAD-PSI) regulates afferent signaling, we investigated hindlimb PAD-PSI during locomotion in an *in vitro* rat spinal cord–hindlimb preparation. We compared the relation of PAD-PSI, measured as dorsal root potentials (DRPs), to observed ipsilateral and contralateral limb endpoint forces. Afferents activated during stance-phase force strongly and proportionately influenced DRP magnitude in the swinging limb. Responses increased with locomotor frequency. Electrical stimulation of contralateral afferents also preferentially evoked DRPs in the opposite limb during swing (flexion). Nerve lesioning, in conjunction with kinematic results, support a prominent contribution from toe Golgi tendon organ afferents. Thus, force-dependent afferent feedback during stance binds interlimb sensorimotor state to a proportional PAD-PSI in the swinging limb, presumably to optimize interlimb coordination. These results complement known actions of ipsilateral afferents on PAD-PSI during locomotion.

Keywords: presynaptic inhibition; primary afferent depolarization; dorsal root potential; primary afferent; Golgi tendon organ

Introduction

The circuitry required for generating vertebrate locomotion resides in the spinal cord and is called the locomotor central pattern generator (CPG).[1,2] The locomotor CPG is present at birth in both mouse and rat.[3] The rodent neonatal spinal cord can be isolated and maintained *in vitro*,[4] and, in the presence of bath-applied neuroactive substances, this preparation produces a motor pattern consistent with locomotion as recorded from ventral roots, peripheral nerves, or hindlimb muscles.[5]

Studies in the intact animal indicate that sensory cues related to limb posture and loading play significant roles in determining the spatiotemporal features of motor output.[6,7] Sensory feedback particularly regulates the timing of flexor[8,9] and amplitude of extensor activation.[10,11] Despite the overwhelming importance of sensory feedback to locomotor function, many studies on spinal sensorimotor circuitry are performed in the absence of intact feedback. In order to provide greater similarity to adult locomotion using *in vitro* preparations, we recently developed a novel preparation—the *in vitro* neonatal rat spinal cord with hindlimbs intact and pendant (SCHIP). This preparation couples natural sensory feedback and behavioral observability with the neural accessibility of the classic *in vitro* preparations.[12] The spinal cord is positioned dorsal-up so that stepping can be activated by bath application of neurochemicals (Fig. 1D–E) to generate coordinated muscle activity patterns with limb kinematics that compare well to what is seen in the adult.[12–16]

We have since incorporated ventral and dorsal root recordings, as well as intracellular recordings,

doi: 10.1111/nyas.12055

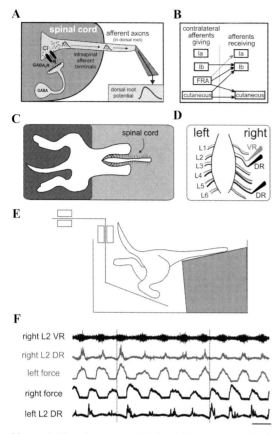

Figure 1. Dorsal root potentials (DRPs) as measures of primary afferent depolarization and experimental setup. (A) GABA release activates GABA$_A$ receptors on intraspinal primary afferent terminals. Afferents possess a high intracellular chloride gradient, leading to chloride efflux. This depolarization travels antidromically as an electrotonic wave, where it can be measured as a DRP at the root entry zone. (B) Previous studies have shown that contralateral afferents can evoke PAD-PSI in ipsilateral afferents. FRA = flexor reflex afferents. (C) Overhead view of the *in vitro* SCHIP preparation with exposed spinal cord and intact hindlimbs. The spinal cord and limbs are maintained in continuously oxygenated artificial cerebrospinal fluid circulated by a peristaltic pump and gravity-fed perfusion system. (D) Recording configuration. DRPs were recorded near the dorsal root (DR) entry zones of L2 and/or L5 DR using glass suction electrodes. Activity in the L2 ventral root (VR) was also recorded. (E) Sagittal view of hindlimb-force platform interaction. Two-dimensional force platforms were constructed to monitor forces exerted by a single hindlimb.[18] (F) Sample recordings of right and left hindlimb vertical ground reaction forces measured during locomotion and displayed relative to the right L2 VR, and right and left DR, activity. Vertical dotted lines mark relative timing of left and right endpoint forces that correspond with DRPs in right and left L2 DRs, respectively. Scale bar is five seconds. Reproduced in part from Ref. 18, with permission.

to relate neural function and behavior in ways not typically possible *in vitro*.[17] A particularly exciting observation regarding afferent feedback concerned the control of transmission in contralateral afferents by a distinctive form of presynaptic inhibition (PSI), called *primary afferent depolarization* (PAD).[18] Such contralateral interlimb control of sensory transmission is barely studied and consequently poorly understood. Yet the magnitude of PAD observed suggests that these control mechanisms are profound.[18]

Here, we first review the role of afferent feedback on the regulation of ongoing locomotion, with subsequent emphasis on afferent PAD-mediated PSI (PAD-PSI). We then present our previously reported results and provide some additional data on a powerful contralateral afferent–evoked PAD-PSI during locomotion. Overall, contralateral stance-phase force feedback may prove to be the most pivotal mechanosensory event encoding sensory gain into the swinging limb during an alternating locomotor gait.

Role of sensory feedback during locomotion

Sensory feedback refines the spatiotemporal features of motor output. Limb extension and loading are primary determinants of phase transition timing.[7–9,19,20] Overall, it appears that a balance between the excitatory stretch and inhibitory load sensory signals determines the exact timing of the stance-to-swing transition. For example, preventing hip extension hinders swing initiation,[8] while stretch or vibration of hip or ankle flexors can alter swing onset timing.[9] Hindlimb flexor stretch-sensitive muscle spindle afferent (Ia afferent) activity during hip extension likely initiates flexion via homonymous and synergistic reflex excitatory feedback. Passive oscillatory hip extensions entrain locomotor speed by altering the duration of stance and the timing of the stance-to-swing transition.[21,22] Loading also controls swing initiation since preventing limb unloading can inhibit flexion generation at the stance-to-swing transition,[19,23] implicating force-sensitive Golgi tendon organ afferent activity (Ib afferents) in loaded extensors.[19,30] The contralateral limb also contributes load-related signals; even when the ipsilateral hip is fully extended, swing will only initiate if the contralateral limb is prepared to accept the load.[8,23]

In addition to timing, sensory feedback regulates the magnitude and duration of extensor activity during stance, particularly at the ankle.[7,24–26] In the cat, the ankle extensor activity magnitude is reduced if the ankle extensor load is reduced, while ankle extensor activity and force production increase if ankle extensors are artificially stretched.[10,24,27] Though responses to length changes are often attributed to Ia muscle spindles, Ib afferents can contribute substantially to ankle extensor activity since Ib feedback onto ankle extensors during locomotion can actually be excitatory, further increasing stance-phase extensor activity and force production.[10,26]

While less well studied, sensory feedback can also influence swing-phase flexor activity. Resisting hip flexion or stimulation of group I and II afferents in flexor nerves during swing enhances flexor activity,[28–30] as does stimulation of toe flexors.[31] More generally, stimulation of peripheral nerves during fictive locomotion can reset or entrain centrally generated rhythms in a task- and phase-dependent manner.[29–35]

Regulation of sensory inflow by presynaptic inhibition

Precisely because sensory feedback exerts such powerful influence over motor output, it must be tightly regulated. During locomotion, sensory inputs help modify muscle timing and magnitude to meet environmental demands. While both presynaptic and postsynaptic inhibition regulate the effectiveness of sensory transmission onto central circuits,[36] their actions are quite different. PSI of intraspinal primary afferent terminals occurs even before the first afferent synapse, so it is well positioned to regulate sensory actions on spinal neurons. Although postsynaptic changes are certainly important and coexist with PSI effects, PSI is a highly selective and effective way to gate and/or redirect afferent actions.[36] PSI can occur via activation of metabotropic or ionotropic receptors. Ionotropic receptor–mediated PSI is a special form of PSI in afferents, typically thought to be caused by activation of $GABA_A$ receptors. Since primary afferents retain high intracellular chloride, $GABA_A$ receptor activation leads to a chloride efflux initiating PAD. PAD is thought to reduce transmitter release by inactivating sodium and calcium channels, and/or by electrical shunting. In this way, PAD reduces the central actions of incoming sensory events.[37,38] Since PAD travels electrotoni-

cally back out the dorsal root toward the periphery, it can be measured experimentally as a dorsal root potential (DRP), and its magnitude monitored as a measure of the relative strength of presynaptic inhibition (Fig. 1A).

PAD-PSI encompasses all-ensuing descriptions of PSI

Over the past five decades, researchers have characterized the many sources of both afferent and descending PAD-PSI onto group I (Ia and Ib) and II muscle, and cutaneous afferents, both at rest or during fictive locomotion. Most have focused on ipsilateral interactions as provided in a recent review.[36] Since we examined actions from contralateral afferents, ipsilateral actions are not described subsequently.

Contralateral contributions to PAD-PSI. Very little attention has been given to contralateral afferent–evoked PAD-PSI. A small number of early studies found that stimulation of group I and flexor reflex afferents produced a contralateral, along with the larger ipsilateral, DRP (Fig. 1B).[39–41] Most contralateral PAD-PSI involves Ib Golgi tendon organ afferents as both the source and receiving afferents.[39] In the pentobarbital anesthetized cat, Ib afferents evoked PAD-PSI of contralateral Ib afferents, but Ia afferents neither gave nor received contralateral PAD-PSI, mirroring the largely ipsilateral Ia afferent reflex patterns.[39,42] In the presence of L-DOPA, higher threshold flexor reflex afferents (typically group III, not II) could evoke inhibition of contralateral Ia afferents.[41] No studies have demonstrated Ia-evoked crossed inhibition under any conditions.

Studies of pedaling on a stationary bike further affirm that sensory inputs from the contralateral limb affect ipsilateral sensory transmission and motor output, particularly of flexors.[43,44] While contralateral movement–related feedback may play a role in ipsilateral sensory regulation, no recent work has investigated crossed PAD-PSI pathways during movements that require interlimb coordination like locomotion.

PAD-PSI during locomotion. While much is known about afferent-evoked PAD-PSI at rest, less is known about PAD-PSI during behavior. It has been difficult to study both centrally evoked and afferent-evoked PAD-PSI during nonfictive locomotion,[45] but PAD-PSI is clearly quite active and phase-dependent during movement.[46,47] DRPs and intra-axonal PAD are rhythmic during fictive locomotion, confirming actions independent of rhythmic afferent feedback (see Refs. 48 and 49). Locomotor-related PAD-PSI is typically maximal during the flexion phase in the majority of group I and II muscle as well as cutaneous afferents.[49,50] While afferent-evoked PAD-PSI appears to be more effective than locomotor-related PAD-PSI,[51] these circuitries interact since locomotor circuits modulate the effectiveness of sensory-evoked PAD-PSI in a phase- and muscle-dependent manner.[45,52]

Thus, the spinal locomotor circuitry and recruited afferents interact to create dynamic patterns of presynaptic sensory regulation. An important goal therefore is to understand these interactions during behaviorally relevant natural patterns of afferent activity. The SCHIP allowed us to mechanically isolate the spinal cord from the limbs to provide sufficient stability for study of contralateral and ipsilateral PSI-PAD during nonfictive locomotion, while manipulating the mechanics and neural system in other ways was not possible.

Methods

The development of SCHIP and the related results obtained have previously been described.[17,18,53] Briefly, locomotion was induced with *N*-methyl-D-aspartate and serotonin, with or without dopamine added to the artificial cerebrospinal fluid solution. Activity in the lumbar L2 or L5 ventral roots was recorded as a monitor of spinal motor output. Bursting activity in these roots corresponds to flexor and extensor locomotor phases, respectively.[14] DRPs were recorded with glass suction electrodes placed *en passant* near the entry zones of dorsal roots L2 and/or L5. DRPs are a measure of PAD-PSI, and DRP amplitude increases indicate increases in PAD-PSI. Limb endpoint forces were measured using two 2D force platforms, one for each hindlimb. A description of the preparation and obtained recordings has previously been presented[18] and is summarized in Figure 1C–F. Throughout the following Results section, ipsilateral refers to the side of the recorded DRP (ipsi-DRP), and contralateral refers to the side contralateral to the DRP, unless otherwise stated.

Results

Figure 1F shows the typical locomotor patterns of L2 DRPs recorded from DRs in relationship to L2 ventral root (VR) and ipsilateral and contralateral forces. Note that contralateral foot contact forces predicted timing of the ipsilateral DRPs (left force on right DRP, right force on left DRP). The DRP amplitude also appeared to scale with contralateral force. Indeed, when L2 DRP area, mean amplitude, or peak amplitude are plotted against the corresponding values for the ipsilateral and contralateral forces in each cycle, significant correlations are only observed with contralateral force ($n = 10/10$ for area, 9/10 for peak, and 8/10 for mean). None showed a significant positive correlation with ipsilateral force magnitude (Fig. 2A).[18]

Within a step cycle, the contralaterally evoked ipsilateral DRP occurred during the early ipsilateral flexion phase in all animals. In order for the contralateral limb to evoke or influence DRP magnitude, and thus the amount of PAD-PSI on the ipsilateral limb, contralateral force must precede the onset of the ipsilateral L2 DRP during each cycle. Figure 2B shows the phasing of the ipsilateral and contralateral force onset relative to DRP onset for a typical bout of locomotion. Note that contralateral force immediately preceded the ipsilateral DRP, while ipsilateral force timing was more variable and out-of-phase. The mean onset of contralateral force always just preceded the onset of the DRP ($n = 10/10$). Overall, we observed that the timing of PAD-PSI is tightly coupled to contralateral limb endpoint force, but not ipsilateral force or motor output timing.

When the locomotor frequency increased, the magnitude of the DRP also increased (Fig. 2C). The relationship between L2 DRP area and locomotor frequency was generally well fit with an exponential curve, suggesting that sensory feedback is progressively depressed as the locomotor frequency increased ($n = 10/10$ with $R^2 > 0.85$; Fig. 2D). In comparison, there was no relationship between DRP magnitude and any hindlimb kinematic variable compared (contralateral ankle, knee, and hip range of motion, area under the angular trajectory, magnitude of concurrent ipsilateral flexion and extension).[18]

Several perturbations further confirmed the source of the ipsilateral DRP (Fig. 3). To establish the

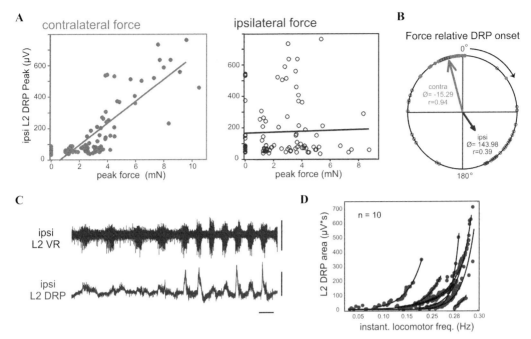

Figure 2. Ipsilateral L2 DRP strength scales with contralateral force and locomotor frequency. (A) Linear regression relating DRP peak to contralateral (left panel; gray dots) and ipsilateral peak force (middle panel; open circles) from a representative animal. Each point represents a single cycle ($n = 95$ cycles). DRP peak was strongly correlated with contralateral ($R = 0.83$, $P < 0.0001$), but not ipsilateral, force ($R = 0.03$; $P = 0.79$). Note that when no contralateral force occurs (gray points on the y-axis), the DRP is small or zero, but the absence of ipsilateral force (right panel; open circle points on the y-axis) does not affect DRP magnitude. (B) Phase relationship of contralateral and ipsilateral force relative to DRP onset. 0° represents the DRP onset, and the cycle progresses clockwise from 0° (in-phase) to 180° (out-of-phase) to 360°/0°. Northwest points precede DRP onset; northeast points lag. Arrow length represents the concentration (r) about the mean angle (Ø). Note that contralateral force consistently precedes DRP onset, while ipsilateral force is typically out of phase with the DRP and contralateral force. (C) In this animal, a perturbation induced a rapid increase in frequency and force in the middle of a locomotor bout. As a result, L2 DRP amplitude rapidly increased, highlighting the increase in PAD-PSI with locomotor frequency and force. Scale bars: five seconds, 200 µV. (D) DRP area versus instantaneous locomotor frequency fitted by exponential curves for 10 animals. Note that DRP area increases with increasing locomotor frequency. Reproduced from Ref. 18, with permission.

essential requirement for the contralateral endpoint force in ipsilateral DRP generation, ipsilateral and contralateral plates were separately removed (Fig. 3A). While ipsilateral DRPs were largely unchanged when the ipsilateral plate was removed ($n = 6/7$), DRPs were clearly reduced or abolished during contralateral plate removal ($n = 7/8$). These changes were not associated with changes in locomotor frequency. To confirm that contralateral afferents were responsible for evoking the force-related DRPs, contralateral lumbar L3–L5 dorsal roots were cut, and DRPs were abolished or reduced in number and consistency in three of four experiments (Fig. 3B). Surprisingly, cutting the contralateral plantar nerve to remove most plantar surface cutaneous afferents did not affect ipsilateral DRP generation ($n = 3/3$; Fig. 3C), suggesting that cutaneous afferents are

not primarily involved in the generation of force-dependent DRPs. To explore this further, we observed that during locomotion, if the foot contacted the plate proximal to the metatarsophalangeal joint (no toe contact), DRPs were reduced or absent ($n = 3$; Fig. 3D). This strongly implicates toe afferents as being primarily responsible for the crossed DRP.[18]

Previously, ipsilateral afferent stimulation-evoked PAD was shown to be maximal during the flexion phase of fictive locomotion in the majority of low-threshold afferents.[49,50] Here, we undertook the first experiments that test whether a similar phase dependent organization applies to contralateral afferents. To demonstrate that the amplitude of afferent stimulation–evoked crossed DRPs is locomotor phase dependent, we stimulated contralateral dorsal roots at five times the intensity

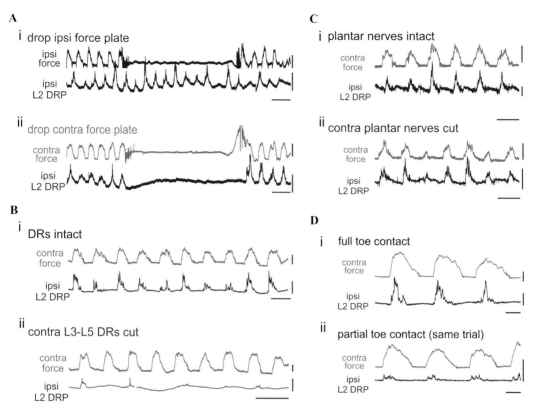

Figure 3. Response to contralateral plate removal and dorsal root rhizotomy and plantar nerve transection. (A) Ipsilateral DRPs are shown with ipsilateral force (Ai) and contralateral force (Aii). Note the lack of locomotor frequency change before, during, or after plate removal. (Ai) When the ipsilateral plate was removed, the ipsilateral DRP was largely unchanged. (Aii) When the contralateral plate was removed, reducing contralateral limb loading, the ipsilateral L2 DRP was nearly abolished. The DRPs returned as soon as the contralateral plate was restored. Scale bars are 10 mN, 200 μV, and 10 seconds. (B) L2 DRP relative to contralateral force during locomotion with dorsal roots intact. (Bi) Ipsilateral L2 DRP relative to contralateral force during locomotion with dorsal roots intact. (Bii) Following rhizotomy of contralateral L3–5 dorsal roots, the DRP was significantly reduced and inconsistent despite high contralateral forces. The frequency increase seen is due to the addition of 2 μM NMDA to induce locomotion without intact roots. Scale bars are 10 mN, 400 μV, and five seconds. (C) Ipsilateral L2 DRP and contralateral force before and after plantar nerve transection. (Cii) DRP persists following transection of the contralateral medial and lateral plantar nerves and continues to scale with contralateral force. Scale bars are 20 mN, 800 μV, and five seconds. (D) Contralateral toe contact and observed DRPs. (Di) L2 DRP with full contralateral toe contact. (Dii) Later in the same locomotor bout, the toe moved forward, so only a small portion contacted the plate. This resulted in reduced amplitude DRPs. Scale bars are 400 μV, 5 mN, and 2 seconds. Reproduced from Ref. 18, with permission.

of afferent volley detection threshold (5T) at 30- to 60-s intervals. We examined evoked responses throughout the locomotor cycle either with the force plate positioned to optimize limb contact at stance (Fig. 4A) or to minimize foot contact (Fig. 4B). In both cases, electrical stimulation of right limb afferents during the onset of right limb extension generated a strong DRP on the opposite side, presumably due to activation of afferents normally recruited via limb loading. In comparison, afferent stimulation at the end of extensor phase, or during the flexor phase, did not reliably evoke crossed

DRPs ($n = 3/3$). Overall, the crossed PAD generated by ground contact would coincide temporally with ipsilateral PAD, previously reported to be maximal during ipsilateral flexion.[49,50] Complementary bilateral actions could serve to reinforce afferent-evoked PAD-PDI, perhaps via actions on common interneurons.

Discussion

Summary

We recorded DRP activity as a measure of PAD-PSI and compared the spatiotemporal dependence

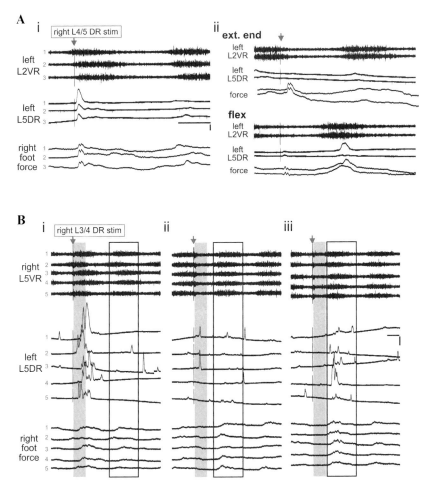

Figure 4. The amplitude of afferent stimulation–evoked crossed DRPs is locomotor phase dependent. Shown are results from two animals (A & B). Right (contralateral) lumbar dorsal roots were stimulated at 5T. Simultaneous individual ventral root (VR), DR, and right force recordings are matched in vertical progression in all panels as indicated with numbers on left records in A and B. Arrows at top identify stimulus onset for each panel. (Ai) Electrical stimulation of right DRs near onset of right extension produced large crossed DRPs that corresponded with force plate interactions (extensor phase timing is inferred from left flexor–related L2VR motor activity). (ii) During the end of the extensor phase (ext. end), electrical stimulation could still produce comparable plate contact force increases (i), but did not recruit DRPs, while stimulation during right limb flexion (flex) evoked only small DRPs and was not associated with any force increases. (B) In this experiment, the force plate was positioned to minimize foot contact during right limb extension. Evoked responses were then separated into three epochs (i–iii) to compare the relative crossed DRP amplitudes during various phases of locomotion (locomotor phase timing is inferred from right extensor–related L5VR motor activity). Green vertical highlighting identifies a 1-s period poststimulation. Vertical boxed regions denote subsequent bouts of right extensor activity. (i) Stimulation of right afferents during onset of right limb extension evoked a strong DRP in the opposite limb, presumably due to activation of afferents normally recruited via limb loading. In comparison, afferent stimulation at the end of extensor (ii) or during the flexor phase (iii) did not reliably evoke crossed DRPs. (iii) Flexor-phase afferent stimulation appeared to enhance the subsequent extensor phase (boxed region), such that limb extension, now generated larger foot contact forces with correspondingly large evoked crossed DRPs. Scale bars are 400 μV (for DR), one second.

of DRP patterns on ipsilateral and contralateral limb force and kinematics. We performed mechanical perturbations on each limb to isolate the influence of the individual limbs and to distinguish between movement- and force-related feedback.

Because both central circuits and sensory feedback can influence PAD-PSI, we also considered the dependence of PAD-PSI on locomotor phase, motor output, and performed deafferentations to distinguish central and sensory sources. We found that

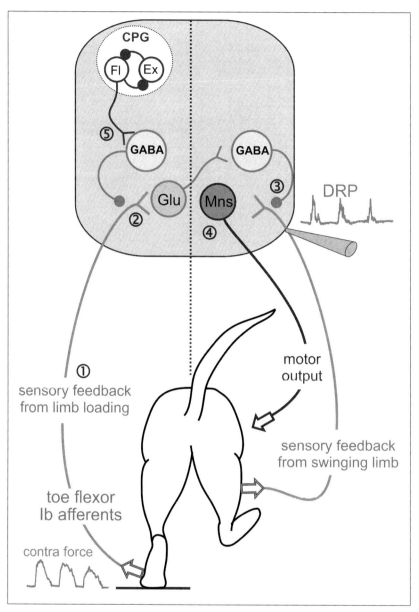

Figure 5. Proposed circuitry underlying the contralateral presynaptic inhibition (PAD-PSI) evoked by contralateral limb loading. Sensory events and dorsal roots are shown in gray. Motor output is shown in black. A simplified representation of the central pattern generator (CPG) is also shown. Identities of events are numbered 1–5. (1) Contralateral limb loading and toe contact activate Ib afferents that then enter contralateral lumbar dorsal roots. The magnitude of afferent activation scales with force. (2) Since the majority of identified commissural interneurons receiving Ib input are glutamatergic,[63] these afferents likely activate glutamatergic commissural interneurons (Glu). (3) Glu interneurons subsequently activate GABAergic interneurons (GABA), producing crossed PAD-PSI; Glu interneurons are recorded as a DRP at ipsilateral lumbar dorsal roots. (4) By blocking sensory inputs and/or closing sensory pathways, PAD-PSI may affect ipsilateral motor output during the swing/flexion phase. (5) Expression of crossed PAD-PSI is locomotor phase dependent and depressed during the flexion phase.

the mechanics of the contralateral limb, particularly limb loading, played a pivotal role in regulating ipsilateral sensory inflow via contralateral afferent-evoked PAD-PSI. One possible purpose for contralaterally generated force-sensitive presynaptic inhibition is to preserve swing by inhibiting unwanted afferent inputs that counteract flexion.

Figure 5 provides a proposed circuitry to explain the underlying PAD-PSI, and the numbers below correspond to sites of action in this figure. We hypothesize that contralateral limb loading activates Ib Golgi tendon organs in extrinsic toe flexors (1), whose afferents project via lumbar dorsal roots to activate first-order glutamatergic excitatory commissural interneurons[63] (2) that subsequently activate last-order GABAergic interneurons (3) responsible for generating the observed PAD-PSI. The net effect is to inhibit sensory input from the ipsilateral swinging limb during flexor motor output (4). The timing is consistent with observations of ipsilateral afferents also producing peak ipsilateral PAD-PSI during limb flexion.[49,50] The crossed PAD-PSI pathway is likely under control from the locomotor CPG (5) since the magnitude of its interlimb actions are suppressed when the limb is not in stance (Fig. 4).

Identity of afferent populations recruited during ground contact and undergoing PAD-PSI

We observed that the crossed flexion-phase DRPs scaled with contralateral afferent signaling. The identity of the afferent modalities responsible is herein considered. First, cutaneous mechanoceptors on the paw plantar surface are not likely to be responsible since removing most of its innervation by cutting the plantar nerve did not reduce the DRPs. However, a cutaneous afferent contribution cannot be excluded, since a small cutaneous receptive field on digits two and three remains after plantar nerve section.[54,55] Plantar nerve section also largely denervates intrinsic toe muscles, arguing against their contribution to contralateral PAD-PSI.

While the overall loading of muscles during ground contact undoubtedly contributes to PAD-PSI, toe contact clearly contributed strongly to the evoked DRPs. Extrinsic toe extensors, such as extensor digitorum longus and extensor hallucis longus (toe extensors/ankle flexors), are unlikely involved as they are most active during swing, and their Ia and

Ib afferents tend to fire in early swing rather than stance.[56,57] In contrast, extrinsic toe flexors, particularly flexor hallucis longus (FHL, toe flexors/ankle extensors), are active during stance with other ankle extensors.[58] Flexor digitorum longus (FDL) can be coactivated with FHL and other ankle extensors in the rat,[14] and recordings from afferents in the moving cat show that FHL and FDL group Ia and Ib afferents are both active during stance, particularly at toe contact.[56,57,59] This makes their peak firing well timed to evoke the observed patterns of PAD-PSI. As some joint afferents signal midrange joint angles, and toe joint afferents are involved in proprioception, joint afferents may also contribute to the PAD-PSI seen.[60]

Overall, based on the known contralateral PAD-PSI pathways reviewed earlier, our conclusion is that Ib afferents from FHL and FDL seem the most likely afferent source of contralateral force-sensitive PAD-PSI.[39] Indeed, increased contralateral limb endpoint force would likely increase Ib firing as ankle extensor/toe flexor activity increases,[10,56] readily explaining the scaling of PAD-PSI with contralateral force. Because most quadrupeds walk digitigrade,[61] the toe muscles are also well positioned to sense ground stability, and unexpected toe and ankle perturbations during stance, making their contribution to contralateral sensory regulation positionally appropriate, along with a contribution from toe muscles.

The afferents possibly receiving contralateral PAD-PSI include Ia, Ib, cutaneous, and joint afferents. Ib afferents are again the most likely since they are known to receive crossed PAD-PSI from contralateral Ib afferents.[39] In contrast, cutaneous and Ia afferents have not been shown to receive crossed PAD-PSI from low-threshold muscle afferents.[39,62]

Overall, this work has identified a stance force-encoded crossed PAD-PSI modulated by locomotor phase. While a prominent contribution from Ib afferents is suggested, additional studies are required to more precisely identify recruited afferents. In comparison, nothing is currently known about the identity of the ipsilateral afferents receiving the crossed input to generate PAD-PSI. Identification of these afferents, interposed interneuronal pathways, and interactions, both with ipsilaterally evoked PAD-PSI and the locomotor CPG, are needed to more fully understand the interaction between sensory and locomotor systems.

Acknowledgments

This work was funded by NIH NS45248, NIH NS65949, and NSF 0745164 to S.H.; NIH AR054760 to Y.H.C.; and NSF GRFP fellowship to H.B.H.

Conflicts of interest

The authors declare no conflicts of interest.

References

1. Brown, T.G. 1911. The intrinsic factors in the act of progression in the mammal. *Proc. R. Soc. Lond.* **84:** 308–319.

2. Grillner, S. 1981. Control of locomotion in bipeds, tetrapods, and fish. In *Handbook of Physiology—The Nervous System II*, vol. II. V.B. Brooks, ed: 1179–1236. Williams and Wilkins. Baltimore.

3. Kiehn, O. 2006. Locomotor circuits in the mammalian spinal cord. *Annu. Rev. Neurosci.* **29:** 279–306.

4. Nishimaru, H. & N. Kudo. 2000. Formation of the central pattern generator for locomotion in the rat and mouse. *Brain Res. Bull.* **53:** 661–669.

5. Smith, J.C., J.L. Feldman & B.J. Schmidt. 1988. Neural mechanisms generating locomotion studied in mammalian brain stem-spinal cord in vitro. *FASEB J.* **2:** 2283–2288.

6. Rossignol, S. 2006. Plasticity of connections underlying locomotor recovery after central and/or peripheral lesions in the adult mammals. *Philos. Trans. R. Soc. Lond. B Biol. Sci.* **361:** 1647–1671.

7. Pearson, K.G., J.E. Misiaszek & K. Fouad. 1998. Enhancement and resetting of locomotor activity by muscle afferents. *Ann. N.Y. Acad. Sci.* **860:** 203–215.

8. Grillner, S. & S. Rossignol. 1978. On the initiation of the swing phase of locomotion in chronic spinal cats. *Brain Res.* **146:** 269–277.

9. Hiebert, G.W. *et al.* 1996. Contribution of hind limb flexor muscle afferents to the timing of phase transitions in the cat step cycle. *J. Neurophysiol.* **75:** 1126–1137.

10. Donelan, J.M. & K.G. Pearson. 2004. Contribution of force feedback to ankle extensor activity in decerebrate walking cats. *J. Neurophysiol.* **92:** 2093–2104.

11. Hiebert, G.W. & K.G. Pearson. 1999. Contribution of sensory feedback to the generation of extensor activity during walking in the decerebrate cat. *J. Neurophysiol.* **81:** 758–770.

12. Hayes, H.B., Y.H. Chang & S. Hochman. 2009. An in vitro spinal cord–hindlimb preparation for studying behaviorally relevant rat locomotor function. *J. Neurophysiol.* **101:** 1114–1122.

13. Juvin, L., J. Simmers & D. Morin. 2005. Propriospinal circuitry underlying interlimb coordination in mammalian quadrupedal locomotion. *J. Neurosci.* **25:** 6025–6035.

14. Kiehn, O. & O. Kjaerulff. 1996. Spatiotemporal characteristics of 5-HT and dopamine-induced rhythmic hindlimb activity in the in vitro neonatal rat. *J. Neurophysiol.* **75:** 1472–1482.

15. Klein, D.A. & M.C. Tresch. 2010. Specificity of intramuscular activation during rhythms produced by spinal patterning systems in the in vitro neonatal rat with hindlimb attached preparation. *J. Neurophysiol.* **104:** 2158–2168.

16. Klein, D.A., A. Patino & M.C. Tresch. 2010. Flexibility of motor pattern generation across stimulation conditions by the neonatal rat spinal cord. *J. Neurophysiol.* **103:** 1580–1590.

17. Hochman, S. *et al.* 2012. Enabling techniques for in vitro studies on mammalian spinal locomotor mechanisms. *Front. Biosci.* **17:** 2158–2180.

18. Hayes, H.B., Y.H. Chang & S. Hochman. 2012. Stance-phase force on the opposite limb dictates swing-phase afferent presynaptic inhibition during locomotion. *J. Neurophysiol.* **107:** 3168–3180.

19. Duysens, J. & K.G. Pearson. 1980. Inhibition of flexor burst generation by loading ankle extensor muscles in walking cats. *Brain Res.* **187:** 321–332.

20. Whelan, P.J., G.W. Hiebert & K.G. Pearson. 1995. Stimulation of the group I extensor afferents prolongs the stance phase in walking cats. *Exp. Brain Res.* **103:** 20–30.

21. Andersson, O. & S. Grillner. 1983. Peripheral control of the cat's step cycle. II: Entrainment of the central pattern generators for locomotion by sinusoidal hip movements during "fictive locomotion." *Acta Physiol. Scand.* **118:** 229–239.

22. Kriellaars, D.J. *et al.* 1994. Mechanical entrainment of fictive locomotion in the decerebrate cat. *J. Neurophysiol.* **71:** 2074–2086.

23. Pang, M.Y. & J.F. Yang. 2000. The initiation of the swing phase in human infant stepping: importance of hip position and leg loading. *J. Physiol.* **528:** 389–404.

24. Hiebert, G.W. & K.G. Pearson. 1999. Contribution of sensory feedback to the generation of extensor activity during walking in the decerebrate Cat. *J. Neurophysiol.* **81:** 758–770.

25. Juvin, L., J. Simmers & D. Morin. 2007. Locomotor rhythmogenesis in the isolated rat spinal cord: a phase-coupled set of symmetrical flexion extension oscillators. *J. Physiol.* **583:** 115–128.

26. Pearson, K.G. & D.F. Collins. 1993. Reversal of the influence of group Ib afferents from plantaris on activity in medial gastrocnemius muscle during locomotor activity. *J. Neurophysiol.* **70:** 1009–1017.

27. Gorassini, M.A. *et al.* 1994. Corrective responses to loss of ground support during walking. I. Intact cats. *J. Neurophysiol.* **71:** 603–610.

28. Lam, T. & K.G. Pearson. 2001. Proprioceptive modulation of hip flexor activity during the swing phase of locomotion in decerebrate cats. *J. Neurophysiol.* **86:** 1321–1332.

29. Perreault, M.C. *et al.* 1995. Effects of stimulation of hindlimb flexor group II afferents during fictive locomotion in the cat. *J. Physiol.* **487:** 211–220.

30. Quevedo, J. *et al.* 2000. Group I disynaptic excitation of cat hindlimb flexor and bifunctional motoneurones during fictive locomotion. *J. Physiol.* **525:** 549–564.

31. Stecina, K., J. Quevedo & D.A. McCrea. 2005. Parallel reflex pathways from flexor muscle afferents evoking resetting and flexion enhancement during fictive locomotion and scratch in the cat. *J. Physiol.* **569:** 275–290.

32. Conway, B.A., H. Hultborn & O. Kiehn. 1987. Proprioceptive input resets central locomotor rhythm in the spinal cat. *Exp. Brain Res.* **68:** 643–656.

33. Iizuka, M., O. Kiehn & N. Kudo. 1997. Development in neonatal rats of the sensory resetting of the locomotor rhythm induced by NMDA and 5-HT. *Exp. Brain Res.* **114:** 193–204.

34. Kiehn, O., M. Iizuka & N. Kudo. 1992. Resetting from low threshold afferents of N-methyl-D-aspartate-induced locomotor rhythm in the isolated spinal cord-hindlimb preparation from newborn rats. *Neurosci. Lett.* **148:** 43–46.

35. Quevedo, J. *et al.* 2005. Stumbling corrective reaction during fictive locomotion in the cat. *J. Neurophysiol.* **94:** 2045–2052.

36. Rudomin, P. & R.F. Schmidt. 1999. Presynaptic inhibition in the vertebrate spinal cord revisited. *Exp. Brain Res.* **129:** 1–37.

37. Eccles, J.C., P.G. Kostyuk & R.F. Schmidt. 1962. Central pathways responsible for depolarization of primary afferent fibres. *J. Physiol.* **161:** 237–257.

38. Eccles, J.C. 1964. Presynaptic inhibition in the spinal cord. *Prog. Brain Res.* **12:** 65–91.

39. Devanandan, M.S., B. Holmqvist & T. Yokota. 1965. Presynaptic depolarization of group I muscle afferents by contralateral afferent volleys. *Acta Physiol. Scand.* **63:** 46–54.

40. Gossard, J.P. & S. Rossignol. 1990. Phase-dependent modulation of dorsal root potentials evoked by peripheral nerve stimulation during fictive locomotion in the cat. *Brain Res.* **537:** 1–13.

41. Jankowska, E., S. Lund & A. Lundberg. 1966. The effect of DOPA on the spinal cord 4. Depolarization evoked in the contralateral terminals of contralateral Ia afferent terminals by volleys in the flexor reflex afferents. *Acta Physiol. Scand.* **68:** 337–341.

42. Jankowska, E. 1992. Interneuronal relay in spinal pathways from proprioceptors. *Progr. Neurobiol.* **38:** 335–378.

43. Alibiglou, L. *et al.* 2009. Bilateral limb phase relationship and its potential to alter muscle activity phasing during locomotion. *J. Neurophysiol.* **102:** 2856–2865.

44. Ting, L.H. *et al.* 1998. Bilateral integration of sensorimotor signals during pedaling. *Ann. N.Y. Acad. Sci.* **860:** 513–516.

45. Menard, A., H. Leblond & J.P. Gossard. 1999. The modulation of presynaptic inhibition in single muscle primary afferents during fictive locomotion in the cat. *J. Neurosci.* **19:** 391–400.

46. Yakhnitsa, I.A., A.I. Pilyavskii & N.V. Bulgakova. 1988. Phase-dependent changes in dorsal root potential during actual locomotion in rats. *Neirofiziologiya* **20:** 333–340.

47. Beloozerova, I.N. & S. Rossignol. 2004. Antidromic discharges in dorsal roots of decerebrate cats. II: studies during treadmill locomotion. *Brain Res.* **996:** 227–236.

48. Duenas, S.H. & P. Rudomin. 1988. Excitability changes of ankle extensor group Ia and Ib fibers during fictive locomotion in the cat. *Exp. Brain Res.* **70:** 15–25.

49. Gossard, J.P., J.M. Cabelguen & S. Rossignol. 1991. An intracellular study of muscle primary afferents during fictive locomotion in the cat. *J. Neurophysiol.* **65:** 914–926.

50. Gossard, J.P., J.M. Cabelguen & S. Rossignol. 1989. Intraaxonal recordings of cutaneous primary afferents during fictive locomotion in the cat. *J. Neurophysiol.* **62:** 1177–1188.

51. Gossard, J.P. 1996. Control of transmission in muscle group IA afferents during fictive locomotion in the cat. *J. Neurophysiol.* **76:** 4104–4112.

52. Menard, A., H. Leblond & J.P. Gossard. 2003. Modulation of monosynaptic transmission by presynaptic inhibition during fictive locomotion in the cat. *Brain Res.* **964:** 67–82.

53. Hayes, H.B., Y.H. Chang & S. Hochman. 2009. An in vitro spinal cord-hindlimb preparation for studying behaviorally relevant rat locomotor function. *J. Neurophysiol.* **101:** 1114–1122.

54. Bouyer, L.J. & S. Rossignol. 2003. Contribution of cutaneous inputs from the hindpaw to the control of locomotion. I: intact cats. *J. Neurophysiol.* **90:** 3625–3639.

55. Holmberg, H. & J. Schouenborg. 1996. Developmental adaptation of withdrawal reflexes to early alteration of peripheral innervation in the rat. *J. Physiol.* **495:** 399–409.

56. Loeb, G.E. & J. Duysens. 1979. Activity patterns in individual hindlimb primary and secondary muscle spindle afferents during normal movements in unrestrained cats. *J. Neurophysiol.* **42:** 420–440.

57. Prochazka, A. & M. Gorassini. 1998. Ensemble firing of muscle afferents recorded during normal locomotion in cats. *J. Physiol.* **507:** 293–304.

58. O'Donovan, M.J. *et al.* 1982. Actions of FDL and FHL muscles in intact cats: functional dissociation between anatomical synergists. *J. Neurophysiol.* **47:** 1126–1143.

59. Prochazka, A., R.A. Westerman & S.P. Ziccone. 1976. Discharges of single hindlimb afferents in the freely moving cat. *J. Neurophysiol.* **39:** 1090–1104.

60. Ferrell, W.R. 1980. The adequacy of stretch receptors in the cat knee joint for signalling joint angle throughout a full range of movement. *J. Physiol.* **299:** 85–99.

61. Cunningham, C.B. *et al.* 2010. The influence of foot posture on the cost of transport in humans. *J. Exp. Biol.* **213:** 790–797.

62. Baldissera, F., H. Hultborn & M. Illert. 1981. Integration in spinal neuronal systems. In *Handbook of Physiology—The Nervous System*, vol. II. V.B. Brooks, ed: 509–595. Williams and Wilkins. Baltimore.

63. Bannatyne, B.A. *et al.* 2009. Excitatory and inhibitory intermediate zone interneurons in pathways from feline group I and II afferents: differences in axonal projections and input. *J. Physiol.* **587:** 379–399.

Ann. N.Y. Acad. Sci. ISSN 0077-8923

ANNALS OF THE NEW YORK ACADEMY OF SCIENCES

Issue: *Neurons, Circuitry, and Plasticity in the Spinal Cord and Brainstem*

Motor primitives and synergies in the spinal cord and after injury—the current state of play

Simon F. Giszter and Corey B. Hart

Department of Neurobiology and Anatomy, Drexel University College of Medicine, Philadelphia, Pennsylvania

Address for correspondence: Simon F. Giszter, Department of Neurobiology and Anatomy, Drexel University College of Medicine, 2900 Queen Lane, Philadelphia, PA 19129. sgiszter@drexelmed.edu

Modular pattern generator elements, also known as burst synergies or motor primitives, have become a useful and important way of describing motor behavior, albeit controversial. It is suggested that these synergy elements may constitute part of the pattern-shaping layers of a McCrea/Rybak two-layer pattern generator, as well as being used in other ways in the spinal cord. The data supporting modular synergies range across species including humans and encompass motor pattern analyses and neural recordings. Recently, synergy persistence and changes following clinical trauma have been presented. These new data underscore the importance of understanding the modular structure of motor behaviors and the underlying circuitry to best provide principled therapies and to understand phenomena reported in the clinic. We discuss the evidence and different viewpoints on modularity, the neural underpinnings identified thus far, and possible critical issues for the future of this area.

Keywords: motor primitives; synergies; modularity; spinal cord; pattern generation; stroke; spinal cord injury

Introduction

Here we review ideas and experimental data related to modularity, in particular motor primitives and muscle synergies, and recent work suggesting the direct clinical relevance of these ideas and data. We will highlight gaps to be filled by the research community, which are likely; when filled, we can produce new therapeutic information and strategies.

Modular systems have components that may be separated and recombined. Modularity thus implies a compositional set of building blocks with different possible arrangements.[1,2] Recently, interest in motor modularity has grown greatly.[3,4] However, the area remains controversial, and competing perspectives are presently available on the topic. For example, there are several levels of analysis—kinematic, muscle, and neural—supporting modularity, and several different sets of ideas about the origins of modularity in each of these. Depending on the perspective, modularity can be seen as a benefit or an impediment to motor control,[19] as a fundamental neural structural feature, as an epiphenomenon of other more fundamental mechanisms, and as of-fering both clinical opportunities and hindrances. However, it is becoming clear that modularity perspectives at the worst offer a concise shorthand to describe clinical changes and pathology,[5–7] and for this reason alone it is important to gain a better insight into modularity. Our understanding of the processes supporting modular motor control and the spinal segmental implementation of modularity remains sketchy,[8–12] but filling out this understanding will be the only way to resolve the controversies and leverage modular descriptions to clinical benefit.

Definitions and types of modules

Modularity is apparent at several levels of analysis in motor control. Table 1 summarizes many of the levels and descriptions used, and some examples of biomimetic robotic uses of modularity for comparison. Here we unpack this description from the table in terms of kinematic, motor pattern, kinetic/force, and neural types of modularity.

Kinematic descriptions

Early kinematic analyses undertaken by ethologists supported modular constructions, although these

Table 1. Modularity terminology and relationships among descriptions and units

Marr level of description	Increasingly finer resolution elements of action composition and control $\longrightarrow$					Type of data and description
Task Behavioral task/kinematic task/task planning	Behavior (e.g., hunt or groom)	Task: (run to capture); [a]**kinematic oscillation**	Subtask: (swing, stance, foot placement, reach to grasp)	Stroke/kinematic segmentation unit = **kinematic stroke** or **kinematic primitive**		*Kinematics*
Algorithms Controls/ algorithms (kinetics necessary)	Timing, rhythm & sequence	[d]Force pattern for subtask, possible coarticulation	Force pattern for individual kinematic stroke, possible coarticulation	Unit building block for force patterns. **Force-field primitives** = kinetic primitives		*Kinetics*
	[b]**Rhythm generator system**	[c]**Pattern shaping** under rhythm generator, descending and sensory control influences or		[g]**Burst synergies = motor primitives = force-field primitive**		*Muscle patterns*
	Rhythm generator system	**Pattern-shaping system**	[e]**Time-varying synergies**	Unitary burst (or "pulse") of a synchronous muscle synergy		
Implementation Implementations/ circuitry	Neural oscillator or state chaining system	[f]**Neural switching systems** or emergent pattern shaping and sequencing processes		Separated neural **premotor drives and burst generation**		*Neural support*
	< = > robotic **Dynamical movement primitive (oscillator primitives) or finite state machine**	< = > robotic **Coupling or selection and switching policy,** e.g., finite state machine with time outs	< = > robotic **Kinematic primitive and PID controller**	< = > robotic **Kinetic primitive or multijoint PD control unit**		*Sample equivalent idea in biomimetic robotics*

NOTES: Missing from the table are the synergy formulations as optimal controlled subspaces, and controlled/uncontrolled manifolds. The missing terms and associated models of modularity are harder to relate simply and directly to pattern generation and spinal neural circuits in testable hypotheses. In the table, specific relationships or areas need significant work and are important for understanding new data on modularity in stroke and spinal cord injury.

[a]The relations of rhythmic and discrete motion at kinematic and production descriptions remain an area of ambiguity.

[b]As may be expected, rhythm/timing generation mechanisms and circuit location remain insufficiently specified.

[c]The structure of pattern shaping remains difficult and controversial. The separation from rhythm generation is controversial. The composition and modularity of pattern is disputed. Modularity can be considered emergent from network dynamics or based on switching and incorporation of explicit modules.

[d]Efficient kinematic and kinetic action may involve coarticulation of modules in sequences, and mechanisms should support this contextual adjustment for increasing skill levels in tasks.

[e]Modules in pattern shaping could be fixed sequences of several bursts, also called time-varying synergies, which are explicitly related to kinematic modules such as strokes or cycles. Alternatively, pattern-shaping modules could be single bursts that are not uniquely or closely related to kinematic strokes and cycles (e.g., unit bursts/synchronous synergies that instead relate directly to units of biomechanical modularity, such as unitary force fields, and to neural drive; see footnote g, below).

[f]The structure of neural switching and pattern sequencing is very poorly understood but is probably crucial to understanding the processes of coarticulation, merging, splitting, and sequence changes after neural damage and during therapy.

[g]Neural underpinning of unit bursts and drive structure are also part of the key to these clinically meaningful processes.

ethological descriptions generally faltered as more detailed analysis of execution was considered.[17] Nonetheless, Fentress and Golani,[13,14] and more recently Whishaw,[15,16] have used such kinematic modular analyses to good effect. Kinematic execution modularity has provided the most insight in work on human reach kinematics and its segmentation. As elements of the segmentation, unitary kinematic strokes were noted by Viviani and Terzuolo.[18] Explanations for the observed stroke properties were first explored by Hogan and Flash.[20,21] In general, the motions of limb endpoints exhibit straight paths, and have unimodal bell-shaped tangential velocity profiles. These profiles and the resultant stroke segmentations were consistent with kinematic optimization and were well predicted by an endpoint minimization of jerk.[20, 21] For point to point motions, these features are preserved across loads and in different species and environments.[12, 22, 23] However, likely owing to features of biological limb design, the kinematic strokes observed are also consistent with various other kinetic and task optimizations. These other optimizations include minimum torque change and minimized signal-dependent noise at the muscle.[24] Superposition of collections of kinematic strokes by the central nervous system (CNS) account for various learning, correction, and rehabilitation phenomena.[23, 25, 27] Flash and colleagues continue to develop better descriptions for similar analyses of more complex 3D kinematics.[26]

Pattern generation and pattern structure descriptions

Rhythmic motions have repeating forms that can be considered modular. Modular and repeatable controls for rhythmic motions have been evaluated at both kinematic and motor pattern levels. Evaluations at the kinematic level include the research of Schaal et al.[28, 29] and Sternad and Hogan.[30, 33] Schaal, Ijspeert, and colleagues also suggested dynamic primitives in robotics.[31, 32] Pattern generation in the nervous system was clearly established by seminal work of Wilson, Grillner, and their successors in both vertebrate and invertebrate animals,[34] and this suggested that central motor pattern was fundamental.[35] Examples of modularity at the pattern generator levels would be unit burst generators for hip flexion and hip extension as examined by Stein et al. in turtles.[36, 39] The CNS can organize modular patterns underlying behaviors (usually rhythmic) independent of any feedback or patterned input. Identifying the structure of pattern generator circuitry is an ongoing process. Although the neural elements constituting the structure of patterned generators are well identified and modeled in simpler vertebrate circuits,[36, 40] the topic is significantly less well understood in complex mammalian systems. The biological central pattern generators (CPGs) could represent several types of engineer-style controllers. Neural networks in the CNS could implement dynamic system limit-cycle oscillators. These might be implemented as half-center oscillators, as in the original models of Brown and succeeding work. Alternatively, the CPG might be a kind of finite state machine, one that cycles through

its states in the absence of rhythmic inputs. Various pacemaking and burst-generating mechanisms in neurophysiology, e.g., respiration,[41] arguably show some features of such a system. Recent data from the paralyzed decerebrate cat supports a hierarchical hybrid CPG in mammalian hindlimb locomotion: a rhythmic or clocking system layer (constructing timing features and perhaps state sequencing) exerts control on a pattern-shaping layer.[42–48] This hybrid framework is also consistent with component motor primitive/synergy elements (see below) and unit burst elements.[36–39]

Muscle synergies

To discover pattern and modularity in unparalyzed animals and humans behaving more broadly, statistical analysis of motor patterns has been used.[7, 49–51] In animals ranging from the frog to the intact or injured human being, and in both rhythmic and nonrhythmic behaviors, application of statistical decomposition techniques show a remarkably similar breakdown of muscle activation patterns into modular muscle groups (or synergies or drive motor primitives). Other nonstatistical physiological methods also support the modules and analysis results.[52, 53] These muscle groups or synergies are activated as unitary synchronous muscle bursts or pulses, either in sequence or simultaneously, across various test paradigms.[51, 54–64] They are adjusted and adapted by the CNS in response to the task conditions by selecting among the different synergies, and by changing overall burst amplitudes and onset timings. Dimensionality reduction of the full range of muscle activations that are possible, down to a significantly smaller set of muscle synergies, seems to be the rule. The data are thus consistent with the separate sequencing and control of a small set of premotor drives in each task. The precise structure of the drives has been disputed. Some authors have favored synchronous unit bursts (equivalent to neurophysiological unit burst generators [9, 51–53, 56, 65]), while others have favored time sequences of muscle activations, with the whole sequence acting as an atom or unit of the motor patterns.[60, 61, 63] Pulses of synergistic activity (or unit burst generators) that act as fundamental building blocks of patterns at spinal levels, added or deleted as units, seem to be favored by most data from both fictive, semi-intact, and intact animal behaviors.[8, 9, 36, 42, 52, 53]

Linking muscle synergy to biomechanics—unit bursts as force-field primitives

Muscles are intrinsically viscoelastic. The precise mechanical properties vary with activation, history of activation, and shortening or lengthening history. Muscles act in concert via soft tissues, ligament and tendon systems and their moment arms, through the skeletal linkage and its Jacobean properties, to generate forces that can be described as position- and velocity-dependent fields. The simultaneous (synchronous) activation of a collection of muscles produces a well-defined viscoelastic force field in the limb that modulates with time. This field depends on the initial limb state. In this way, force-field descriptions can capture the action of muscle synergies (Fig. 1 and Refs. 66–74). Muscles in a synchronous burst synergy have fixed ratios to each other, matching the ratios of c_is in Figure 1 (for example, see Fig. 2). These force-field descriptions show the property of linear-vector superposition, which can be demonstrated when fields are combined through electrical stimulation of the spinal cord, or during natural drive coactivations.[52,53,65,71,75] Feedback systems in the spinal cord do not disrupt these viscoelastic force-field structures, but rather seem to support them.[53]

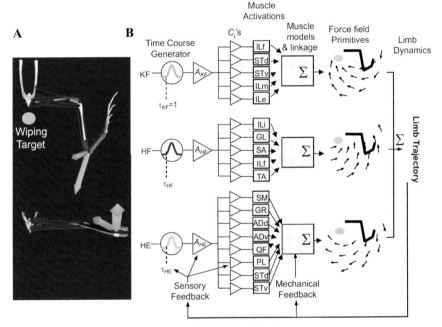

Figure 1. Viscoelastic force-field simulation design. Simulating hindlimb wiping with a detailed model of the spinal primitives. (A) The 13 hindlimb muscles forming the biomechanical model are shown as red lines. Colored arrows mark the force directions of the three force primitives at a fixed limb position during the isometric wiping response: KF (knee flexor primitive), light purple; HE (hip extensor primitive), green; HF (hip flexor primitive), dark purple. (B) The framework used to simulate wiping (left to right): each primitive had a time-course generator, representing the premotor drive burst that put out a normalized waveform (peak = 1.0) at time τ. The variable A scaled this waveform, which was then distributed to each of the muscles within the primitive. Each muscle had a muscle-specific variable (C) that scaled the excitation wave form. The synergy muscle groups generate contractile forces (MF) that are transmitted through the limb to produce an isometric endpoint force (at one position) or force field FF (when forces are measured across a range of positions). Normalized force fields produced by each primitive are shown in the far right. When the model limb is freed to move, MFs drive the motion of the model. MF values are in turn regulated by the limb motion (i.e., the force–velocity and force–length properties of muscle and stress–strain properties of in-series connective tissue alter MF forces). In this version model, sensory feedback from muscles potentially regulate τ, A, or C. ILf, iliofibularis; STd, semitendinosus dorsal branch; STv, semitendinosus ventral branch; ILm, iliacus median; Ile, iliacus externus; ILi, iliacus internus; GL, gluteus; SA, sartorius; TA, tibialis anticus; SM, semimembranosus; GR, gracilis; ADd, adductor dorsal; ADv, adductor ventralis; QF, quadratus femoris; PL, peroneus longus. Reproduced from Ref. 83, with permission.

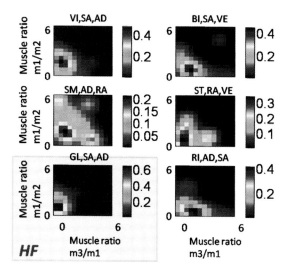

VI,SA,AD — BI,SA,VE — SM,AD,RA — ST,RA,VE — GL,SA,AD — RI,AD,SA

HF

Muscle ratio m1/m2

Muscle ratio m3/m1

Figure 2. Probability of muscle ratios in observed spinal EMG data. Examining reproducibility of frog primitives. Stability of muscle proportionality ratios in the six main primitives are observed across frogs and across behaviors. The action of the spinal cord across the tested behaviors was to recruit the muscles in fixed ratios and thereby couple muscles so as to generate specific force-field primitives and associated preflex responses. Reflex actions (i.e., feedback effects) modulated these primitives, not individual muscles (see Ref. 53), and thus acted on their component muscles as groups. HF, hip flexor synergy. Muscle abbreviations (nomenclature of Ecker, with equivalent Abbott and Lombard): VI, vastus internus (= iliacus internus); VE, vastus externus (= iliacus externus); BI, biceps (= iliofibularis); AD, adductor magnus; SA, sartorius; RA, rectus anticus (= quadratus femoris); GL, gluteus. Reproduced from Ref. 83, with permission.

A synergy-driven force field is a predictable pattern, whether the synergy is activated as a single primitive, as part of a coordinated reflex, or in a voluntary pattern. A "force-field motor primitive" in biomechanics can be associated with a muscle synergy, and corresponds 1:1 to the specific premotor drive bursts (or primitive pulses) occurring in the motor pattern. In effect, the actions of the multiple simultaneously pulsed drives and their associated muscle recruitment effects can be summarized as a sum of the individual muscle biomechanical force effects:

$$F(q, \dot{q}, t) = \sum_i A_i a_i(t)\Phi_i(q, \dot{q}), \quad (1)$$

where q is a configuration vector, F is the total limb force field, A_i is the amplitude of the activation of the ith force-field motor primitive, $a_i(t)$ is the

normalized activation time course of the ith primitive, and Φ_i is the normalized viscoelastic field associated with the ith premotor drive and its collection of associated muscles. This summation has been demonstrated experimentally.[68–71] Since each field Φ_i consists of the sum of individual muscle fields, this framework naturally extends from an initially constrained low degree of freedom synergies/motor primitives in reflex and pattern generation through to the fullest voluntary use of spinal cord capabilities of the motor system using individuated muscles that are driven independently from one another.

To generate flexible behaviors and action using the individual premotor drive pulses, the pulses may be repeated and staggered in different combinations:

$$F(q, \dot{q}, t) = \sum_i A_i a(t + \tau_i)\Phi_i(q, \dot{q}), \quad (2)$$

where τ_i represents the time shift of the ith drive pulse. The point of this brief mathematical excursion is that a matching between Newton's laws, a compact control framework, and synergy bursts as motor pattern elements is possible and, taken together, this suggests a simple compositional scheme for movement. This framework ably represents spinal-generated behaviors and reflexes[53,83] (see Figs. 1–3). It also maps cleanly onto the inferred hierarchical control of the pattern-shaping systems suggested by McCrea, Rybak, and others.

Support of motor modularity across species

The kinematic stroke modularity seen in human reaching[25] is also exhibited in octopus reaching[12,76] in a marine environment using very different effector and neural control circuitry. This suggests fundamental aspects of physics, mechanics, and task control may be a basis for the structure of kinematic strokes. Similarly, basic postures and stereotypic motions are exhibited in swimming turns, escape (C-start turns), and other kinematic controls.[77,78] Rhythmic kinematics of locomotion in aquatic, terrestrial, and aerial media all support common compositional features, thus leading Koditschek and Full to the notion of *templates* and *anchors*.[79] Templates represent the lowest order (lowest dimensional) description of the essence of the physical task and its requirements. Anchors are specific hypotheses regarding the neuromuscular and biomechanical controls used for implementation of the template. In their scheme, motor primitives of any kind (kinematic or kinetic) would be the anchors that allow low dimensional and efficient physical task performance.

Motor patterns and muscle/pattern modularity across species

Features of the motor patterns, stereotypic bursting, and synergies supporting both stereotypic and more flexible behaviors are common from invertebrates to humans. Examples of common features have been reviewed and remarked on extensively.[80] Recent work from Lacquaniti, Ivanenko, Dominici, and colleagues has shown the similarity of rhythmic pattern construction in guinea fowl, rat, cat and human, and across human development, linking rhythm generation and primitives across species.[64]

Interpreting modular structures—task, algorithm, and implementation

Modularity as exhibited across species in kinematic, kinetic, and motor pattern features could be considered in several ways. The neuroscientist David Marr[81] suggested considering nested levels of task, algorithm, and implementation in analyzing CNS function in behavior. This scheme is useful in modularity because it highlights the points of convergence and divergence in different accounts of modularity. At the task level, modularity may simply represent task features that must be managed as constraints or task affordances in the real world: most locomotion and physical engines employ cyclic features, etc., and often have, in the parlance of Full and Koditschek, similar templates. At the algorithmic level, if task solutions are modular, how is this accomplished computationally? What are the computational anchors? How flexible and general are the algorithms employed in different biological systems, in different tasks, and at different levels of the neuraxis? Control theory as a field has developed numerous principled and powerful ways of constructing and managing controls that often converge on modular solutions, and the field continues to advance rapidly. Complex issues such as problems of degrees of freedom are proving more tractable than originally supposed, though still by no means paper tiger problems that can be ignored.[84] Optimization methods in control lead to modular solutions.[85–89] It is natural to wonder how much of the biological circuitry represents the implementation of a set of powerful, flexible, and principled control algorithms.[90–97] Alternatively, how much of biological circuitry represents hacks, heuristics, and the historical baggage from evolution? Engineers all recognize that real-world systems are a compromise between what is ideal and what is possible, given the manufacturing, design, material, and other costs available. We do not know how best to balance this cost as we think about biology and motor evolution, ontogeny, and learned interactions. Data and simulation results are often consistent with both perspectives.[98,99] It seems that algorithmic flexibility, generality, and learning time must be weighed against the evolutionary pressure to function rapidly (e.g., the wildebeest calf, hatchling sea turtle). The pack-and-go cost of added neural hardware for flexibility in weight and energy consumption must be balanced against other options for using the mass and energy. Some of these kinds of trade-offs are observed in modern consumer technologies such as mobile phones, which also must survive in the marketplace in competition with other choices. Not every phone type is equally smart, or similarly controlled, and each has its niche. Although some researchers view modules as built into the neural structure of the CNS by evolution,[8] others see them as arising from an optimal control implementation, operating on the anatomical affordances in the musculoskeletal system. In this last view, modularity arises as an optimal strategy and division of labor by the optimal controller, which parcels task dimensions into a controlled and an uncontrolled manifold (i.e., different subspaces) with the controlled task operating in low dimensions.[94–96]

Some behavioral needs that can always be anticipated are partly embedded in the mechanical design of limbs. *De novo* neural motor solutions in each generation may waste time and lose access to scarce resources, and be selected against. Further, in biological systems, all requirements for function are not available a priori in early development. Future control needs must be anticipated or learning must be channeled to support the critical functions needed later, beyond the current locally optimal learning. Developmental research supports much prestructuring of circuits before feedback use.[103–106]

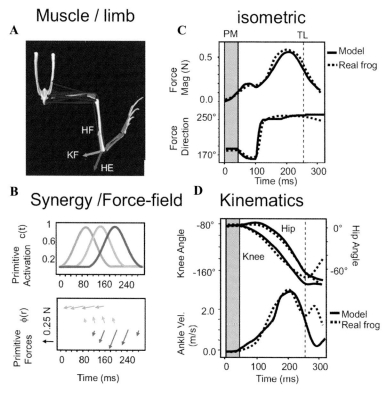

Figure 3. Competence of force-field primitives to replicate force and kinematic behaviors. Simulation results using a framework of primitives. Simulating wiping forces and kinematics with the primitive framework and model frog. (A) Model structure. (B) Activation of muscles as synergies. (C) The isometric force pattern produced by the model frog (solid lines) closely matched the force pattern recorded experimentally (dotted lines). (D) After making minor adjustments to the isometric motor pattern (amplitude scaling the ensemble down slightly), the model frog also reproduced the free limb kinematics of the experimental frog. Without downscaling, the forces were too strong, indicating potential feedback adjustment in isometric conditions. The top row shows hip and knee angles. The bottom row shows ankle velocity. Dashed line marks the time of target limb contact in the real frog. The gray area (PM) represents the 40-ms premovement period between EMG onset and motion onset that is observed in real frogs.

Nonetheless, subsequent adaptation and extensibility of behaviors are also needed, and automaton-like behavior is clearly insufficient.[100–102] Optimization and adaptation in adults is essential. Modularity, if well crafted by evolution, should support both early function and later adjustment to conditions and new affordances.

How might the circuitry and algorithms suggested by these different views arise and be supported in the spinal cord? Are they built-in by evolution, learned in child development, or developed *de novo* in task solutions? In Marr's scheme,[81] the implementation details of neural systems, where available, may help resolve these conundrums. Current data suggest that different perspectives can be correct, depending on species, age, task, and the specific CNS circuit regions that are examined.

Implementation of spinal primitives and modules—spinal structures

Several lines of investigation support spinal circuits explicitly geared toward modularity. The competence of modular systems is largely established. From the perspective of synergy-based modularity, work in frog, cat, and monkey bears strongly on these issues. In our lab, Hart has demonstrated with single unit extracellular recordings in the frog spinal cord that interneurons with mono- or di-synaptic projections to motor pools replicate the patterns and weightings of synergy drives.[107] Other interneurons showed a range of features that further support the idea that the frog reflex circuitry is organized to drive and control onset timing, amplitude, and combinations of synergy and primitives in spinal reflexes (Ref. 107 and unpublished data). Recordings by Berkowitz and colleagues in turtles support similar shared and specialized neural structures that might match with synergy/primitives analyses of motor pattern.[108–110] Behavioral state-dependent control and altered output patterns from a spinal circuit in this case are expected to arise from rearranging the onset timing and combinations of synergy primitives, not from radical alteration of these drive circuits. Within such a framework, duplication of similar projecting drive circuit neurons for specialized use in various contexts could occur. Specialized drive systems that contribute to reaching were characterized in the work of Alstermark and colleagues on the C3–C4 interneuron systems (reviewed by Alstermark and Isa[111]). The precise premotor drive and motoneuron projection patterns of these neurons are not sufficiently characterized to show whether there are truly discrete sets of modular drive projections, but the roles and even genetic developmental processes of the C3–C4 systems in mice, cats, and monkeys are now beginning to yield to modern methods.[112] The work of Seki and colleagues, recording in monkey spinal cord, also supports possible modular groupings in upper limb controls.[113] Taken together, these studies all represent modern extensions of the modular ideas initiated by Hall, Brown, and Sherrington.[121] In mouse CPG, the extensive data from the combination of physiological and genetic dissection of the interneuron systems may in the future be related to ideas of modular spinal organization[107] and to hierarchical pattern-generator organization.[42,43] The explicit linkages of developmental mechanisms, genetics, and resulting interneuron form and projection details remain to be determined but will surely yield to this powerful strategy.

Learning and subsuming spinal modular structure through development

A scaffold of basic infrastructure for modular control and pattern generation is likely to be common across tetrapods. Through development this may be put into use very rapidly, or more gradually shaped by learning. Dominici and colleagues have recently performed an analysis and comparison of human and cross-species development of locomotion.[64] Their data support the idea that the basic modular pattern generation of tetrapods is, in humans, gradually specialized for bipedalism through devel-

opment. The quadrupedal infrastructure is refined: some is preserved and some is modified to satisfy the novel biomechanical constraints of bipedalism. Their account is very much in keeping with a hybrid perspective of evolutionarily embedded bootstrap circuitry, onto which novel extensions and optimizations are layered.[8,114] In some instances the older circuits are subsumed to form elements of the new control, but in others instances these may be reshaped or replaced to satisfy the novel needs. This is consistent with the points emphasized by Valero-Cuevas and colleagues, namely that neural modularity constraints will both support and limit biomechanical and behavioral control options, not unlike limb design and the tendon anatomy in the hand.[94–96] The fullest flexibility of behavior is achieved by transcending these constraints where it is feasible to do so. The degree of departure from the early infrastructure in an adult individual will determine the relative importance and therapeutic value of the evolutionarily older circuitry. Various groups are now examining how modularity and spinal modularity affect the motor learning of able-bodied individuals.[61] Older tetrapod circuitry could potentially either aid or hinder learning and rehabilitation after some kinds of injury. Depending on the final adult movement compositionality in able-bodied individuals, it is conceivable that the older circuitry could reemerge as pathology.[8] However, despite this risk, when thoracic and lumbar circuits are isolated from descending controls in complete spinal cord injury, there may be few options in the clinic but to engage the older and now isolated spinal circuits for therapeutic purposes. It is becoming clear that recruiting these isolated spinal circuits in fact often offers great promise.[115–117] Nonetheless, it is important to understand the end state of spinal circuitry after development and motor learning in humans to best use this promise. Taken together, this all suggests an important role for spinal modularity and pattern generator elaboration in the understanding of disease processes and trauma, and in the design of effective clinical rehabilitation. Newer studies of human clinical outcomes using modularity analyses support this perspective.

Clinical ramifications of modularity

Recently, the modularity perspectives developed in animal models have begun to be tested in clinical settings. These studies have involved both the upper and lower limb. More is known about animal pattern generation in the lower limb, and work from various groups supports modularity and pattern generation contributions in humans.[7,54,55] Clark, Neptune, Ting, Zajac, and Kautz have shown that, following stroke, the modularity reported by themselves and others in able-bodied individuals can collapse to still lower dimensions.[7] This collapse looks like a merging of the healthy modules or synergies. The degree of collapse may be related to the degree of deficit. With rehabilitation, the merged synergies may again separate themselves and the degree of this redifferentiation may relate to the degree of rehabilitation improvement. These data provide a compelling account of the pathology and recovery processes with clear links to the hierarchical pattern generation and synergy literature. Bizzi and colleagues have examined stroke in the upper limb.[5,6] The C3-C4 systems of Alstermark and the overall modularity described by Drew, Krouchev, and colleagues in cats[50] may be relevant. By comparing the unaffected and affected arms, and the muscle pattern structures recorded in reaching movement tasks in each, they have shown that the same modularity or premotor drive structures often exist in both limbs. However, drive use is disordered in pattern in the affected limb. Beyond this, they also observed merging, as was seen in the lower limb by Clark and colleagues. This merging process could be directly correlated with the degree of deficit on clinical scales. Such merging is seen in animal models in deafferentation or other manipulations.[82] The data from both upper- and lower-limb human studies suggest that the basic synergy circuits are embedded in the spinal cord and brainstem. Taken together, the Cheung/Bizzi and Clark/Neptune/Ting data suggest that loss of cortical controls causes initial failures in the temporally differentiated control of spinal interneuron systems for synergies in humans, but that temporally differentiated controls can be reestablished. The Cheung /Bizzi work also showed a further process at work in late chronic stroke. This involved the splitting of some synergies in the affected arm from those that were observed in the unaffected arm (i.e., splitting in the affected limb of the modules observed in the healthy and unaffected limb). The splitting process they observed was interpreted as a possible compensatory reorganization of brainstem and spinal controls that was only possible on a timescale of several or more years. The finer subdivision of modules, as opposed to module merging and separation, seems to take very long periods of concerted effort by patients, suggesting a robustness of modules and a need for prolonged practice or difficult identification of latent circuits to create these subdivisions. Stroke recovery also involves changes in kinematic modules or elements, though these take a different form,[23] with kinematic merging indicating increasing function and skill. What spinal mechanisms support such processes? Transient addition, deletion, merging, and switching processes in modular pattern elements are part of the normal operation of spinal cord reflex and pattern generating systems as underscored by work in the turtle by Stein,[118,119] Berkowitz,[120] and the data in the frog discussed previously.[52,53,82,107] It is likely the pathological merging observed in trauma relates closely to these low-level mechanisms and their neural support. Understanding splitting, merging, and control of component synergy drives and bursts may thus form a key to more principled rehabilitation. Animal models of injury and rehabilitation that exhibit similar merging and splitting phenomena through injury and recovery are likely to provide such understanding.

Conclusions—future issues and known unknowns

Current data support a well-defined spinal infrastructure in tetrapods that is layered onto earlier trunk locomotor swimming systems, to augment or supplant them.[114] This infrastructure seems to have both rhythm generation and separated modular pattern-shaping layers. The modular shaping may consist of specific premotor drive systems, and these are likely to participate, albeit in modified fashion, in adult human movements. Activating the isolated lumbar circuits to restore some level of stepping in severe spinal cord injury recruits all of these systems. Our knowledge of the modular drive systems is still very sketchy. Understanding the organization and operations in the pattern-shaping part of the CPG hierarchy is likely to have strong clinical significance. In injury, modular mergings or timing collapses of synergies occur. Later, on timescales of several years, splitting of normal synergies may be possible. We do not know in detail how merging, splitting, and rhythm to pattern-shaping interactions operate at the circuit level, what neural systems they employ, or how best to support these circuits in a rehabilitative or other intervention. These latter data are likely to arise more quickly in basic animal models, using neural recordings in reduced preparations and fictive and genetic methods. The resulting knowledge of how the building blocks of pattern interact and support one another in the context of descending and other controls is essential to developing principled integrative therapies and interventions for the clinic.

Acknowledgments

This study is supported by NIH NS40412, NS54894, and NSF CRCNS IIS-0827684.

Conflicts of interest

The authors declare no conflicts of interest.

References

1. Callebaut, W. & D. Rasskin-Gutman, Eds. 2005. *Modularity: Understanding the Development and Evolution of Natural Complex Systems.* Cambridge, MA: MIT Press.

2. Wagner, G.P., M. Pavlicev & J.M. Cheverud. 2007. The road to modularity. *Nat. Rev. Genet.* **8:** 921–931.

3. Schouenborg, J. 2002. Modular organization and spinal somatosensory imprinting. *Brain Res. Rev.* **40:** 80–91.

4. Slotine, J.J. & W. Lohmiller. 2001. Modularity, evolution, and the binding problem: a view from stability theory. *Neural. Netw.* **14:** 137–145.

5. Cheung, V.C., L. Piron, M. Agostini, *et al.* 2009. Stability of muscle synergies for voluntary actions after cortical stroke in humans. *Proc. Natl. Acad. Sci. USA* **106:** 19563–19568.

6. Cheung, V.C., A. Turolla, M. Agostini, *et al.* 2012. Muscle synergy patterns as physiological markers of motor cortical damage. *Proc. Natl. Acad. Sci. U.S.A.* **109:** 14652–14656.

7. Clark, D.J., L.H. Ting, F.E. Zajac, *et al.* 2010. Merging of healthy motor modules predicts reduced locomotor performance and muscle coordination complexity post-stroke. *J. Neurophysiol.* **103:** 844–857.

8. Giszter, S.F., C.B. Hart & S.P. Silfies. 2010. Spinal cord modularity: evolution, development, and optimization and the possible relevance to low back pain in man. *Exp. Brain Res.* **200:** 283–306.

9. Giszter S, V. Patil & C. Hart. 2007. Primitives, premotor drives, and pattern generation: a combined computational and neuroethological perspective. *Prog. Brain Res.* **165:** 323–346.

10. Giszter, S.F. 2008. Motor primitives. In *Encyclopedia of Neuroscience.* L.R. Squire Ed.: 102–1040. Academic Press. Oxford.

11. Tresch, M.C. & A. Jarc. 2009. The case for and against muscle synergies. *Curr. Opin. Neurobiol.* **19:** 601–607.

12. Flash, T. & B. Hochner. 2005. Motor primitives in vertebrates and invertebrates. *Curr. Opin. Neurobiol.* **15:** 660–666.

13. Fentress, J.C. 1973. Development of grooming in mice with amputated forelimbs. *Science* **179:** 704–705.

14. Golani, I. & J.C. Fentress. 1985. Early ontogeny of face grooming in mice. *Dev. Psychobiol.* **18:** 529–544.

15. Sacrey, L.A., J.M. Karl & I.Q. Whishaw. 2012. Development of rotational movements, hand shaping, and accuracy in advance and withdrawal for the reach-to-eat movement in human infants aged 6–12 months. *Infant Behav. Dev.* **35:** 543–560.

16. Whishaw, I.Q., S.G. Travis, S.W. Koppe, *et al.* 2010. Hand shaping in the rat: conserved release and collection vs. flexible manipulation in overground walking, ladder rung walking, cylinder exploration, and skilled reaching. *Behav. Brain Res.* **206:** 21–31. doi: 10.1016/j.bbr.2009.08.030

17. McFarland, D.J. & A. Houston. 1981. *Quantitative Ethology: The State Space Approach.* Pitman.

18. Viviani, P. & C. Terzuolo. 1982. Trajectory determines movement dynamics. *Neuroscience* **7:** 431–437.

19. Bernstein, N. 1967. *The Co-ordination and Regulation of Movements.* Pergamon Press. Oxford.

20. Hogan, N. 1984. An organizing principle for a class of voluntary movements. *J. Neurosci.* **4:** 2745–2754.

21. Flash, T. & N. Hogan. 1985. The coordination of arm movements: an experimentally confirmed mathematical model. *J. Neurosci.* **5:** 1688–1703.

22. Burdet, E. & T.E. Milner. 1998. Quantization of human motions and learning of accurate movements. *Biol. Cybern.* **78:** 307–318.

23. Rohrer, B., S. Fasoli, H.I. Krebs, *et al.* 2002. Smoothness during stroke recovery. *J. Neurosci.* **22:** 8297–8304.

24. Wolpert, D.M., Z. Ghahramani & J.R. Flanagan. 2001. Perspectives and problems in motor learning. *Trends Cogn. Sci.* **5:** 487–494.

25. Sosnik, R., B. Hauptmann, A. Karni & T. Flash. 2004. When practice leads to co-articulation: the evolution of geometrically defined movement primitives. *Exp. Brain Res.* **156:** 422–438.

26. Polyakov, F., E. Stark, R. Drori, *et al.* 2009. Parabolic movement primitives and cortical states: merging optimality with geometric invariance. *Biol. Cybern.* **100:** 159–184.

27. Santello, M., M. Flanders & J.F. Soechting. 1998. Postural hand synergies for tool use. *J. Neurosci.* **18:** 10105–10115.

28. Schaal, S., D. Sternad, R. Osu & M. Kawato. 2004. Rhythmic arm movement is not discrete. *Nat. Neurosci.* **7:** 1136–1143. [Erratum in: *Nat Neurosci.* 2004 Nov; 7: 1279.]

29. Schaal, S., A. Ijspeert & A. Billard. 2003. Computational approaches to motor learning by imitation. *Philos. Trans. R. Soc. Lond. B. Biol. Sci.* **358:** 537–547.

30. Hogan, N. & D. Sternad. 2007. On rhythmic and discrete movements: reflections, definitions and implications for motor control. *Exp. Brain Res.* **181:** 13–30.

31. Ijspeert, A., J. Nakanishi & S. Schaal. 2003. Learning attractor landscapes for learning motor primitives. In *Advances in Neural Information Processing Systems 15.* S. Becker, S. Thrun & K. Obermayer, Eds.: 1547–1554. MIT Press. Cambridge, MA.

32. Ijspeert, A., J. Nakanishi, P. Pastor, *et al.* 2013. Dynamical movement primitives: learning attractor models for motor behaviors. *Neural Comput.* **25:** 328–373.

33. Hogan, N. & D. Sternad. 2012. Dynamic primitives of motor behavior. *Biol. Cybern.* **106:** 727–739.

34. Grillner, S. 2003. The motor infrastructure: from ion channels to neuronal networks. *Nat. Neurosci. Rev.* **4:** 673–686.

35. Kiehn, O., J. Hounsgard & K.T. Sillar. 1997. Basic building blocks of vertebrate CPGs. In *Neurons, Networks and Motor Behavior.* P.S.G. Stein, S. Grillner, A.I. Selverston & D.G. Stuart, Eds.: 47–60. MIT Press. Cambridge, MA.

36. Stein, P.S. 2005. Neuronal control of turtle hindlimb motor rhythms. *J. Comp. Physiol. A. Neuroethol. Sens. Neural Behav. Physiol.* **191:** 213–229.

37. Stein, P.S. 2008. Motor pattern deletions and modular organization of turtle spinal cord. *Brain Res. Rev.* **57:** 118–124.

38. Stein, P.S. & S. Daniels-McQueen. 2002. Modular organization of turtle spinal interneurons during normal and deletion fictive rostral scratching. *J. Neurosci.* **22:** 6800–6809.

39. Stein, P.S. & S. Daniels-McQueen. 2004. Variations in motor patterns during fictive rostral scratching in the turtle: knee-related deletions. *J. Neurophysiol.* **91:** 2380–2384.

40. Sillar, K.T., D. Combes, S. Ramanathan, *et al.* 2008. Neuromodulation and developmental plasticity in the locomotor system of anuran amphibians during metamorphosis. *Brain Res. Rev.* **57:** 94–102.

41. Rybak, I.A., A.P. Abdala, S.N. Markin, *et al.* 2007. Spatial organization and state-dependent mechanisms for respiratory rhythm and pattern generation. *Prog. Brain Res.* **165:** 201–220.

42. McCrea, D.A. & I.A. Rybak. 2007. Modeling the mammalian locomotor CPG: insights from mistakes and perturbations. *Prog. Brain Res.* **165:** 235–253.

43. McCrea, D.A. & I.A. Rybak. 2008. Organization of mammalian locomotor rhythm and pattern generation. *Brain Res. Rev.* **57:** 134–146.

44. Lafreniere-Roula, M. & D.A. McCrea. 2005. Deletions of rhythmic motoneuron activity during fictive locomotion and scratch provide clues to the organization of the mammalian central pattern generator. *J. Neurophysiol.* **94:** 1120–1132.

45. Quevedo, J., K. Stecina, S. Gosgnach & D.A. McCrea. 2005. Stumbling corrective reaction during fictive locomotion in the cat. *J. Neurophysiol.* **94:** 2045–2052.

46. Rybak, I.A., N.A. Shevtsova, M. Lafreniere-Roula & D.A. McCrea. 2006a. Modelling spinal circuitry involved in locomotor pattern generation: insights from deletions during fictive locomotion. *J. Physiol.* **577**(Pt 2): 617–639.

47. Rybak, I.A., K. Stecina, N.A. Shevtsova & D.A. McCrea. 2006b. Modelling spinal circuitry involved in locomotor pattern generation: insights from the effects of afferent stimulation. *J. Physiol.* **577**(Pt 2): 641–658.

48. Abbas, J.J. & H.J. Chizeck. 1995. Neural network control of functional neuromuscular stimulation systems: computer simulation studies. *IEEE Trans. Biomed. Eng.* **42:** 1117–1127.

49. Tresch, M.C., V.C. Cheung & A. d'Avella. 2006. Matrix factorization algorithms for the identification of muscle synergies: evaluation on simulated and experimental data sets. *J. Neurophysiol.* **5:** 2199–2212.

50. Krouchev N., J.F. Kalaska & T. Drew. 2006. Sequential activation of muscle synergies during locomotion in the intact cat as revealed by cluster analysis and direct decomposition. *J. Neurophysiol.* **96:** 1991–2010.

51. Hart, C.B. & S.F. Giszter. 2004. Modular premotor drives and unit bursts as primitives for frog motor behaviors. *J. Neurosci.* **24:** 5269–5282.

52. Kargo, W.J. & S.F. Giszter. 2000. Rapid corrections of aimed movements by combination of force-field primitives. *J. Neurosci.* **20:** 409–426.

53. Kargo, W.J. & S.F. Giszter. 2008. Individual premotor drive pulses, not time-varying synergies, are the units of adjustment for limb trajectories constructed in spinal-cord. *J. Neurosci.* **28:** 2409–2425.

54. Olree, K.S. & C.L. Vaughan. 1995. Fundamental patterns of bilateral muscle activity in human locomotion. *Biol. Cybern.* **73:** 409–414.

55. Cappellini, G., Y.P. Ivanenko, R.E. Poppele & F. Lacquaniti. 2006. Motor patterns in human walking and running. *J. Neurophysiol.* **95:** 3426–3437.

56. Ting, L.H. 2007. Dimensional reduction in sensorimotor systems: a framework for understanding muscle coordination of posture. *Prog. Brain Res.* **165:** 299–321.

57. Ting, L.H. & J.M. Macpherson. 2005. A limited set of muscle synergies for force control during a postural task. *J. Neurophysiol.* **93:** 609–613.

58. Torres-Oviedo, G., J.M. Macpherson & L.H. Ting. 2006. Muscle synergy organization is robust across a variety of postural perturbations. *J. Neurophysiol.* **96:** 1530–1546.

59. Torres-Oviedo, G. & L.H. Ting. 2007. Muscle synergies characterizing human postural responses. *J. Neurophysiol.* **98:** 2144–2156.

60. d'Avella, A. & E. Bizzi. 2005. Shared and specific muscle synergies in natural motor behaviors. *Proc. Natl. Acad. Sci. U.S.A.* **102:** 3076–3081.

61. d'Avella, A., A. Portone, L. Fernandez & F. Lacquaniti. 2006. Control of fast-reaching by muscle synergy combinations. *J. Neurosci.* **26:** 7791–7810.

62. d'Avella, A., P. Saltiel & E. Bizzi. 2003. Combinations of muscle synergies in the construction of a natural motor behavior. *Nat. Neurosci.* **6:** 300–308.

63. Muceli, S., A.T. Boye, A. d'Avella & D. Farina. 2010. Identifying representative synergy matrices for describing muscular activation patterns during multidirectional reaching in the horizontal plane. *J. Neurophysiol.* **103:** 1532–1542.

64. Dominici, N., Y.P. Ivanenko, G. Cappellini, *et al.* 2011. Locomotor primitives in newborn babies and their development. *Science* **334:** 997–999.

65. Giszter, S.F. & W.J. Kargo. 2000. Conserved temporal dynamics and vector superposition of primitives in frog wiping reflexes during spontaneous extensor deletions. *Neurocomputing* **32–33:** 775–783.

66. Bizzi, E., F.A. Mussa-Ivaldi & S. Giszter. 1991. Computations underlying the execution of movement: a biological perspective. *Science* **253:** 287–291.

67. Giszter, S.F., F.A. Mussa-Ivaldi & E. Bizzi. 1993. Convergent force fields organized in the frog spinal cord. *J. Neurosci.* **13:** 467–491.

68. Mussa-Ivaldi, F.A. 1992. From basis functions to basis fields: using vector primitives to capture vector patterns. *Biol. Cybern.* **67:** 479–489.

69. Mussa-Ivaldi, F.A. & E. Bizzi. 2000. Motor learning through the combination of primitives. *Philos. Trans. R. Soc. Lond. B. Biol. Sci.* **355:** 1755–1769.

70. Mussa-Ivaldi, F.A. & S.F. Giszter. 1992. Vector field approximation: a computational paradigm for motor control and learning. *Biol. Cybern.* **67:** 491–500.

71. Mussa-Ivaldi, F.A., S.F. Giszter & E. Bizzi. 1994. Linear combination of primitives in vertebrate motor control. *Proc. Natl. Acad. Sci. U.S.A.* **91:** 7534–7538.

72. Lemay M.A., & W.M. Grill. 2004. Modularity of motor output evoked by intraspinal microstimulation in cats. *J. Neurophysiol.* **91:** 502–514.

73. Boyce, V.S. & M.A. Lemay. 2009. Modularity of endpoint force patterns evoked using intraspinal microstimulation in treadmill trained and/or neurotrophin-treated chronic spinal cats. *J. Neurophysiol.* **101:** 1309–1320.

74. Tresch, M.C. & E. Bizzi. 1999. Responses to spinal microstimulation in the chronically spinalized rat and their relationship to spinal systems activated by low threshold cutaneous stimulation. *Exp. Brain Res.* **129:** 401–416.

75. Tresch, M.C., P. Saltiel & E. Bizzi. 1999. The construction of movement by the spinal cord. *Nature Neurosci.* **2:** 162–167.

76. Sumbre, G., G. Fiorito, T. Flash & B. Hochner. 2006. Octopuses use a human-like strategy to control precise point-to-point arm movements. *Curr. Biol.* **16:** 767–772.

77. Fetcho, J.R. & D.L. McLean. 2010. Some principles of organization of spinal neurons underlying locomotion in zebrafish and their implications. *Ann. N.Y. Acad. Sci.* **1198:** 94–104.

78. Ryczko, D., R. Dubuc & J.M. Cabelguen. 2010. Rhythmogenesis in axial locomotor networks: an interspecies comparison. *Prog. Brain Res.* **187:** 189–211.

79. Full, R.J. & D.E. Koditschek. 1999. Templates and anchors: neuromechanical hypotheses of legged locomotion on land. *J. Exp. Biol.* **202**(Pt 23): 3325–3332.

80. Grillner, S. & T.M. Jessell. 2009. Measured motion: searching for simplicity in spinal locomotor networks. *Curr. Opin. Neurobiol.* **19:** 572–586.

81. Marr, D. 1983. *Vision: A Computational Investigation into the Human Representation and Processing of Visual Information.* New York: WH Freeman. Reprinted 2010. MIT Press. Cambridge, MA.

82. Kargo, W.J. & S.F. Giszter. 2000. Afferent roles in hindlimb wiping reflex: free limb kinematics and motor patterns. *J. Neurophysiol.* **83:** 1480–1501.

83. Kargo, W.J., A. Ramakrishnan, C.B. Hart, *et al.* 2010. A simple experimentally-based model using proprioceptive regulation of motor primitives captures adjusted trajectory formation in spinal frogs. *J. Neurophysiol.* **103:** 573–590.

84. Loeb, G.E. 2000. Overcomplete musculature or underspecified tasks? *Motor Control* **4:** 81–83; discussion 97–116.

85. Loeb, G.E., J. He & W.S. Levine. 1989. Spinal cord circuits: are they mirrors of musculoskeletal mechanics? *J. Mot. Behav.* **21:** 473–491.

86. Loeb, G.E., W.S. Levine & J. He. 1990. Understanding sensorimotor feedback through optimal control. *Cold Spring Harb. Symp. Quant. Biol.* **55:** 791–803.

87. Sanger, T.D. 2000. Human arm movements described by a low-dimensional superposition of principal components. *J. Neurosci.* **20:** 1066–1072.

88. Sanger, T.D. 1994. Optimal unsupervised motor learning for dimensionality reduction of nonlinear control systems. *IEEE Trans. Neural Netw.* **5:** 965–973.

89. Chabra, M. & R.A. Jacobs. 2006. Properties of synergies arising from a theory of optimal motor behavior. *Neural Comput.* **18:** 2320–2342.

90. Todorov, E. 2004. Optimality principles in sensorimotor control. *Nat. Neurosci.* **7:** 907–915.

91. Todorov, E. & Z. Ghahramani. 2003. Unsupervised learning of sensory-motor primitives. In *Proc., 25th Annual International Conference of the IEEE Engineering in Medicine and Biology Society.* IEEE. Cancun, Mexico.

92. Todorov, E. & M.I. Jordan. 2002. Optimal feedback control as a theory of motor coordination. *Nat. Neurosci.* **5:** 1226–1235.

93. Todorov, E., W. Li & X. Pan. 2005. From task parameters to motor synergies: a hierarchical framework for approximately-optimal control of redundant manipulators. *J. Robot Syst.* **22:** 691–710.

94. Valero-Cuevas, F.J. 2009. A mathematical approach to the mechanical capabilities of limbs and fingers. *Adv. Exp. Med. Biol.* **629:** 619–633.

95. Valero-Cuevas, F.J., M. Venkadesan & E. Todorov. 2009. Structured variability of muscle activations supports the minimal intervention principle of motor control. *J. Neurophysiol.* **102:** 59–68.

96. Valero-Cuevas, F.J., J.W. Yi, D. Brown, *et al.* 2007. The tendon network of the fingers performs anatomical computation at a macroscopic scale. *IEEE Trans. Biomed. Eng.* **54**(Pt 2): 1161–1166.

97. Liu, D. & E. Todorov. 2007. Evidence for the flexible sensorimotor strategies predicted by optimal feedback control. *J. Neurosci.* **27:** 9354–9368.

98. Lockhart, D.B. & L.H. Ting. 2007. Optimal sensorimotor transformations for balance. *Nat. Neurosci.* **10:** 1329–1336.

99. Berniker, M., A. Jarc, E. Bizzi & M.C. Tresch. 2009. Simplified and effective motor control based on muscle synergies to exploit musculoskeletal dynamics. *Proc. Natl. Acad. Sci. U.S.A.* **106:** 7601–7606.

100. Grau, J.W., E.D. Crown, A.R. Ferguson, *et al.* 2006. Instrumental learning within the spinal cord: underlying mechanisms and implications for recovery after injury. *Behav. Cogn. Neurosci. Rev.* **5:** 191–239.

101. Wolpaw, J.R. 2007. Spinal cord plasticity in acquisition and maintenance of motor skills. *Acta Physiol. (Oxf.)* **189:** 155–169.

102. Pang, M.Y., T. Lam & J.F. Yang. 2003. Infants adapt their stepping to repeated trip-inducing stimuli. *J. Neurophysiol.* **90:** 2731–2740.

103. Clarac, F., F. Brocard & L. Vinay. 2004. The maturation of locomotor networks. *Prog. Brain Res.* **143:** 57–66.

104. Frank, E. & B. Mendelson. 1990. Specification of synaptic connections mediating the simple stretch reflex. *J. Exp. Biol.* **153:** 71–84.

105. Mendelson, B. & E. Frank. 1991. Specific monosynaptic sensory-motor connections form in the absence of

patterned neural activity and motoneuronal cell death. *J. Neurosci.* **11:** 1390–1403.

106. Haverkamp, L.J. 1986. Anatomical and physiological development of the Xenopus embryonic motor system in the absence of neural activity. *J. Neurosci.* **6:** 1338–1348.

107. Hart, C.B. & S.F. Giszter. 2010. Neural underpinnings of motor primitives. *J. Neurosci.* **30:** 1322–1336.

108. Berkowitz, A. 2008. Physiology and morphology of shared and specialized spinal interneurons for locomotion and scratching. *J. Neurophysiol.* **99:** 2887–2901.

109. Berkowitz, A. & Z.Z. Hao. 2011. Partly shared spinal cord networks for locomotion and scratching. *Integr. Comp. Biol.* **51:** 890–902.

110. Berkowitz, A., A. Roberts & S.R. Soffe. 2010. Roles for multifunctional and specialized spinal interneurons during motor pattern generation in tadpoles, zebrafish larvae, and turtles. *Front. Behav. Neurosci.* **4:** 36.

111. Alstermark, B. & T. Isa. 2012. Circuits for skilled reaching and grasping. *Annu. Rev. Neurosci.* **35:** 559–578.

112. Kinoshita, M., R. Matsui, S. Kato, *et al.* 2012. Genetic dissection of the circuit for hand dexterity in primates. *Nature* **487:** 235–238.

113. Takei, T. & K. Seki. 2010. Spinal interneurons facilitate coactivation of hand muscles during a precision grip task in monkeys. *J. Neurosci.* **30:** 17041–17050.

114. Rauscent, A., D. Le Ray, M.J. Cabirol-Pol, *et al.* 2006. Development and neuromodulation of spinal locomotor networks in the metamorphosing frog. *J. Physiol. Paris* **100:** 317–327.

115. Harkema, S., Y. Gerasimenko, J. Hodes, *et al.* 2011. Effect of epidural stimulation of the lumbosacral spinal cord on voluntary movement, standing, and assisted stepping after motor complete paraplegia: a case study. *Lancet* **377:** 1938–1947.

116. Edgerton, V.R. & S. Harkema. 2011. Epidural stimulation of the spinal cord in spinal cord injury: current status and future challenges. *Expert Rev. Neurother.* **11:** 1351–1353.

117. Edgerton, V.R., R.D. Leon, S.J. Harkema, *et al.* 2001. Retraining the injured spinal cord. *J. Physiol.* **533**(Pt 1): 15–22.

118. Earhart, G.M. & P.S. Stein. 2000. Scratch-swim hybrids in the spinal turtle: blending of rostral scratch and forward swim. *J. Neurophysiol.* **83:** 156–165.

119. Earhart, G.M. & P.S. Stein. 2000. Step, swim and scratch motor patterns in the turtle. *J. Neurophysiol.* **84:** 2181–2190.

120. Hao, Z.Z., L.E. Spardy, E.B. Nguyen, *et al.* 2011. Strong interactions between spinal cord networks for locomotion and scratching. *J. Neurophysiol.* **106:** 1766 1781.

121. Sherrington, C.S. 1961. *The Integrative Action of the Nervous System.* Yale University Press. New Haven, CT.

Ann. N.Y. Acad. Sci. ISSN 0077-8923

ANNALS OF THE NEW YORK ACADEMY OF SCIENCES
Issue: *Neurons, Circuitry, and Plasticity in the Spinal Cord and Brainstem*

A dual spinal cord lesion paradigm to study spinal locomotor plasticity in the cat

Marina Martinez and Serge Rossignol

Department of Physiology, Université de Montréal, Montréal, Québec, Canada

Address for correspondence: Dr. Serge Rossignol, Department of Physiology, Groupe de Recherche sur le Système Nerveux Central, Faculty of Medicine, Université de Montréal, P.O. Box 6128, Station Centre-Ville, Montréal, Québec, Canada H3C 3J7. serge.rossignol@umontreal.ca

After a complete spinal cord injury (SCI) at the lowest thoracic level (T13), adult cats trained to walk on a treadmill can recover hindlimb locomotion within 2–3 weeks, resulting from the activity of a spinal circuitry termed the *central pattern generator* (CPG). The role of this spinal circuitry in the recovery of locomotion after partial SCIs, when part of descending pathways can still access the CPG, is not yet fully understood. Using a dual spinal lesion paradigm (first hemisection at T10 followed three weeks after by a complete spinalization at T13), we showed that major changes occurred in this locomotor spinal circuitry. These plastic changes at the spinal cord level could participate in the recovery of locomotion after partial SCI. This short review describes the main findings of this paradigm in adult cats.

Keywords: central pattern generator; plasticity; training; spinal cord injury; locomotion

Introduction

Hindlimb locomotion is controlled by a tripartite system that involves dynamic interactions among a spinal neuronal network, supraspinal structures, and peripheral sensory inputs.[1–5] Following incomplete spinal cord injury (SCI), this tripartite interaction is perturbed, but animals can recover a surprising degree of locomotor performance, which probably involves widespread plastic changes within remaining ascending and descending pathways,[6–9] segmental spinal reflexes,[10] and the spinal cord itself.[11,12] While many studies emphasize the compensatory role of supraspinal structures in functional recovery after partial SCI, there is little evidence indicating how plastic changes within the spinal circuitry itself can account for some of the locomotor recovery after SCI. This review focuses on locomotor recovery and plasticity after partial SCI by describing some fundamental concepts that emerged from complete SCIs in cats. Adaptation within the spinal locomotor circuitry after partial SCI is then discussed.

Complete spinal lesions and the role of the CPG

In many species, some recovery of hindlimb locomotion can be observed after a complete spinal cord section.[1,2] Since locomotion can be expressed in kittens spinalized 7–14 days after birth, that is, before they gain their optimal walking capacities, one can suggest that this behavior is genetically programmed.[13–15] After a complete SCI at the level of the last thoracic segment (T13), cats can recover hindlimb locomotion that closely resembles a normal locomotor pattern by using electrical stimulation,[16,17] locomotor training,[18,19] pharmacological agents,[20–24] or a combination of the above.[25] This, per se, confirms the concept of spinal generation of locomotion by an intrinsic spinal circuitry at the lombosacral level, the central pattern generator (CPG). Since descending and ascending pathways are completely destroyed after a complete SCI, some intrinsic changes must have occurred in the CPG to allow the re-expression of hindlimb locomotion. After a complete SCI, the excitability of motoneurons,[26,27]

doi: 10.1111/j.1749-6632.2012.06823.x

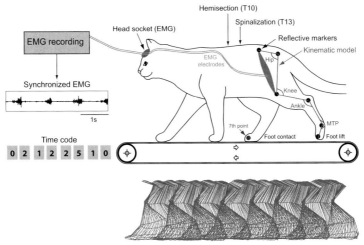

Figure 1. General methodology for the study of locomotion in cats. The animal is placed on the treadmill (arrows within the belt indicate the direction of movement) and is allowed to walk with all four limbs. Pairs of EMG wires are implanted into various muscles (only one pair is represented here) and soldered to a multipin connector cemented to the skull. The EMG signal was filtered and digitized and was then displayed on a computer. The left panel illustrates the EMG trace of the sartorius (hip flexor and knee extensor). Reflective markers are placed at various points on the limb, and the angle measurements are taken in the indicated orientations. For each video field (16.7 milliseconds between fields), the coordinates of the reflective markers are obtained and the hindlimb movement is reconstructed as indicated in the kinematic model (MTP = metatarsophalangeal joint). A seventh point is put on the right hindpaw to obtain kinematic data for this limb and information relative to coordination between hindlimbs. From such data, the swing and stance phases of each cycle can be reconstructed as shown in stick figures and duty cycles. The foot contact and lift are also measured to determine cycle length and duration and also to synchronize EMG events when needed. A digital time code (SMPTE) is used to synchronize video and EMG recordings.

cutaneous reflexes,[28] and receptors of various neurotransmitter systems[29] are changed.

Recovery from partial spinal lesions

Considering that incomplete SCIs are more frequent than complete SCIs in humans, we asked whether the spinal mechanisms observed after complete SCIs also apply to hindlimb locomotion generated after partial SCI. To accurately evaluate the locomotor pattern after spinal lesions, the most objective method consists of using chronic electromyographic recordings combined with synchronized video recordings of movements (Fig. 1). This method permits a powerful quantitative comparison of locomotor performances before and after spinal lesions, as well as its evolution over time. To evaluate the role of descending pathways in locomotor control, studies have investigated the consequences of lesions targeting specific tracts at the spinal level.

Dorsal/dorsolateral spinal tracts. After the dorsal/dorsolateral SCI's bilateral effect on the cortico- and rubrospinal fibers, cats can recover voluntary

quadrupedal locomotion while maintaining their weight and equilibrium after a 3- to 10-day period. They, however, have some long-lasting deficits such as foot drag (resulting from impaired coupling between hip and knee flexors) and they lose the capacity for anticipatory modifications when negotiating obstacles placed on the treadmill.[30] Despite these defects, the cats do walk remarkably well with all four limbs on the treadmill, suggesting that the cortico- and rubrospinal tracts do not appear critical for the control of treadmill locomotion but are important for skilled locomotion.

Ventral/ventrolateral spinal tracts. Vestibulo- and reticulospinal pathways traveling in the ventral part of the cord are known to be, respectively, involved in postural control and initiation of locomotion. After small ventral/ventrolateral lesions, cats can walk voluntarily at speeds of up to 0.7 m/s with all four limbs one to three days after the lesion.[31] With larger lesions, sparing only part of dorsal columns, cats behave initially as complete spinal cats. However, with regular treadmill training, all cats could regain voluntary quadrupedal locomotion but, with the largest lesions, the animals could not walk faster than 0.4 m/s even after several weeks of training. The important observation here is that, even with extensive bilateral lesions to ventral pathways, cats could initiate and maintain voluntary quadrupedal locomotion, suggesting that the ventral tracts do not appear essential for triggering hindlimb locomotion.

Hemisections. In contrast to bilateral dorsal or ventral lesions of the cord, unilateral hemisections damage the dorsal end ventral tracts on one side, while the other side remains intact. The functional consequences of such lesions mainly depend on the extent of the lesion. Indeed, we showed in cats hemisected at thoracic level 10 that the smaller the lesion the faster the locomotor recovery.[11,32] During the first two to three days after a nearly complete hemisection on one side, the hindlimb ipsilateral to the lesion shows flaccid paresis and animals adopt a tripod gait, such that they require help for body support and stabilization. One week after hemisection, voluntary quadrupedal locomotion reappeared despite limping of the hindlimb and inconsistent plantar foot contact on the side of the lesion. Stepping activity greatly improves during the first three weeks posthemisection, but some deficits were shown to persist at this postoperative time. To minimize the time spent on the effected hindlimb, the support time was reduced on the side of the lesion while the swing phase was increased. Concomitantly, the burst duration of extensors decreased while that of the flexors increased on the side of the lesion.[12] By plotting the relationships between the cycle and its subphases (stance and swing) at various treadmill speeds, we showed that the hemisection had induced profound changes in the intrinsic structure of the cycle on both sides and had thus altered the neural control of locomotion.[33] Moreover, forelimb/hindlimb coordination on the side of the lesion and left/right hindlimb coupling was altered and cats exhibited an asymmetrical gait.[12] The ipsilateral descending reflexes (scratch, vestibular drop reflex) were lost after hemisection and skilled locomotion such as ladder walking was also impaired durably on the side of the lesion.[34–36] Altogether, although a gradual recovery of walking occurs after a spinal hemisection, some deficits associated with dorsal (postural deficits and altered coordination)

and ventral lesions (impaired skilled locomotion) persist.

From the studies on partial spinal lesions described previously, we can conclude that although there are some deficits due to the interruption of specific tracts, animals can in most cases regain a functional locomotor pattern. Several nonexclusive mechanisms occurring at different levels of the neuraxis probably participate in locomotor recovery after a partial SCI such as physiological and/or anatomical reorganizations within remnant pathways could somehow take over the functions of the damaged spinal cord.[37–43] However, an alternative interpretation is that after a partial SCI, the CPG below the lesion assumes a greater role in generating the hindlimb locomotor pattern. In this case, the plastic changes in descending pathways would serve to reorganize functionally the spinal cord and modify new input–output characteristics of the spinal CPG. This interpretation requires demonstrating that changes occur within the CPG after a partial lesion, which is discussed next.

Role of the CPG after partial spinal lesions: the dual lesion paradigm

A dual lesion paradigm has been developed[11] to study the changes occurring within the lumbar spinal locomotor circuitry (CPG) after a partial spinal lesion. In this paradigm, a unilateral hemisection of the cord is first performed at T10–11, well above the CPG located in the lumbo-sacral segments, and is followed, several weeks later, by a second complete spinal lesion two spinal segments below the first one at T13 (Fig. 2) (i.e., at the level where the complete spinal lesions were usually made[18,25,44]) to isolate the spinal CPG from its supraspinal influences. The main idea of this paradigm is that if intrinsic changes occurred within the spinal cord itself during locomotor recovery after the initial hemisection, these changes could probably be retained and expressed very early after a second and complete spinalization a few segments below. In the first study with this paradigm, cats were trained to walk on the treadmill during the interim between the two lesions and the second complete lesion was performed only when the locomotor performances of the cats reached a plateau of locomotor performance. The first major finding was that within 24 h (i.e., the first testing session) following complete spinalization, cats could reexpress

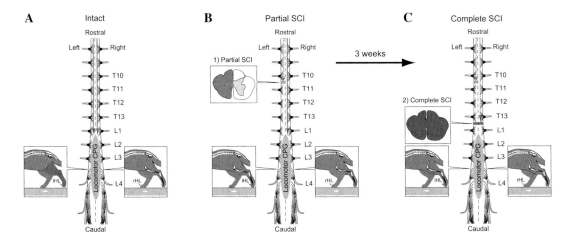

Figure 2. The dual lesion paradigm. (A) In intact cats, locomotion is controlled and modulated by descending inputs (red arrows) from supraspinal structures that interact with the spinal CPG (in gray) to coordinate the left (lHL, in red) and right (rHL, in green) hindlimb activity. (B) After a unilateral hemisection targeting the left part of the spinal cord at T10, descending inputs are disconnected from the CPG on the left side (red dotted arrows). While the right hindlimb remained actively controlled by the remnant supraspinal fibers, the left hindlimb is mainly under the control of the partially deafferented CPG. (C) Three weeks after the hemisection, a complete spinal section is performed at T13, removing all the supraspinal inputs to the CPG, such that the left and right hindlimbs are mainly under the control of the CPG completely devoid of supraspinal inputs.

hindlimb locomotion with plantar foot placement at high speeds (up to 1.0 m/s) without any pharmacological stimulation.[11,32] It should again be recalled that under normal circumstances the expression of spinal locomotion requires two to three weeks of intense treadmill training in spinal cats.[18] This result indicates that the hemisection had induced changes within the spinal circuitry below the lesion, such that it was already primed to reexpress locomotion after the complete SCI. As these cats were trained to walk during the interim between the two lesions and that this interim period varied among cats, it was important to standardize the conditions required to induce plastic changes in the spinal locomotor circuitry after hemisection.

Could such intraspinal changes occur spontaneously after the first partial lesion—that is, in the absence of training and within a short period of time between the two lesions? To address this question, the interim period between the two lesions was limited to a short period of three weeks in a cohort of 11 cats.[12] To limit self-training that could influence locomotor recovery after hemisection, cats were kept in individual cages and their locomotor performances were recorded only once a week on the treadmill. Such housing condition limits sensorimotor experience and contrasts with the daily

treadmill training procedure used in previous studies.[11,32] Interestingly, 24 h after spinalization, 6/11 cats (55%) expressed a bilateral hindlimb locomotion, 3/11 expressed a unilateral pattern on the side of the previous hemisection, and 2/11 were not able to walk. These results indicate that the spinal circuitry below the hemisection can reorganize spontaneously and within a relatively short period of time after hemisection.

This result raised the question of the type of reorganizations occurring within the spinal locomotor circuitry after hemisection. To determine how the spinal circuits reorganize after hemisection, we investigated, over the entire dual lesion paradigm, the evolution of specific locomotor parameters mainly controlled by the spinal locomotor circuitry.[33] We measured the various subcomponents of the step cycle (swing, stance) (see, for instance, Fig. 3A) and plotted the relationships between the cycle and its subphases at various treadmill speeds. First, the results showed that after a hemisection, the intrinsic structure of the cycle is altered in both hindlimbs, such that the cycle period can change by adapting the duration of each subphase according to the speed. Second, removing all supraspinal inputs by spinalization revealed that the changes observed after hemisection were retained for a long time but

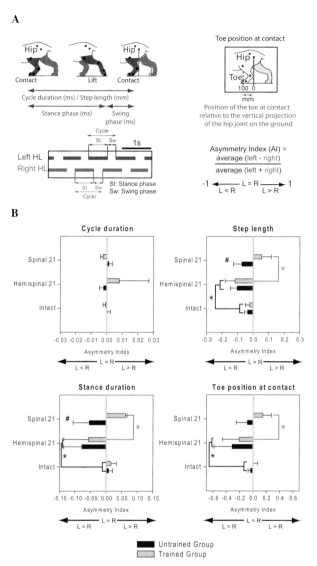

Figure 3. Effect of locomotor training on hindlimb kinematics in completely spinalized cats previously submitted to a spinal hemisection on the left side. (A) Illustrations of the kinematic measures. *Step cycle duration* is the time between two successive contacts of the same foot on the treadmill, whereas the *stance duration* refers to the time between foot contact and toe off that initiates the swing phase. *Step length* is the distance traveled during the stance and swing phase of a complete step cycle. *Toe position relative to the hip at contact* was determined by calculating the mean position (in mm) of the toe relative to the vertical projection of the hip joint on the ground at contact. As the first spinal lesion (the hemisection on the left side) is known to induce an asymmetrical walking pattern, the locomotor parameters described previously were compared between hindlimbs by calculating an *asymmetry index* (AI). An AI of 0 indicates a perfect symmetry, and in normal conditions, the AIs obtained in all parameters are equal to 0 ± 0.05. The AI gives information on the direction of asymmetry. When a parameter is greater on the left than on the right side, the AI will be positive and vice versa. For example, in the case of a step length AI < 0, the step of the right hindlimb will be longer than the left. (B) Evolution of the asymmetry indices calculated for cycle duration, step length, stance duration, and toe position at contact in cats trained to locomote after spinalization and untrained cats over the entire dual lesion paradigm. In the intact state, the two groups of cats exhibited a symmetrical walking pattern (AIs $= 0 \pm 0.05$). Twenty-one days after a hemisection on the left side of the cord, both groups displayed an asymmetrical walking pattern due to the deficits of the left hindlimb (side of the lesion). Note that both groups were comparable before the spinalization and the beginning of training. Twenty-one days after spinalization, the asymmetrical walking pattern displayed by all cats after hemisection was retained in untrained cats, while it reversed in trained cats. Statistical differences between delays are indicated by the symbol (*). Statistical differences between groups are indicated by the symbol (#). L, left; R, right.

only on the side of the previous hemisection, thus demonstrating a durable asymmetrical reorganization at the spinal level resulting from the previous partial spinal lesion. In addition, the asymmetrical walking pattern observed after hemisection was also retained by cats after the complete SCI[45] (Fig. 3B), such that the hindlimb on the side of the previous hemisection kept its deficits after spinalization. Moreover, some changes that occurred in simple cutaneous reflexes after the partial lesion were also retained following spinalization.[46] These carry-over of changes that result from the previous partial SCI indicated that the spinal cord has an intrinsic capacity to be imprinted by past experiences.

However, can a spinal cord previously modified by past experience (such as a hemisection) again adapt to new demands, or are these changes immutable? To address this question, the performance of cats subjected to treadmill training after the complete SCI were compared with that of untrained cats during the entire dual lesion paradigm.[45] While without training the asymmetrical locomotion induced by the previous unilateral hemisection was retained for three weeks after the subsequent spinalization, training cats after spinalization reversed the asymmetrical locomotor pattern induced, suggesting that new plastic changes occurred within the spinal cord in response to training (Fig. 3B). Moreover, training cats after spinalization was shown to improve the locomotor performance of the hindlimb previously affected by the hemisection. These results demonstrated that a spinal cord previously modified by past experience (such as after a hemisection) can remarkably adapt to new demands imposed by locomotor training and suggest that locomotor training can be beneficial in the spinal state regardless of the previous experience of the spinal cord.

Considering the role of locomotor training in shaping intraspinal plasticity in cats previously submitted to spinal hemisection, can the locomotor deficits and asymmetrical intraspinal changes observed after hemisection be compensated by locomotor training? In the most recent study, we evaluated the effect of treadmill training on locomotor plasticity after hemisection. We compared the functional locomotor outcome of eight cats trained to locomote for three weeks after hemisection, and eight untrained cats that served as control. The activity-induced intraspinal plasticity was then as-

sessed by comparing the locomotor performances of both groups of cats 24 h after the subsequent spinalization. We confirmed that a three-week period of locomotor training after hemisection had a beneficial role in the recovery of voluntary quadrupedal locomotion. Furthermore, locomotor training enhanced plasticity in the spinal cord below the lesion because 100% of the trained cats reexpressed a high level of bilateral hindlimb locomotion immediately after spinalization compared to 60% of untrained cats. This study highlighted the beneficial role of locomotor training on facilitating adaptive plastic changes within the spinal circuitry and in promoting locomotor recovery after hemisection.

From these studies using a dual spinal lesion paradigm, we demonstrated that (1) the spinal locomotor circuitry has an intrinsic potential of plasticity; (2) this plasticity most likely participates in the recovery of locomotion after partial SCIs; (3) the spinal cord is a flexible integration center capable of adapting and shaping its motor circuits in order to optimize function with available inputs (i.e., descending and sensory); and (4) locomotor training has a powerful effect on spinal plasticity and acts positively on spinal circuits by reestablishing kinematic parameters approaching the normal state.

Clinical implications

For years, several concepts relative to the notion of a spinal locomotor CPG have been obtained from cat models of SCI. This review highlights the notion that the spinal locomotor CPG plays a prominent role in the recovery of locomotion after complete but also after incomplete SCIs. The question of whether such notions apply to humans, especially when such a strong role is considered for a spinal cord CPG, has been debated. Although the existence of a spinal CPG in humans cannot be directly and stringently demonstrated as in animal models, there are a number of indications that the isolated spinal cord of humans also contains rhythmogenic capabilities.[47–50] Consequently, targeting intrinsic spinal circuits together with remaining descending and sensory pathways should be a prime approach for rehabilitative strategies in humans with incomplete SCI, which is the concept of locomotor training.[22]

Acknowledgments

This work was funded by a Canada Research Chair on the Spinal Cord, the SensoriMotor Rehabilitation

Research Team (ERRSM) as part of the Regenerative Medicine, and the Nanomedicine Strategic Initiative of the Canadian Institute for Health Research (CIHR), as well as by an individual grant from the CIHR to S.R. M.M was funded by postdoctoral fellowships from ERRSM, FQRNT, and CIHR.

Conflicts of interest

The authors declare no conflicts of interest.

References

1. Grillner, S. 1981. Control of locomotion in bipeds, tetrapods, and fish. In *Handbook of Physiology. The Nervous System II*. J.M. Brookhart & V.B. Mountcastle, Eds.: 1179–1236. Bethesda, MD.

2. Rossignol, S. 1996. Neural control of stereotypic limb movements. In *Handbook of Physiology, Section 12. Exercise: Regulation and Integration of Multiple Systems*. L.B. Rowell & J.T. Sheperd, Eds.: 173–216. Oxford University Press. New York.

3. Kiehn, O. 2006. Locomotor circuits in the mammalian spinal cord. *Annu. Rev. Neurosci.* **29:** 279–306.

4. Rossignol, S., R. Dubuc & J.P. Gossard. 2006. Dynamic sensorimotor interactions in locomotion. *Physiol. Rev.* **86:** 89–154.

5. Rossignol, S. & A. Frigon. 2011. Recovery of locomotion after spinal cord injury: some facts and mechanisms. *Annu. Rev. Neurosci.* **34:** 413–440.

6. Ghosh, A., E. Sydekum, F. Haiss, *et al.* 2009. Functional and anatomical reorganization of the sensory-motor cortex after incomplete spinal cord injury in adult rats. *J. Neurosci.* **29:** 12210–12219.

7. Martinez, M., M. Delcour, M. Russier, *et al.* 2010. Differential tactile and motor recovery and cortical map alteration after C4-C5 spinal hemisection. *Exp. Neurol.* **221:** 186–197.

8. Martinez, M. & S. Rossignol. 2011. Changes in CNS structures after spinal cord lesions implications for BMI. *Prog. Brain Res.* **194:** 191–202.

9. Martinez, M., J.M. Brezun, Y. Zennou-Azogui, *et al.* 2009. Sensorimotor training promotes functional recovery and somatosensory cortical map reactivation following cervical spinal cord injury. *Eur. J. Neurosci.* **30:** 2356–2367.

10. Frigon, A. & S. Rossignol. 2009. Partial denervation of ankle extensors prior to spinalization in cats impacts the expression of locomotion and the phasic modulation of reflexes. *Neuroscience* **158:** 1675–1690.

11. Barrière, G., H. Leblond, J. Provencher & S. Rossignol. 2008. Prominent role of the spinal central pattern generator in the recovery of locomotion after partial spinal cord injuries. *J. Neurosci.* **28:** 3976–3987.

12. Martinez, M., H. Delivet-Mongrain, H. Leblond & S. Rossignol. 2011. Recovery of hindlimb locomotion after incomplete spinal cord injury in the cat involves spontaneous compensatory changes within the spinal locomotor circuitry. *J. Neurophysiol.* **106:** 1969–1984.

13. Grillner, S. 1973. Locomotion in the spinal cat. In *Control of Posture and Locomotion*. R.B. Stein, K.G. Pearson, R.S. Smith & J.B. Redford, Eds.: 515–535. Plenum Press. New York.

14. Forssberg, H., S. Grillner, J. Halbertsma & S. Rossignol. 1980. The locomotion of the low spinal cat: II. *Interlimb Coordination Acta Physiol. Scand.* **108:** 283–295.

15. Forssberg, H., S. Grillner & J. Halbertsma. 1980. The locomotion of the low spinal cat: I. Coordination within a hindlimb. *Acta Physiol. Scand.* **108:** 269–281.

16. Barthélemy, D., H. Leblond, J. Provencher & S. Rossignol. 2006. Non-locomotor and locomotor hindlimb responses evoked by electrical microstimulation of the lumbar cord in spinalized cats. *J. Neurophysiol.* **96:** 3273–3292.

17. Barthélemy, D., H. Leblond & S. Rossignol. 2007. Characteristics and mechanisms of locomotion induced by intraspinal microstimulation and dorsal root stimulation in spinal cats. *J. Neurophysiol.* **97:** 1986–2000.

18. Barbeau, H. & S. Rossignol. 1987. Recovery of locomotion after chronic spinalization in the adult cat. *Brain Res.* **412:** 84–95.

19. Lovely, R.G., R.J. Gregor, R.R. Roy & V.R. Edgerton. 1990. Weight-bearing hindlimb stepping in treadmill-exercised adult spinal cat. *Brain Res.* **514:** 206–218.

20. Barbeau, H., C. Julien & S. Rossignol. 1987. The effects of clonidine and yohimbine on locomotion and cutaneous reflexes in the adult chronic spinal cat. *Brain Res.* **437:** 83–96.

21. Barbeau, H. & S. Rossignol. 1990. The effects of serotonergic drugs on the locomotor pattern and on cutaneous reflexes of the adult chronic spinal cat. *Brain Res.* **514:** 55–67.

22. Barbeau, H. & S. Rossignol. 1994. Enhancement of locomotor recovery following spinal cord injury. *Curr. Opin. Neurol.* **7:** 517–524.

23. Chau, C., H. Barbeau & S. Rossignol. 1998. Effects of intrathecal α_1- and α_2-noradrenergic agonists and norepinephrine on locomotion in chronic spinal cats. *J. Neurophysiol.* **79:** 2941–2963.

24. Giroux, N., C. Chau, H. Barbeau, *et al.* 2003. Effects of intrathecal glutamatergic drugs on locomotion. II. NMDA and AP-5 in intact and late spinal cats. *J. Neurophysiol.* **90:** 1027–1045.

25. Chau, C., H. Barbeau & S. Rossignol. 1998. Early locomotor training with clonidine in spinal cats. *J. Neurophysiol.* **79:** 392–409.

26. Hounsgaard, J., H. Hultborn, J. Jespersen & O. Kiehn. 1988. Bistability of alpha-motoneurones in the decerebrate cat and in the acute spinal cat after intravenous 5-hydroxytryptophan. *J. Physiol.* **405:** 345–367.

27. Rossignol, S. & A. Frigon. 2011. Recovery of locomotion after spinal cord injury: some facts and mechanisms. *Annu. Rev. Neurosci.* **34:** 413–440.

28. Frigon, A. & S. Rossignol. 2008. Adaptive changes of the locomotor pattern and cutaneous reflexes during locomotion studied in the same cats before and after spinalization. *J. Physiol.* **586:** 2927–2945.

29. Giroux, N., S. Rossignol & T.A. Reader. 1999. Autoradiographic study of α_1-, α_2-Noradrenergic and Serotonin $_{1A}$ receptors in the spinal cord of normal and chronically transected cats. *J. Comp. Neurol.* **406:** 402–414.

30. Jiang, W. & T. Drew. 1996. Effects of bilateral lesions of the dorsolateral funiculi and dorsal columns at the level of the low thoracic spinal cord on the control of locomotion in the adult cat: I. Treadmill walking. *J. Neurophysiol.* **76:** 849–866.

31. Brustein, E. & S. Rossignol. 1998. Recovery of locomotion after ventral and ventrolateral spinal lesions in the cat: I. Deficits and adaptive mechanisms. *J. Neurophysiol.* **80:** 1245–1267.

32. Barrière, G., A. Frigon, H. Leblond, *et al.* 2010. Dual spinal lesion paradigm in the cat: evolution of the kinematic locomotor pattern. *J. Neurophysiol.* **104:** 1119–1133.

33. Martinez, M., H. Delivet-Mongrain, H. Leblond & S. Rossignol. 2012. Incomplete spinal cord injury promotes durable functional changes within the spinal locomotor circuitry. *J. Neurophysiol.* **108:** 124–134.

34. Murray, M. & M.E. Goldberger. 1974. Restitution of function and collateral sprouting in the cat spinal cord: the partially hemisected animal. *J. Comp. Neurol.* **158:** 19–36.

35. Kato, M. 1992. Walking of cats on a grid: performance of locomotor task in spinal intact and hemisected cats. *Neurosci. Lett.* **145:** 129–132.

36. Helgren, M.E. & M.E. Goldberger. 1993. The recovery of postural reflexes and locomotion following low thoracic hemisection in adult cats involves compensation by undamaged primary afferent pathways. *Exp. Neurol.* **123:** 17–34.

37. Fouad, K., V. Pedersen, M.E. Schwab & C. Brosamle. 2001. Cervical sprouting of corticospinal fibers after thoracic spinal cord injury accompanies shifts in evoked motor responses. *Curr. Biol.* **11:** 1766–1770.

38. Weidner, N., A. Ner, N. Salimi & M.H. Tuszynski. 2001. Spontaneous corticospinal axonal plasticity and functional recovery after adult central nervous system injury. *Proc. Natl. Acad. Sci. USA* **98:** 3513–3518.

39. Raineteau, O., K. Fouad, F.M. Bareyre & M.E. Schwab. 2002. Reorganization of descending motor tracts in the rat spinal cord. *Eur. J. Neurosci.* **16:** 1761–1771.

40. Schucht, P., O. Raineteau, M.E. Schwab & K. Fouad. 2002. Anatomical correlates of locomotor recovery following dorsal and ventral lesions of the rat spinal cord. *Exp. Neurol.* **176:** 143–153.

41. Bareyre, F.M., M. Kerschensteiner, O. Raineteau, *et al.* 2004. The injured spinal cord spontaneously forms a new intraspinal circuit in adult rats. *Nat. Neurosci.* **7:** 269–277.

42. Ballermann, M. & K. Fouad. 2006. Spontaneous locomotor recovery in spinal cord injured rats is accompanied by anatomical plasticity of reticulospinal fibers. *Eur. J. Neurosci.* **23:** 1988–1996.

43. Ghosh, A., F. Haiss, E. Sydekum, *et al.* 2010. Rewiring of hindlimb corticospinal neurons after spinal cord injury. *Nat. Neurosci.* **13:** 97–104.

44. Bélanger, M., T. Drew, J. Provencher & S. Rossignol. 1996. A comparison of treadmill locomotion in adult cats before and after spinal transection. *J. Neurophysiol.* **76:** 471–491.

45. Martinez, M., H. Delivet-Mongrain, H. Leblond & S. Rossignol. 2012. Effect of locomotor training in completely spinalized cats previously submitted to a spinal hemisection. *J. Neurosci.* **32:** 10961–10970.

46. Frigon, A., G. Barriere, H. Leblond & S. Rossignol. 2009. Asymmetric changes in cutaneous reflexes after a partial spinal lesion and retention following spinalization during locomotion in the cat. *J. Neurophysiol.* **102:** 2667–2680.

47. Dimitrijevic, M.R., Y. Gerasimenko & M.M. Pinter. 1998. Evidence for a spinal central pattern generator in humans. *Ann. N.Y. Acad. Sci.* **860:** 360–376.

48. Harkema, S.J. 2008. Plasticity of interneuronal networks of the functionally isolated human spinal cord. *Brain Res. Rev.* **57:** 255–264.

49. Nadeau, S., G. Jacquemin, C. Fournier, *et al.* 2010. Spontaneous motor rhythms of the back and legs in a patient with a complete spinal cord transection. *Neurorehabil. Neural Repair* **24:** 377–383.

50. Calancie, B., N. Alexeeva, J.G. Broton & M.R. Molano. 2005. Interlimb reflex activity after spinal cord injury in man: strengthening response patterns are consistent with ongoing synaptic plasticity. *Clin. Neurophysiol.* **116:** 75–86.

Ann. N.Y. Acad. Sci. ISSN 0077-8923

ANNALS OF THE NEW YORK ACADEMY OF SCIENCES
Issue: *Neurons, Circuitry, and Plasticity in the Spinal Cord and Brainstem*

The effects of endocannabinoid signaling on network activity in developing and motor circuits

Peter Wenner

Department of Physiology, Emory University School of Medicine, Atlanta, Georgia

Address for correspondence: Peter Wenner, Department of Physiology, Emory University School of Medicine, 615 Michael Street, Atlanta, GA 30322-1047. pwenner@emory.edu

Endocannabinoid signaling typically mediates a form of synaptic plasticity in which a postsynaptic cell acts retrogradely to reduce vesicle release from presynaptic terminals impinging on that cell. In the embryonic spinal cord, endocannabinoids inhibit spontaneously released glutamatergic vesicles in both a brief and ongoing tonic manner. Together these endocannabinoid-mediated forms of synaptic regulation appear to play an important role in regulating the frequency of a form of spontaneous network activity (SNA) that is expressed in the embryonic spinal cord. Because of the importance of SNA to the maturation of the developing network, fetal exposure to drugs that influence endocannabinoid signaling may have profound effects on spinal maturation. In this review, endocannabinoid signaling in the embryonic spinal cord is described and compared to signaling in the mature lamprey spinal cord as well as in the developing hippocampal network, which expresses a form of SNA.

Keywords: endocannabinoids; activity; network

Introduction

Strong depolarizations and metabotropic signaling trigger the synthesis and release of endocannabinoids (2-arachidonoyl glycerol or 2-AG, anandamide or AEA), which retrogradely activate a presynaptic G protein–coupled cannabinoid receptor (CB1) and thus inhibit vesicle release.[1–4] In the embryonic spinal cord, endocannabinoids inhibit spontaneous glutamatergic vesicle release and act to reduce the frequency of spinal spontaneous network activity (SNA) *in vitro* and *in vivo*.[5] Endocannabinoids have also been reported to influence a form of SNA that is expressed in the developing hippocampus.[6] Here we compare endocannabinoid signaling in two different networks at stages when GABA is depolarizing and supports the expression of SNA.[7–9] The comparison demonstrates similar roles for endocannabinoids in distinct networks at this early stage of development. Endocannabinoids also influence network activity in the mature spinal cord.[3,10–12] Comparisons are also made among the roles of endocannabinoid signaling in the developing and mature spinal network, in order to investigate how such signaling may transition from nascent to adult circuitry.

Embryonic movements are generated by spontaneous episodic bursts of spiking activity in the developing spinal network.[13] This spontaneous network activity (SNA) results from the highly excitable nature of the spinal synaptic circuitry, as both GABAergic and AMPAergic currents are depolarizing and excitatory at this early developmental stage.[7,14,15] SNA is expressed in most developing networks, suggesting a fundamental importance of this form of activity in the maturation of these neuronal circuits.[16,17] In the embryonic spinal cord, SNA has been shown to be important for the proper development of the limbs and motoneuron axon pathfinding.[13,18] Further, we have recently shown that SNA is important in the regulation of synaptic strength through a form of homeostatic synaptic plasticity.[19] While it is becoming clear that SNA is important for several aspects of development, it is less clear what regulates the expression of this activity. Previous reports have shown that both spontaneous and evoked synaptic responses were depressed in spinal neurons after an episode of

doi: 10.1111/nyas.12068

SNA, but then recovered in the interepisode interval (IEI).[5, 20, 21] This modulation of synaptic strength seems to be important in regulating the frequency of episodes of SNA. The postepisode depression and recovery of glutamatergic synaptic strength appears to be at least partially mediated by changes in presynaptic vesicle release.

The possibility that endocannabinoids play roles in this SNA-dependent glutamatergic synaptic modulation was tested in our recent study.[5] Endocannabinoid-mediated synaptic plasticity typically involves a triggering stimulus such as a strong depolarization and/or metabotropic signaling, both of which occur during episodes of SNA. We predicted that during these bursts of activity, endocannabinoids would be postsynaptically released, activate CB1 receptors on presynaptic terminals, and thereby inhibit glutamatergic vesicle release. Once the endocannabinoids were cleared from the extracellular space during the inter-episode interval, the endocannabinoid-mediated inhibition would be relieved. This negative feedback system could have explained the SNA-dependent glutamatergic modulation seen in the embryonic spinal cord. Although this particular prediction proved to be incorrect, we did find evidence for significant endocannabinoid signaling in embryonic motoneurons.

Endocannabinoid signaling in the embryonic spinal cord

First, it was confirmed that CB1 receptors were expressed in the embryonic spinal cord using Western blots.[5] Bath application of various endocannabinoid modulators was used to test the influence of endocannabinoid signaling in the isolated embryonic day 10 (E10) spinal cord preparation of the chick.[5] Activation of CB1 receptors with an agonist (WIN 55,212-2) reduced glutamatergic miniature postsynaptic current (mPSC) frequency in spinal motoneurons (see Fig. 1 legend for description of mPSC isolation). On the other hand, CB1 activation had no effect on glutamatergic mPSC amplitude or the amplitude or frequency of GABAergic mPSCs (Fig. 1A–C). The results suggest that endocannabinoid signaling can reduce spontaneous glutamatergic vesicle release (mEPSC frequency) in a fairly specific fashion. Conversely, when the CB1 receptor was blocked with an inverse agonist (AM 251), mEPSC frequency was significantly increased; like the agonist, the in-

verse antagonist had no effect on glutamatergic mPSC amplitude or GABAergic mPSCs (Fig. 1D–F). This result suggested that ongoing glutamatergic vesicle release was tonically inhibited by CB1 activation. Tonic activation of CB1 receptors could be mediated by an ongoing release of endocannabinoids or through a constitutively active CB1 receptor independent of endocannabinoid activation. Bath application of drugs (AM404, LY 2183240) that lead to a buildup of endogenously released endocannabinoids reduced mEPSC frequency, suggesting that endocannabinoids were constitutively released and thus responsible for the ongoing activation of CB1 receptors.[5] It has recently been proposed that AEA may mediate the tonic suppression of GABAergic vesicle release in the hippocampus.[22]

In addition to this tonically active suppression of spontaneous glutamatergic release, it was discovered that glutamatergic mPSC frequency was also temporarily reduced following strong, one-minute depolarizations of the motoneuron (Fig. 1G–H). The depolarization-induced enhancement of endocannabinoid signaling only lasted for the first minute after the depolarization. This transient inhibition of mPSC frequency through endocannabinoid signaling has been described previously in several different systems.[22–25]

The results described previously suggest that endocannabinoid signaling could regulate spontaneous glutamatergic vesicle release (mEPSC frequency). To determine whether endocannabinoid signaling mediated the reduction of glutamatergic mPSC frequency observed following an episode of SNA, mEPSCs were recorded in motoneurons in the presence of a CB1 inverse agonist. Surprisingly, blockade of CB1 signaling had no effect on the modulation of mEPSC frequency by episodes of SNA, although the overall frequency was increased.[5] The finding demonstrated that the SNA-evoked modulation of mEPSC frequency was not mediated by endocannabinoids.

The discussion thus far has only considered the role of endocannabinoids on spontaneous vesicle release, but most of the work on endocannabinoid signaling focuses on its inhibition of evoked (action potential–dependent) vesicle release. Therefore, it remained possible that modulation of evoked responses in the IEI could be mediated by endocannabinoids. Glutamatergic-evoked responses were measured through extracellular recordings of

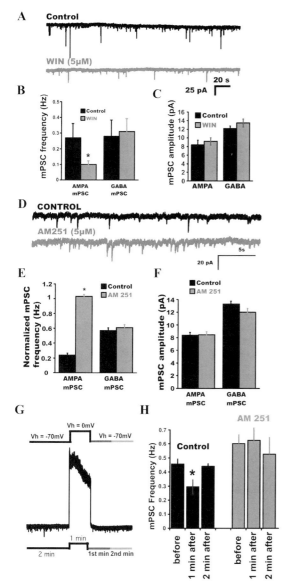

the ventral root. These recordings measure the electrotonically degraded motoneuron response following stimulation of inputs. The dorsal root (sensory neurons) or ventrolateral funiculus (VLF, interneurons) were alternately stimulated and pharmacologically isolated glutamatergic responses were measured (Fig. 2A). Following stimulation of the VLF there was a clear SNA-dependent depression and recovery of the ventral root response in the IEI. Surprisingly, CB1 inhibition had no effect on amplitude modulation or average amplitude of the VLF-evoked response (Fig. 2B). Similarly, blocking CB1 receptors had no effect on the amplitude of the dorsal root–evoked glutamatergic responses[20] (Fig. 2C). The results suggest that endocannabinoids in the embryonic spinal cord regulate spontaneous release of glutamatergic vesicles, but do not influence evoked release or the SNA-dependent modulation of glutamatergic currents. How is it possible that endocannabinoids regulate spontaneous vesicle release, but not evoked release? The biochemical cascades that mediate spontaneous and evoked release in the E10 embryonic motoneuron appear to be distinct, as evoked release was found to be calcium dependent, while spontaneous release was not affected by calcium levels.[5] Therefore, it is likely that CB1 activation targets the calcium-independent pathway for spontaneous release independent of the evoked-release pathway.

The regulation of spontaneous glutamatergic vesicle release by endocannabinoid signaling suggests that there may be a special role for mEPSCs. Previously, it had been established that episodes of SNA were triggered by spontaneously occurring postsynaptic potentials, similar in size to

Figure 1. Effects of endocannabinoid signaling on spontaneous vesicle release in embryonic motoneurons. (A) Representative traces of AMPA mPSCs from a motoneuron before (black trace) and 20 minutes after bath addition of the CB1 receptor agonist WIN (5 μM, gray trace). (B–C) Summary of the effects of WIN (1–10 μM) on AMPA and GABA mPSC frequency (B) and amplitude (C). AMPA and GABA mPSCs were isolated on the basis of their faster and slower kinetics, respectively. Glycinergic mPSCs have not been detected in motoneurons at this stage in the chick embryo, and NMDA mPSCs are not observed at the −70 mV holding potential used in the study.[19] (D) Representative traces of mPSCs from an E10 motoneuron before (black trace) and after bath application of the CB1 inverse agonist AM 251 (5 μM, gray trace). (E–F) Summary of the effects of AM 251 on mPSC frequency (E) and amplitude (F). (G) Schematic representation of the protocol followed for depolarization-induced

suppression of mEPSC frequency. AMPA mPSC frequency was measured in the presence of TTX for two minutes, holding the cell at −70 mV, and delivering a 70 mV step to 0 mV for one minute. AMPA mPSC frequency was then measured in the first and second minute after depolarization. (H) Summary of the effect of 0 mV step on AMPA mPSC frequency in the first and second minute following depolarization. A significant reduction in AMPA mPSC frequency was observed in the first, but not second, minute in control conditions (black bars). Bath application of AM 251 (gray bars) prevented this form of plasticity. Baseline values for mEPSC frequency were quite variable, and this variability can also be seen in E. Further, the data in E were collected more recently, using a patch clamp amplifier with better noise characteristics than that used for the data of H. Figure is modified from Ref. 5.

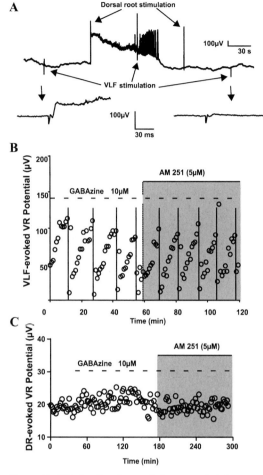

Figure 2. CB1 signaling does not influence evoked glutamatergic responses in embryonic motoneurons. (A) Ventral root recording showing an episode of SNA and the alternation of dorsal root and VLF stimulation. Expanded traces of ventral root potentials evoked by VLF stimulation before and after an episode of SNA are shown below. (B) Plot of the amplitudes of VLF-evoked glutamatergic responses in motoneurons, as measured from the ventral root, in the presence of a GABA_A antagonist, gabazine. The responses modulate in an episode-dependent manner, dropping after an episode (vertical lines) and recovering in the IEI. Application of AM 251 had no effect on these responses. (C) DR-evoked responses were not modulated by SNA and were unaffected following bath application of AM251. Figure is modified from Ref. 5.

mEPSCs, which drove motoneurons to fire action potentials.[26] Therefore, it was possible that reducing mEPSCs would lead to a reduction in the frequency of episodes of SNA. To test this possibility, CB1 signaling was blocked in an *in vitro* spinal cord preparation, and this increased SNA frequency (Fig. 3A). More importantly, when CB1 signaling was blocked in the living embryo by *in ovo* injection

of the CB1 inverse agonist, an increase in embryonic movements was observed (Fig. 3B). The findings are consistent with the idea that tonic CB1 activation reduces the frequency of mEPSCs in motoneurons, thereby reducing the likelihood of triggering an episode of SNA. Alternatively, or in addition to this mechanism, the increase in SNA frequency could result from CB1 activation, influencing release through different mechanisms in other parts of the circuitry.

Overall, these results demonstrate two forms of endocannabinoid signaling. The first is likely associated with large depolarizations that occur during SNA, but are only short lived. The second form of endocannabinoid signaling appears to be independent of SNA, acting to tonically suppress spontaneous, but not evoked, glutamatergic vesicle release. The inhibition of spontaneous glutamatergic release appears to continually suppress the frequency of SNA *in vitro* and *in vivo*.

Endocannabinoid signaling in the mature spinal cord

Studies in the lamprey demonstrate the importance of endocannabinoid signaling in synaptic modulation and the consequent effect on rhythmic activity in the mature spinal cord.[3,27] El Manira and collaborators demonstrated that CB1 receptor signaling acts to increase the burst frequency of locomotor-like activity (LLA) by suppressing inhibitory glycinergic inputs to motor- and interneurons while enhancing excitatory glutamatergic inputs.[10–12] The modulation of rhythmic inhibitory and excitatory synaptic drive in motoneurons occurred in the presence of ongoing network activity associated with LLA.[11,12] Endocannabinoid-dependent signaling was also demonstrated in the absence of LLA; evoked monosynaptic inhibitory currents were shown to be under tonic suppression by CB1 signaling.[10] The endocannabinoid 2-AG is thought to be released following metabotropic glutamate receptor (mGluR) activation of the postsynaptic cell, and to subsequently act to inhibit glycinergic and enhance glutamatergic inputs.[12] Presumably, glycinergic inhibition occurs in the classical manner, where 2-AG activates CB1 receptors on the presynaptic terminal, leading to reduced calcium entry into the terminal and consequently reduced action potential–dependent vesicle release. It is less clear how endocannabinoids enhance glutamatergic

                                                                              Ann. N.Y. Acad. Sci. 1279 (2013) 135–142 © 2013 New York Academy of Sciences.

A

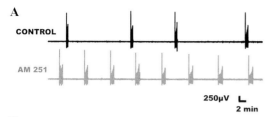

B

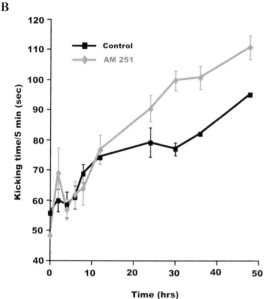

Figure 3. Endocannabinoid signaling reduces SNA in the chick embryo spinal cord *in vitro* and *in vivo*. (A) Example traces of ventral root recordings showing the expression of episodes of SNA before (control, black) and after the addition of a CB1 antagonist (gray). (B) Plot of the average embryonic kicking time *in ovo* in 5-min observation periods during the two-day treatment with vehicle (DMSO black) and 5 μM AM 251 dissolved in DMSO (gray). Limb movements started to increase in AM 251-treated embryos 12 hours after injection, and remained more frequent than controls during the rest of the two-day long treatment. Figure is modified from Ref. 5.

transmission, but it is dependent on nitric oxide signaling for its expression.[11]

Certain features of endocannabinoid signaling in the embryonic chick and the adult lamprey spinal cord were similar. In both systems, endocannabinoids modulate glutamatergic transmission and can be tonically active when the network is inactive.[5,10–12] On the other hand, the differences in endocannabinoid signaling between the immature and adult cord are striking.[3,5,10–12,27] CB1 signaling increases burst frequency in the adult and decreases it in the embryo. In the mature cord, endocannabinoid signaling has a profound effect

on inhibitory transmission (glycinergic). In the embryo, endocannabinoids do not influence GABAergic mPSCs, although it is possible that they influence evoked GABAergic responses, which remain untested.[5] In the chick embryo, endocannabinoids reduce spontaneously released glutamate vesicles via a calcium-independent mechanism, but have no effect on evoked release. Alternatively, in the lamprey spinal cord, endocannabinoids enhance action potential-evoked glutamatergic transmission.

The mechanisms that lead to the release of endocannabinoids in embryonic and mature spinal cord show some similarities, but differences may also exist. In the embryonic cord, endocannabinoids are briefly (seconds) released following the kind of depolarization that would occur during SNA, but it is not clear whether this happens in the lamprey. In addition to this depolarization-dependent release in the embryo, there is a constitutive release whose regulation is not understood, but is known to be independent of SNA. In the mature lamprey spinal cord there also appears to be a constitutive release of endocannabinoids that is independent of LLA, but there is also an mGluR-dependent release that is associated with LLA.[10]

The results of studies on endocannabinoid signaling in developing and mature spinal networks raise several interesting questions. Is there an mGluR-dependent modulation of endocannabinoid signaling in embryos as there is in the adult lamprey cord? Is there a depolarization-induced release of endocannabinoids in lamprey spinal neurons as there is in the embryo? Are the differences in these two systems bridged during development? It may be that endocannabinoids inhibit glycinergic transmission in the adult through the later expression of CB1 receptors on glycinergic terminals. Several studies suggest a partial transition from GABAergic to glycinergic neurotransmission in the ventral cord during spinal maturation.[28–31] It would be interesting to know whether the expression of CB1 receptors in inhibitory interneurons occurs at the same time as the GABA to glycine transition or the depolarizing to hyperpolarizing transition in chloride currents that starts slightly earlier in development. The development of endocannabinoid-mediated inhibition of evoked release is likely to come about because the signaling machinery downstream of the CB1 receptor is different in adult glycinergic terminals (release dependent on calcium) than in embryonic glutamatergic terminals (release is calcium-independent). Clearly, if these transitions were made during maturation they could at least partially explain how endocannabinoids change roles from inhibiting network activity in the embryo to exciting network activity in the adult.

It is possible that some of the differences are species specific; however, it is clear that embryonic and mature spinal networks go through a profound reconfiguration as excitatory GABAergic transmission in the embryo is converted to inhibitory glycinergic transmission later in development. Whether or not there is a transition in endocannabinoid signaling from embryonic to adult spinal networks, it will be important to understand how such signaling influences spinal network activity in the period between embryo and adult, as the neonatal-isolated spinal cord preparation has become a favored model system for locomotor studies.[32–34]

Endocannabinoid signaling in the developing hippocampus

SNA occurs in many networks during the period that GABA is depolarizing and excitatory.[16,35] As in the spinal cord, SNA in other circuits occurs episodically and is driven, in part, by depolarizing GABAergic currents. Giant depolarizing potentials (GDPs) are the hippocampal equivalent of spinal SNA, and occur in the rodent through the first neonatal week.[8,36] Interestingly, endocannabinoids have been implicated in the expression of GDPs.[6] In the CA1 region of the P5–6 hippocampal slice, strong depolarizations in pyramidal cells or interneurons produce an inhibition of evoked GABAergic inputs onto these cells. This depolarization-induced suppression of GABAergic currents last for seconds and is mediated by the release of endocannabinoids from the postsynaptic cell, which activate CB1 receptors on GABAergic presynaptic terminals; this reduces action potential-mediated vesicle release. While cellular depolarizations appear to cause short-lived CB1 activation, the authors also found evidence for an ongoing or tonic CB1 activation, which appears to be a developmental phenomenon as it is not observed in the more mature hippocampus.[24] Because GABA is depolarizing and excitatory at this stage of development, endocannabinoid-mediated suppression of the GABAergic drive resulted in less frequent GDPs. This was shown both *in vitro* and *in vivo*.[6]

There are numerous similarities between endocannabinoid signaling in the developing hippocampus and spinal cord.[5,6] Both describe an endocannabinoid-dependent inhibition of the release of vesicles of an excitatory neurotransmitter following strong depolarizations of the postsynaptic neuron. These reductions in transmission are transient, lasting only seconds. Such signaling is likely to occur during network activity in both systems, and could function as a negative feedback system by contributing to the termination of the bouts of activity (GDP in the hippocampus, SNA in the spinal cord). In addition to this depolarization-induced short-term suppression of transmission, there also appears to be a tonic signaling of the CB1 receptor in both systems. In the embryonic spinal cord this baseline endocannabinoid signaling is dependent on the tonic release of endocannabinoids but independent of SNA, as it occurs in the complete absence of spiking activity following TTX application (Fig. 1H, compare "before" responses in the absence and presence of AM 251). It is not clear whether tonic endocannabinoid signaling in the neonatal hippocampus is dependent on network or spiking activity. In both systems, CB1 antagonists increase neurotransmission, whereas drugs that inhibit the clearance of endogenously released endocannabinoids reduce neurotransmission. Finally, in both of these developing circuits, endocannabinoid signaling acts to reduce network activity both *in vitro* and *in vivo*, suggesting an important role for endocannabinoid signaling in an ongoing suppression of network activity.

While the general principles of CB1 signaling in the two systems are similar, some of the specific details are different in the spinal and hippocampal tissues. Although endocannabinoid signaling inhibits excitatory neurotransmission in both systems, the actual transmitter that is presynaptically inhibited is different (GABA in the hippocampus, glutamate in the spinal cord). However, because they were not examined in embryonic motoneurons, we cannot rule out the possibility that endocannabinoids actually do influence evoked GABAergic responses. Further, the presynaptic mechanism of action appears to be distinct. In the hippocampus, endocannabinoid signaling inhibits transmission by reducing action potential-dependent vesicle release, which is likely calcium dependent.[6] In the spinal cord, CB1 signaling has no effect on evoked glutamatergic release but reduces spontaneous vesicle release in a calcium-independent manner.[5]

These studies raise many new questions and open avenues for research. It will be important to determine what triggers the tonic CB1 activation that occurs in these systems. It is tempting to speculate that there may be a constitutive release of endocannabinoids. If this is the case then it will be of great interest to recognize how such release is regulated. Endocannabinoids seem to be well positioned to regulate network activity in a diverse array of circuits that express SNA during development. SNA is important in the maturation of synaptic strength, cellular excitability, axon pathfinding, and much more. Therefore, it will be critical to understand how CB1 signaling regulates such activity, so that we can appreciate normal development, but also the consequences of perturbing endocannabinoid signaling in developing networks.

Conflicts of interest

The author declares no conflicts of interest.

References

1. Alger, B.E. 2009. Endocannabinoid signaling in neural plasticity. *Curr. Top. Behav. Neurosci.* **1:** 141–172.
2. Heifets, B.D. & P.E. Castillo. 2009. Endocannabinoid signaling and long-term synaptic plasticity. *Annu. Rev. Physiol.* **71:** 283–306.
3. El Manira, A. & A. Kyriakatos. 2010. The role of endocannabinoid signaling in motor control. *Physiology* **25:** 230–238.
4. Ohno-Shosaku, T. *et al.* 2012. Endocannabinoids and retrograde modulation of synaptic transmission. *Neuroscientist* **18:** 119–132.
5. Gonzalez-Islas, C.A., M.A. Garcia-Bereguiain & P. Wenner. 2012. Tonic and transient endocannabinoid regulation of AMPAergic mPSCs and homeostatic plasticity in embryonic motor networks. *J. Neurosci.* **32:** 13597–13607.
6. Bernard, C. *et al.* 2005. Altering cannabinoid signaling during development disrupts neuronal activity. *Proc. Natl. Acad. Sci. USA* **102:** 9388–9393.
7. Chub, N. & M.J. O'Donovan. 2001. Post-episode depression of GABAergic transmission in spinal neurons of the chick embryo. *J. Neurophysiol.* **85:** 2166–2176.
8. Ben-Ari, Y. *et al.* 1989. Giant synaptic potentials in immature rat CA3 hippocampal neurones. *J. Physiol.* **416:** 303–325.
9. Ben-Ari, Y. *et al.* 2012. Refuting the challenges of the developmental shift of polarity of GABA actions: GABA more exciting than ever! *Front Cell Neurosci.* **6:** 35.
10. Kettunen, P. *et al.* 2005. Neuromodulation via conditional release of endocannabinoids in the spinal locomotor network. *Neuron* **45:** 95–104.
11. Kyriakatos, A. & A. El Manira. 2007. Long-term plasticity of the spinal locomotor circuitry mediated by endocannabinoid and nitric oxide signaling. *J. Neurosci.* **27:** 12664–12674.
12. Song, J., A. Kyriakatos & A. El Manira. 2012. Gating the polarity of endocannabinoid-mediated synaptic plasticity by nitric oxide in the spinal locomotor network. *J. Neurosci.* **32:** 5097–5105.
13. O'Donovan, M.J. *et al.* 1998. Mechanisms of spontaneous activity in the developing spinal cord and their relevance to locomotion. *Ann. N.Y. Acad. Sci.* **860:** 130–141.
14. Wilhelm, J.C. & P. Wenner. 2008. GABAA transmission is a critical step in the process of triggering homeostatic increases in quantal amplitude. *Proc. Natl. Acad. Sci. USA* **105:** 11412–11417.
15. Gonzalez-Islas, C., N. Chub & P. Wenner. 2009. NKCC1 and AE3 appear to accumulate chloride in embryonic motoneurons. *J. Neurophysiol.* **101:** 507–518.
16. Blankenship, A.G. & M.B. Feller. 2010. Mechanisms underlying spontaneous patterned activity in developing neural circuits. *Nat. Rev. Neurosci.* **11:** 18–29.
17. O'Donovan, M.J., N. Chub & P. Wenner. 1998. Mechanisms of spontaneous activity in developing spinal networks. *J. Neurobiol.* **37:** 131–145.
18. Gonzalez-Islas, C. & P. Wenner. 2010. Role of spontaneous activity in the maturation of GABAergic synapses in spinal circuits. In *Developmental Plasticity of Inhibitory Circuitry.* Springer-Verlag. New York.
19. Gonzalez-Islas, C. & P. Wenner. 2006. Spontaneous network activity in the embryonic spinal cord regulates AMPAergic and GABAergic synaptic strength. *Neuron* **49:** 563–575.
20. Fedirchuk, B. *et al.* 1999. Spontaneous network activity transiently depresses synaptic transmission in the embryonic chick spinal cord. *J. Neurosci.* **19:** 2102–2112.
21. Tabak, J., J. Rinzel & M.J. O'Donovan. 2001. The role of activity-dependent network depression in the expression and self-regulation of spontaneous activity in the developing spinal cord. *J. Neurosci.* **21:** 8966–8978.
22. Alger, B.E. 2012. Endocannabinoids at the synapse a decade after the dies mirabilis (29 March 2001): what we still do not know. *J. Physiol.* **590**(Pt 10): 2203–2212.
23. Ohno-Shosaku, T., T. Maejima & M. Kano. 2001. Endogenous cannabinoids mediate retrograde signals from depolarized postsynaptic neurons to presynaptic terminals. *Neuron* **29:** 729–738.
24. Wilson, R.I. & R.A. Nicoll. 2001. Endogenous cannabinoids mediate retrograde signalling at hippocampal synapses. *Nature* **410:** 588–592.
25. Kreitzer, A.C. & W.G. Regehr. 2001. Retrograde inhibition of presynaptic calcium influx by endogenous cannabinoids at excitatory synapses onto Purkinje cells. *Neuron* **29:** 717–727.
26. Wenner, P. & M.J. O'Donovan. 2001. Mechanisms that initiate spontaneous network activity in the developing chick spinal cord. *J. Neurophysiol.* **86:** 1481–1498.
27. El Manira, A. *et al.* 2008. Endocannabinoid signaling in the spinal locomotor circuitry. *Brain Res. Rev.* **57:** 29–36.
28. Berki, A.C., M.J. O'Donovan & M. Antal. 1995. Developmental expression of glycine immunoreactivity and its colocalization with GABA in the embryonic chick lumbosacral spinal cord. *J. Comp. Neurol.* **362:** 583–596.

29. Gao, B.X. & L. Ziskind-Conhaim. 1995. Development of glycine- and GABA-gated currents in rat spinal motoneurons. *J. Neurophysiol.* **74:** 113–121.

30. Gao, B.X., C. Stricker & L. Ziskind-Conhaim. 2001. Transition from GABAergic to glycinergic synaptic transmission in newly formed spinal networks. *J. Neurophysiol.* **86:** 492–502.

31. Allain, A.E. *et al.* 2010. Serotonin controls the maturation of the GABA phenotype in the ventral spinal cord via 5-HT1b receptors. *Ann. N. Y. Acad. Sci.* **1198:** 208–219.

32. Smith, J.C. & J.L. Feldman. 1987. In vitro brainstem-spinal cord preparations for study of motor systems for mammalian respiration and locomotion. *J. Neurosci. Methods* **21:** 321–333.

33. Smith, J.C., J.L. Feldman & B.J. Schmidt. 1988. Neural mechanisms generating locomotion studied in mammalian brain stem-spinal cord in vitro. *Faseb J.* **2:** 2283–2288.

34. Bagust, J. & G.A. Kerkut. 1981. An in vitro preparation of the spinal cord of the mouse. In *Electrophysiology of Isolated Mammalian CNS Preparations.* G.A. Kerkut & H.V. Wheal, Eds.: 337–365. London: Academic Press.

35. O'Donovan, M.J. 1999. The origin of spontaneous activity in developing networks of the vertebrate nervous system. *Curr. Opin. Neurobiol.* **9:** 94–104.

36. Ben-Ari, Y. *et al.* 2007. GABA: a pioneer transmitter that excites immature neurons and generates primitive oscillations. *Physiol. Rev.* **87:** 1215–1284.

Ann. N.Y. Acad. Sci. ISSN 0077-8923

ANNALS OF THE NEW YORK ACADEMY OF SCIENCES
Issue: *Neurons, Circuitry, and Plasticity in the Spinal Cord and Brainstem*

Hypoxia-induced phrenic long-term facilitation: emergent properties

Michael J. Devinney, Adrianne G. Huxtable, Nicole L. Nichols, and Gordon S. Mitchell

Department of Comparative Biosciences, University of Wisconsin–Madison, Madison, Wisconsin

Address for correspondence: Gordon S. Mitchell, Department of Comparative Biosciences, University of Wisconsin, 2015 Linden Drive, Madison, WI 53706. mitchell@svm.vetmed.wisc.edu

As in other neural systems, plasticity is a hallmark of the neural system controlling breathing. One spinal mechanism of respiratory plasticity is phrenic long-term facilitation (pLTF) following acute intermittent hypoxia. Although cellular mechanisms giving rise to pLTF occur within the phrenic motor nucleus, different signaling cascades elicit pLTF under different conditions. These cascades, referred to as Q and S pathways to phrenic motor facilitation (pMF), interact via cross-talk inhibition. Whereas the Q pathway dominates pLTF after mild to moderate hypoxic episodes, the S pathway dominates after severe hypoxic episodes. The biological significance of multiple pathways to pMF is unknown. This review will discuss the possibility that interactions between pathways confer emergent properties to pLTF, including pattern sensitivity and metaplasticity. Understanding these mechanisms and their interactions may enable us to optimize intermittent hypoxia-induced plasticity as a treatment for patients that suffer from ventilatory impairment or other motor deficits.

Keywords: plasticity; motor neuron; intermittent hypoxia; phrenic nerve; pattern sensitivity; metaplasticity

Introduction

The respiratory control system has historically been viewed as fixed and immutable, controlled primarily via negative feedback from sensory receptors.[1] This view was held, in large part, because breathing is an automatic, often subconscious, motor behavior. However, control systems governed by negative feedback alone are frequently unstable due to inappropriate reflex gain.[2] Systems that preserve homeostasis through strong negative feedback loops are vulnerable to insults; robust control systems optimize for worst-case scenarios (e.g., ventilatory failure) and incorporate mechanisms to prevent such failure when the system is challenged. Plasticity is one key property of neural systems, including the respiratory control system, that promotes robust and effective homeostatic regulation.[1] In this brief review, a general definition of respiratory plasticity will be used, namely, a change in future system performance (e.g., breathing or blood gas regulation) based on experience.[1]

Plasticity in the neural system controlling breathing has only been widely appreciated for the past few decades.[1,3–5] Recently, the field of respiratory neuroplasticity has grown considerably; plasticity has been discovered at the neuromuscular,[6] chemoreceptor,[7] spinal,[8] and brainstem[9] levels of respiratory control, and our knowledge concerning these forms of plasticity is increasing at a rapid pace. Here, the focus will be on a single, widely studied model of spinal respiratory plasticity, *phrenic long-term facilitation* (pLTF), characterized by a persistent increase in phrenic motor output lasting hours after a few brief episodes of low oxygen, or acute intermittent hypoxia (AIH; see Refs. 4–6). Considerable progress has been made toward understanding cellular and network mechanisms giving rise to pLTF.[10, 11] For example, one recent discovery showed that multiple distinct cellular cascades give rise to similar phenotypic plasticity.[10] An important question then is, why do these multiple pathways exist? In this brief review, the potential advantages conferred by this complexity will be considered. Specifically, we present the hypotheses that the existence of

doi: 10.1111/nyas.12085

multiple interacting pathways confers two emergent properties of pLTF: pattern sensitivity and metaplasticity.

Phrenic long-term facilitation

Millhorn *et al.* originally demonstrated that episodic carotid sinus nerve stimulation elicits a long-lasting increase in phrenic motor output in anesthetized cats.[3,4] We subsequently demonstrated a similar phenomenon, pLTF, in anesthetized rats following three brief hypoxic episodes.[12,13] Following AIH, pLTF is expressed as a prolonged increase in phrenic nerve burst amplitude lasting several hours after the final hypoxic episode (Fig. 1A). Phrenic LTF is a form of serotonin- and protein synthesis–dependent spinal plasticity[12,14,15] whose induction is independent of increased phrenic nerve activity[16,17] and is a form of neuromodulator-induced plasticity[1] distinct from conventional forms of activity-dependent synaptic plasticity, such as hippocampal long-term potentiation (LTP).[18]

One hallmark of pLTF is pattern sensitivity; pLTF is elicited by intermittent, but not a single period of acute sustained hypoxia (ASH) with the same cumulative duration (Fig. 1B).[19] Although similar pattern sensitivity is shared by many forms of neuroplasticity,[20–24] we have little understanding of how such pattern sensitivity arises in any system. Another property of pLTF that is not yet fully understood is its ability to express metaplasticity (i.e., the ability of prior experience to alter subsequent plasticity; see subsequent section);[1] for example, following cervical dorsal rhizotomy[25] or chronic intermittent hypoxia (CIH),[26] subsequent responses to AIH (i.e., pLTF) are amplified.

Although our understanding of mechanisms giving rise to respiratory plasticity remains incomplete, considerable progress has been made.[10,11,27,28] For example, multiple, distinct cellular cascades have been shown to exist, each capable of eliciting long-lasting phrenic motor facilitation (pMF; a general term describing augmented phrenic burst amplitude that includes pLTF).[10] These pathways interact in interesting and complex ways, possibly increasing flexibility as the respiratory control system responds to diverse challenges throughout life. Knowledge of the diverse mechanisms giving rise to pMF and their implications will advance our understanding of pattern sensitivity and metaplasticity in other forms of neuroplasticity, and will be essential as we begin to

harness the potential of AIH-induced spinal plasticity to treat severe clinical disorders that impair breathing, such as spinal cord injury[29–31] or amyotrophic lateral sclerosis (ALS).[32]

Multiple cellular mechanisms of pLTF: the *Q* and *S* pathways

Phrenic LTF is frequently studied in anesthetized, paralyzed, and ventilated rats—administered a standardized AIH protocol that we refer to as *moderate AIH* (three 5-min hypoxic episodes; PaO$_2$ 35–45 mmHg; 5-min intervals; see Ref. 25). Following moderate AIH, pLTF requires spinal serotonin type 2 (5-HT$_2$) receptor activation,[12,14,15] new synthesis of brain-derived neurotrophic factor (BDNF),[33] activation of the high-affinity tyrosine kinase B (TrkB) receptor,[34] and ERK MAP kinase signaling.[35] Because 5-HT$_2$ receptors are G$_q$ protein–coupled receptors, we refer to this as the *Q pathway* to pMF (Fig. 2A). Pharmacological activation of other G$_q$-coupled receptors (such as α_1 receptors) elicit similar pMF.[10]

A distinct pathway to pMF relies on activation of G$_s$ protein–coupled metabotropic receptors, such as adenosine 2A (A$_{2A}$)[36] and 5-HT$_7$ receptors.[37] This form of pMF requires new synthesis of an immature TrkB isoform (not BDNF) and PI3 kinase/Akt signaling (not ERK).[36] We refer to this as the *S* pathway to pMF, since multiple G$_s$ protein–coupled metabotropic receptors elicit the same mechanism (Fig. 2B).[13] Contrary to our initial expectation, the *S* pathway does not contribute to pLTF following moderate AIH,[38] but dominates pLTF following severe AIH.[39] In fact, the *S* pathway negatively regulates pLTF following moderate AIH, demonstrating inhibitory interactions between pathways;[38] we proposed from these findings that the predominant interaction between pathways is characterized by *cross-talk inhibition*. Although the mechanistic basis of cross-talk inhibition is not fully understood, *S* to *Q* inhibition may require protein kinase A (PKA) activity (Hoffman and Mitchell, unpublished observations).

During severe AIH, the *S* pathway is activated to a greater extent, and dominates pLTF.[39] Rats exposed to severe AIH exhibit pLTF that is phenotypically similar to that in rats exposed to moderate AIH, but via an A$_{2A}$ receptor–dependent (serotonin-independent) mechanism (i.e., *S* pathway).[39] This finding suggests that cross-talk inhibition between

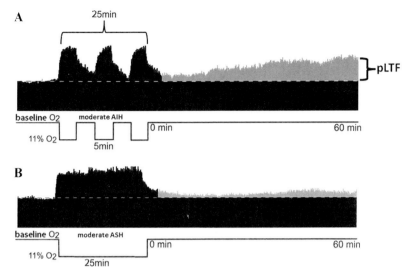

Figure 1. Pattern sensitivity of phrenic long-term facilitation (pLTF). The pLTF is elicited by moderate, acute intermittent hypoxia (AIH), but not by moderate, acute sustained hypoxia (ASH). (A) Representative tracing of phrenic nerve activity from an anesthetized, vagotomized, paralyzed, and pump-ventilated rat exposed to moderate AIH. Following AIH, there is a progressive increase in phrenic nerve burst amplitude, indicating pLTF. (B) Representative tracing from a rat exposed to moderate ASH. Following ASH, there is little increase in phrenic nerve burst amplitude from baseline.

pathways assures dominance of one or the other; the switch appears to be precipitous, since PaO_2 levels above 35 mmHg elicit pLTF via the Q pathway,[35] whereas PaO_2 levels of 30 mmHg or below elicit pLTF via the S pathway.[39] This transition occurs possibly because of relatively greater accumulation of extracellular adenosine during severe hypoxic episodes, shifting the balance toward the S pathway. Once the tipping-point is reached, we hypothesize that the now dominant S pathway suppresses the Q pathway. Although the subordinate pathway does not positively contribute to pLTF, it nevertheless modulates (inhibits) the dominant pathway.[38]

More pathways to pMF

Three other stimuli elicit unique forms of pMF in anesthetized rats. Spinal injections of the growth/trophic factors, vascular endothelial growth factor (VEGF) or erythropoietin (EPO), near the phrenic motor nucleus cause pMF via mechanisms that require both ERK and Akt activation for full expression.[40,41] These hypoxia-sensitive genes might enable phrenic motor plasticity in longer time domains. For example, VEGF- or EPO-induced pMF might play a role in longer term adjustments of

phrenic motor activity during chronic intermittent, or sustained, hypoxia.[40,41]

A completely different cellular cascade gives rise to pMF after brief periods of phrenic inactivity (iPMF) induced by hypocapnia, vagal stimulation, and/or isoflurane.[42] Unlike pLTF, iPMF requires atypical protein kinase C (PKC) activation.[43,44] iPMF might ensure that this critical motor pool is constantly active by, for example, contributing to the preservation of adequate phrenic activity when synaptic inputs are disrupted, such as by spinal injury.

Emergent properties of pLTF

Some properties of pLTF, such as pattern sensitivity and metaplasticity, may be determined by interactions between competing pathways to pMF.[26,45] These emergent properties may determine whether pLTF is expressed after a given stimulus (e.g., different patterns or severity of hypoxia), or its magnitude (e.g., greater magnitude in response to the same AIH).

Pattern sensitivity
Pattern sensitivity is a common feature in many models of neuroplasticity, including models of serotonin-dependent synaptic facilitation.[20,21] pLTF

A

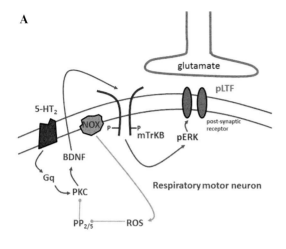

B

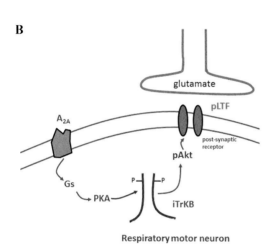

Figure 2. Working models of cellular pathways giving rise to phrenic motor facilitation following AIH. (A) The Q pathway is initiated by activation of the G_q protein–coupled 5-HT$_2$ receptor, leading to protein kinase C (PKC) activation and new synthesis of brain-derived neurotrophic factor (BDNF). BDNF then activates its high-affinity receptor, TrkB. TrkB activation phosphorylates extracellular signal-related kinase (ERK) MAP kinase, which facilitates inputs to phrenic motor neurons by an unknown mechanism (possibly glutamate receptor trafficking). The Q pathway is regulated by protein phosphatases 2A and/or 5, which can inhibit PKC activation. NADPH oxidase–induced ROS formation inhibits these phosphatases, enabling pLTF expression. Thus, NADPH oxidase and PP2A/5 constitute a regulatory cassette for pLTF. (B) The S pathway is induced by the G_s protein–coupled adenosine 2A (A_{2A}) receptor, subsequently activating protein kinase A (PKA). PKA stimulates new synthesis of an immature TrkB isoform, which autoactivates, phosphorylates, and activates Akt. Subsequent to Akt activation, synaptic inputs to phrenic motor neurons are facilitated by an unknown mechanism.

is pattern sensitive, since it is induced by moderate AIH, but not ASH of similar severity and cumulative duration (9–25 min; Fig. 1B).[12] Pattern sensitivity of respiratory plasticity was initially recognized in studies of ventilatory LTF (vLTF) in goats,[46,47] and subsequently observed in rat[48,49] and human vLTF.[50–55] Interestingly, vLTF in humans requires slightly elevated carbon dioxide (CO_2) during AIH.[56] It was recently reported that elevated CO_2 reveals vLTF in humans exposed to 32 min of ASH,[57] although the investigators found that about half of this vLTF was due to drift in ventilation caused by sustained hypercapnia alone. Griffin *et al.*[57] observed a persistent increase in ventilation for 20 min after AIH or ASH, but they did not measure ventilation at later time points necessary to clearly demonstrate LTF. Thus, further studies are needed to confirm that vLTF is pattern sensitive in awake humans.

Although significant progress has been made, the mechanism of pLTF pattern sensitivity remains unknown. The optimal spacing interval has not been well characterized, but is reported to be between 1 and 30 min.[58] Phrenic LTF is relatively insensitive to other characteristics of AIH, such as the severity or duration of hypoxic episodes, in that it is unaffected by the level of PaO_2 from 35 to 60 mmHg,[59,60] or by episode durations between 15 s and 5 min.[61] Similar to AIH-induced pLTF, pMF induced by intraspinal 5-HT requires episodic injections, suggesting that pattern sensitivity occurs at or downstream from 5-HT receptor activation.[16,17,62] Here, we propose three possible mechanisms that could contribute to pLTF pattern sensitivity.

Serotonin receptor desensitization. Serotonin receptors are diverse, with complex signaling and activation requirements; the 5-HT$_2$ class alone includes multiple isoforms with unconventional properties, such as G protein–independent signaling[63] or constitutive activity.[64] One pertinent feature of 5-HT$_2$ receptors is desensitization (Fig. 3A), which is a decreased ligand response after prolonged ligand exposure. In the brain, 5-HT$_2$ receptors desensitize rapidly with persistent elevations in extracellular serotonin via mechanisms that involve receptor internalization and/or functional uncoupling.[65] Receptor internalization occurs when agonist-bound receptors are internalized by clathrin-coated pits into endosomes, where they are sequestered

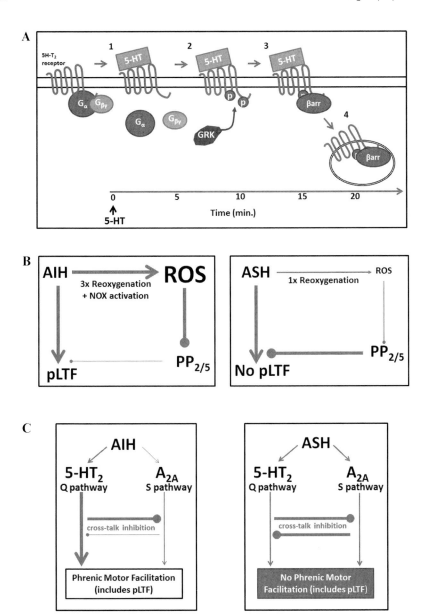

Figure 3. Possible mechanisms giving rise to pLTF pattern sensitivity. (A) 5-HT$_2$ receptor desensitization might occur with prolonged 5-HT release, such as during ASH: (1) 5-HT binds to its receptor, releasing Gα and G$\beta\gamma$ proteins, which activate downstream second messengers; (2) after prolonged agonist exposure ($\sim$10 min), 5-HT$_2$ receptors are phosphorylated by G protein–coupled receptor kinase (GRK), preventing subsequent activation of G-proteins (i.e., functional uncoupling) and promoting β-arrestin (βarr)–binding to the 5-HT$_2$ receptor; (3) β-arrestin–binding prevents further 5-HT$_2$ receptor activation while promoting receptor internalization; (4) receptors are internalized via a clathrin-dependent mechanism, decreasing the number of receptors available to elicit pLTF. (B) Differential reactive oxygen species (ROS) formation during AIH versus ASH may underlie pattern sensitivity. During AIH, ROS is produced during repeated reoxygenation events and NADPH oxidase (NOX) activation. Increased ROS would inhibit protein phosphatases 2 and 5, disinhibiting PKC (or other kinases) and enabling forward progression of pLTF. During ASH, insufficient ROS formation due to a single reoxygenation event may not inhibit the protein phosphatases sufficient to relieve their constraint to pLTF. (C) Inhibitory interactions between the Q and S pathways to pMF may underlie pattern sensitivity of pLTF. (1) During AIH, the Q pathway is dominant because of 5-HT2 receptor activation with relatively little adenosine accumulation. (2) During ASH, greater adenosine release/accumulation may cause sufficient S pathway activation to offset Q pathway activation. Balanced activation of both pathways may prevent any pLTF expression due to balanced cross-talk inhibition.

until reinserted or degraded. Functional uncoupling disrupts G-protein signaling and occurs via receptor phosphorylation, usually by G protein–coupled receptor kinases (GRKs); this phosphorylation causes the protein, arrestin, to bind the receptor, preventing further G-protein activation.[66] *In vitro* studies demonstrate 5-HT$_2$ receptor desensitization after 5–10 min of agonist exposure,[67,68] well within the time frame of a 25-min ASH exposure. Although no studies have assessed 5-HT$_2$ receptor desensitization in the spinal cord, it is a candidate to undermine pLTF during ASH.

Regulation of protein phosphatase activity. Protein phosphatases are inhibitory to many forms of CNS synaptic plasticity;[69,70] the expression of plasticity is often regulated by protein phosphatase activity during the induction phase (Fig. 3B). Phrenic LTF is constrained by constitutive okadaic acid–sensitive protein phosphatases during ASH.[45] Thus, in rats pretreated with spinal okadaic acid, ASH elicits pLTF.[45] Reactive oxygen species (ROS) inhibit protein phosphatases involved in plasticity.[71] Indeed, ROS produced by NADPH oxidase are necessary for pLTF expression,[72] a requirement that is offset by spinal protein phosphatase inhibition with okadaic acid.[73] Thus, NADPH oxidase and okadaic acid–sensitive protein phosphatases (most likely PP2A) appear to constitute a regulatory cassette that determines whether, and how much, pLTF will be expressed (Fig. 3B).[37,64] One likely difference between ASH and AIH in their ability to elicit pLTF could be relative ROS production (and phosphatase inhibition). Although no published studies have directly compared ROS production in response to AIH versus ASH, there is evidence that hypoxia-induced superoxide production occurs predominantly during reoxygenation,[73–75] suggesting that AIH could generate greater ROS than could ASH because of multiple reoxygenations. Further studies are needed to determine how AIH (vs. ASH) stimulates greater ROS production in the cervical spinal cord.

Cross-talk inhibition between *Q* and *S* pathways. Although pLTF shifts to *S* pathway-dependence with severe AIH,[39] it is not known how this shift occurs. We proposed that cross-talk inhibition between pathways (Fig. 3C) enables one pathway to gain an upper hand and, thus, dominate pMF. A key factor is the strength of the initiating stimulus (5-HT$_2$ vs. A$_{2A}$ receptor activation). For example,

during severe AIH, greater adenosine formation/accumulation may increase A$_{2A}$ receptor activation, creating stronger *S* pathway activation and, subsequently, *Q* pathway suppression. The prevailing mechanism of pLTF may result from competition between pathways for dominance (Fig. 3C).

We speculate that, at some levels of hypoxia (e.g., severity, duration), inhibitory interactions could become balanced, creating an impasse where the *S* or *Q* pathways offset one another (i.e., no pLTF). If so, this state is an emergent property of competing pathways to pMF, and could underlie pLTF pattern sensitivity. During longer hypoxic exposures, such as moderate ASH, greater adenosine formation/accumulation[76] may cause balanced activation of the *Q* and *S* pathways (Fig. 3C), thereby obscuring pLTF. This hypothesis is consistent with the study by Griffin *et al.*,[57] since hypercapnia could cause greater serotonin release, shifting the balance toward the *Q* pathway and vLTF expression. Further studies are needed to determine the role of pathway interactions in pLTF pattern sensitivity.

Are other forms of pMF pattern sensitive? Although many studies have focused on pLTF pattern sensitivity following moderate AIH, no studies have examined pattern sensitivity in the other forms of pMF described previously (e.g., *S* pathway, VEGF- and EPO-induced pMF, or iPMF). Studies are needed to determine whether these forms of pMF are also pattern sensitive.

Metaplasticity
Metaplasticity is loosely defined as plastic plasticity, but more specifically as the ability of prior experience to alter subsequent plasticity.[77] Metaplasticity can be expressed as enhanced pLTF triggered by preconditioning of adults with repetitive intermittent hypoxia[26,78,79] or cervical sensory denervation.[25] On the other hand, developmental exposure to CIH suppresses pLTF in adult rats.[80] Phrenic LTF enhancement from preconditioning is particularly interesting because it could increase the success of AIH protocols applied to improve breathing capacity or other motor functions after spinal injury.[8,27–30] Thus, metaplasticity (and its mechanisms) are important considerations for clinical utilization of AIH-induced motor plasticity.

Enhanced pLTF after preconditioning with repetitive intermittent hypoxia. Rats exposed to

CIH (10–12% O_2/air, 2–5 min intervals, 8–12 h/night) exhibit enhanced serotonin-dependent pLTF.[26] Other investigators have reproduced this effect with shorter hypoxic episodes (5–12% O_2, 15-s episodes with 5-min intervals, 8 h/day) to more closely simulate episodic hypoxia during sleep-disordered breathing.[81] In this study, pLTF enhancement was blocked by antioxidant administration,[81] demonstrating ROS dependence. Thus, CIH enhances pLTF, presumably by enhanced Q pathway signaling (i.e., serotonin- and ROS-dependent). Although CIH is a potent stimulus to metaplasticity, it also elicits morbidity, including hypertension, hippocampal apoptosis and cognitive deficits, among others.[82–84] Thus, we developed more subtle protocols of repetitive intermittent hypoxia that elicit pLTF metaplasticity without detrimental effects elicited by CIH.[30,85,86]

Rats exposed to modest protocols of repetitive acute intermittent hypoxia (rAIH) exhibit enhanced pLTF[78,79] and increased respiratory and nonrespiratory somatic motor recovery following cervical spinal injury.[30] Two different rAIH protocols were used in these studies. Daily AIH (ten 5-min episodes of 10.5% O_2, 5-min intervals, 7 days) improved respiratory and forelimb function and increased phrenic burst amplitude in rats with C2 spinal hemisections.[30] In Brown–Norway rats, a strain with low constitutive pLTF and no hypoglossal LTF daily AIH enabled hypoglossal LTF, but an apparent doubling of pLTF was only marginally significant.[85] These results suggest that daily AIH is a modest inducer of metaplasticity in the phrenic motor pool of otherwise normal rats. Rats exposed to AIH three times/week (ten 5-min episodes/day, 3 days/week for 4 weeks) exhibit enhanced pLTF,[78,79] and profound neurochemical plasticity in the phrenic motor nucleus,[86] suggesting that longer, less frequent exposures are more effective at eliciting pLTF metaplasticity. Neither protocol caused hippocampal apoptosis, astrogliosis, or hypertension,[30,85,86] suggesting the ability to elicit respiratory metaplasticity without detectable pathology. Further studies are needed to determine optimal protocols to elicit respiratory metaplasticity, so that AIH-induced plasticity can be harnessed as a means of restoring respiratory and nonrespiratory motor function in clinical disorders that challenge ventilatory control.

What mechanism underlies pLTF metaplasticity following rAIH? In animals exposed to daily AIH, increased spinal BDNF was observed,[85] suggesting Q pathway enhancement. If the Q pathway is enhanced following rAIH, what mechanism underlies this enhancement? One possibility is decreased cross-talk inhibition. For example, decreased S pathway activation (Fig. 4A) could increase Q pathway–dependent pLTF and BDNF synthesis. Alternatively, decreased inhibitory interactions between pathways could enable direct S pathway contributions to pLTF, even with moderate

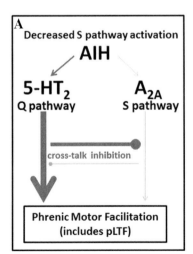
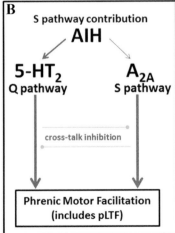
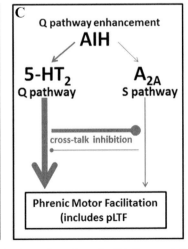

Figure 4. Possible mechanisms of metaplasticity, enhancing pLTF following repetive intermittent hypoxia. (A) Decreased S pathway activation, thereby eliminating S to Q inhibition, could explain metaplasticity in pLTF following rAIH. (B) Reduced cross-talk inhibition, enabling a positive S pathway contribution, could enhance pLTF. (C) Increased Q pathway signaling (hypertrophy), possibly involving increased 5-HT_2 receptors, ROS production, or BDNF synthesis, could explain greater pLTF following rAIH.

AIH, since the pathways would be uncoupled (Fig. 4B). Indeed, greater phospho-Akt levels were observed after AIH exposure three times/week for 10 weeks,[86] consistent with increased *S* pathway activation. Finally, pLTF metaplasticity may arise directly from a more robust *Q* pathway (Fig. 4C); following 10 weeks of AIH three times/week, 5-HT$_{2A}$ receptor, BDNF, TrkB, and phospho-ERK expression are all increased,[86] suggesting hypertrophy of this cellular cascade. Further studies are needed to determine how rAIH causes pLTF metaplasticity.

Other forms of metaplasticity in the phrenic motor system. This brief review was intended to pose rather than answer questions, and is not expansive enough to cover all potential forms of metaplasticity in respiratory motor pools. However, factors known to induce metaplasticity in other models of plasticity, such as hippocampal LTP, include prior activation[77,87,88] and stress.[89] Hippocampal LTP is impaired by stress or glucocorticoids.[89] In the phrenic motor pool, we postulate that stress has two separate effects. Initially, stress might induce pMF, due to norepinephrine release from locus coeruleus neurons followed by activation of α_1 receptors on phrenic motor neurons.[10] On the other hand, if pLTF is similar to other models of plasticity, chronic exposure to stressful stimuli may impair pLTF, possibly through a glucocorticoid-dependent mechanism. Anecdotally, rats subject to stressful stimuli often fail to express pLTF; however, further studies are needed to determine the effects of stress on phrenic motor plasticity, and to confirm or refute these speculations.

Conclusion and significance

Distinct pathways (i.e., *Q* and *S* pathways) give rise to phenotypically similar pMF. However, since the *Q* and *S* pathways interact via cross-talk inhibition, either pathway can dominate and produce pLTF; at a specific balance point, it is possible that these pathways neutralize one another, with equal and opposing inhibition of the other. Thus, cross-talk inhibition may have the ability to confer key emergent properties of pLTF, such as pattern sensitivity. Another emergent property, enhanced pLTF (i.e., metaplasticity), may arise from diminished inhibitory coupling of these pathways, so that both combine to produce enhanced pLTF following AIH. The presence of multiple, competing pathways to pLTF confers flexibility, enabling different manifestations of plasticity as an animal responds to diverse conditions that vary in severity, pattern, and/or duration of hypoxia. Greater understanding of multiple pathways giving rise to respiratory motor plasticity and their interactions could increase our understanding of physiological responses to environmental or pathological changes.

As one example, human subjects at high altitude experience chronic sustained hypoxia, giving rise to homeostatic increases in ventilation; this form of respiratory plasticity is often referred to as ventilatory acclimatization to high altitude.[90,91] What role, if any, do the *Q* and *S* pathways, and their interactions, play in this process? Further studies are needed to determine how pathways interact to produce changes in respiratory control appropriate for the prevailing conditions. As another example, humans, particularly those suffering from obstructive sleep apnea, most frequently experience intermittent hypoxia during sleep. In such cases, does the severity and pattern of hypoxic episodes determine the type and extent of compensatory respiratory plasticity? These differences may also be of considerable relevance as we consider distinctions between ventilatory plasticity versus plasticity in respiratory-related motor output to the upper airways, which may stabilize or destabilize breathing depending on complex interactions between upper airway patency versus chemoreflex gain and apneic threshold. Further studies are needed to answer these questions.

An understanding of complex interactions between mechanisms of respiratory plasticity has considerable importance as we develop strategies to harness intermittent hypoxia-induced motor plasticity to treat clinical disorders that impair breathing and other movements.[31] Although much research has focused on characterizing new forms of plasticity, we are just beginning to understand the significance of interactions between them. These interactions, if understood and controlled, may enable us to optimize therapeutic protocols of rAIH in the treatment of spinal injury, neurodegenerative diseases, and even sleep-disordered breathing.[27]

Acknowledgments

This research is supported by HL080209, HL111598, and HL69064. N.N. is supported by a fellowship from the Francis Family Foundation.

Conflicts of interest

The authors declare no conflicts of interest.

References

1. Mitchell, G.S. & S.M. Johnson. 2003. Neuroplasticity in respiratory motor control. *J. Appl. Physiol.* **94:** 358–374.
2. Doyle, J.C. & M. Csete. 2011. Architecture, constraints, and behavior. *Proc. Natl. Acad. Sci. U.S.A.* **108**(Suppl. 3)**:** 15624–15630.
3. Millhorn, D.E., F.L. Eldridge & T.G. Waldrop. 1980. Prolonged stimulation of respiration by a new central neural mechanism. *Respir. Physiol.* **41:** 87–103.
4. Millhorn, D.E., F.L. Eldridge & T.G. Waldrop. 1980. Prolonged stimulation of respiration by endogenous central serotonin. *Respir. Physiol.* **42:** 171–188.
5. Eldridge, F.L. & D.E. Millhorn. 1986. "Oscillation, gating, and memory in the respiratory control system." In *Handbook of Physiology: Section 3, The Respiratory System II.* P.T. Machklem & J. Mead, Eds.: 93–114. Bethesda, MD: American Physiological Society.
6. Rowley, K.L., C.B. Mantilla & G.C. Sieck. 2005. Respiratory muscle plasticity. *Respir. Physiol. Neurobiol.* **147:** 235–251.
7. Prabhakar, N.R. 2011. Sensory plasticity of the carotid body: role of reactive oxygen species and physiological significance. *Respir. Physiol. Neurobiol.* **178:** 375–380.
8. Dale-Nagle, E.A., M.S. Hoffman, P.M. MacFarlane, *et al.* 2010. Spinal plasticity following intermittent hypoxia: implications for spinal injury. *Ann. N.Y. Acad. Sci.* **1198:** 252–259.
9. Morris, K.F., D.M. Baekey, S.C. Nuding, *et al.* 2003. Invited review: neural network plasticity in respiratory control. *J. Appl. Physiol.* **94:** 1242–1252.
10. Dale-Nagle, E.A., M.S. Hoffman, P.M. MacFarlane & G.S. Mitchell. 2010. Multiple pathways to long-lasting phrenic motor facilitation. *Adv. Exp. Med. Biol.* **669:** 225–230.
11. Mitchell, G.S., T.L. Baker, S.A. Nanda, *et al.* 2001. Invited review: intermittent hypoxia and respiratory plasticity. *J. Appl. Physiol.* **90:** 2466–2475.
12. Bach K.B. & G.S. Mitchell. 1996. Hypoxia-induced long-term facilitation of respiratory activity is serotonin dependent. *Respir. Physiol.* **104:** 251–260.
13. Hayashi, F., S.K. Coles, K.B. Bach, *et al.* 1993. Time-dependent phrenic nerve responses to carotid afferent activation: intact vs. decerebellate rats. *Am. J. Physiol.* **265**(Pt 2)**:** R811–R819.
14. Fuller, D.D., A.G. Zabka, T.L. Baker & G.S. Mitchell. 2001. Phrenic long-term facilitation requires 5-HT receptor activation during but not following episodic hypoxia. *J. Appl. Physiol.* **90:** 2001–2006; discussion 2000.
15. Baker-Herman, T.L. & G.S. Mitchell. 2002. Phrenic long-term facilitation requires spinal serotonin receptor activation and protein synthesis. *J. Neurosci.* **22:** 6239–6246.
16. MacFarlane, P.M. & G.S. Mitchell. 2009. Episodic spinal serotonin receptor activation elicits long-lasting phrenic motor facilitation by an NADPH oxidase-dependent mechanism. *J. Physiol. (Lond.)* **587**(Pt 22)**:** 5469–5481.
17. Bocchiaro, C.M. & J.L. Feldman. 2004. Synaptic activity-independent persistent plasticity in endogenously active mammalian motoneurons. *Proc. Natl. Acad. Sci. U.S.A.* **101:** 4292–4295.
18. Neves, G., S.F. Cooke & T.V.P. Bliss. 2008. Synaptic plasticity, memory and the hippocampus: a neural network approach to causality. *Nat. Rev. Neurosci.* **9:** 65–75.
19. Baker, T.L. & G.S. Mitchell. 2000. Episodic but not continuous hypoxia elicits long-term facilitation of phrenic motor output in rats. *J. Physiol. (Lond.)* **529**(Pt 1)**:** 215–219.
20. Mauelshagen, J., C.M. Sherff & T.J. Carew. 1998. Differential induction of long-term synaptic facilitation by spaced and massed applications of serotonin at sensory neuron synapses of Aplysia californica. *Learn. Mem.* **5:** 246–256.
21. Sutton, M.A., J. Ide, S.E. Masters & T.J. Carew. 2002. Interaction between amount and pattern of training in the induction of intermediate- and long-term memory for sensitization in aplysia. *Learn. Mem.* **9:** 29–40.
22. Kauer, J.A. 1999. Blockade of hippocampal long-term potentiation by sustained tetanic stimulation near the recording site. *J. Neurophysiol.* **81:** 940–944.
23. Nguyen, P.V, S.N. Duffy & J.Z. Young. 2000. Differential maintenance and frequency-dependent tuning of LTP at hippocampal synapses of specific strains of inbred mice. *J. Neurophysiol.* **84:** 2484–2493.
24. Scharf, M.T., N.H. Woo, K.M. Lattal, *et al.* 2002. Protein synthesis is required for the enhancement of long-term potentiation and long-term memory by spaced training. *J. Neurophysiol.* **87:** 2770–2777.
25. Kinkead, R., W.Z. Zhan, Y.S. Prakash, *et al.* 1998. Cervical dorsal rhizotomy enhances serotonergic innervation of phrenic motoneurons and serotonin-dependent long-term facilitation of respiratory motor output in rats. *J. Neurosci.* **18:** 8436–8443.
26. Ling, L., D.D. Fuller, K.B. Bach, *et al.* 2001. Chronic intermittent hypoxia elicits serotonin-dependent plasticity in the central neural control of breathing. *J. Neurosci.* **21:** 5381–5388.
27. Mahamed, S. & G.S. Mitchell. 2007. Is there a link between intermittent hypoxia-induced respiratory plasticity and obstructive sleep apnoea? *Exp. Physiol.* **92:** 27–37.
28. Mitchell, G.S. & J. Terada. 2011. Should we standardize protocols and preparations used to study respiratory plasticity? *Respir. Physiol. Neurobiol.* **177:** 93–97.
29. Vinit, S., M.R. Lovett-Barr & G.S. Mitchell. 2009. Intermittent hypoxia induces functional recovery following cervical spinal injury. *Respir. Physiol. Neurobiol.* **169:** 210–217.
30. Lovett-Barr, M.R., I. Satriotomo, G.D. Muir, *et al.* 2012. Repetitive intermittent hypoxia induces respiratory and somatic motor recovery after chronic cervical spinal injury. *J. Neurosci.* **32:** 3591–3600.
31. Trumbower, R.D., A. Jayaraman, G.S. Mitchell & W.Z. Rymer. 2012. Exposure to acute intermittent hypoxia augments somatic motor function in humans with incomplete spinal cord injury. *Neurorehabil. Neural Repair* **26:** 163–172.
32. Nichols, N.L., G. Gowing, I. Satriotomo, *et al.* 2012. Intermittent hypoxia and stem cell implants preserve breathing capacity in a rodent model of ALS. *Am. J. Respir. Crit. Care Med.* **112:** 1678–1688.

33. Baker-Herman, T.L., D.D. Fuller, R.W. Bavis, *et al.* 2004. BDNF is necessary and sufficient for spinal respiratory plasticity following intermittent hypoxia. *Nat. Neurosci.* **7:** 48–55.

34. Hoffman, M.S., N.L. Nichols, P.M. Macfarlane & G.S. Mitchell. 2012. Phrenic long-term facilitation following acute intermittent hypoxia requires spinal ERK activation but not TrkB synthesis. *J. Appl. Physiol.* **113:** 1184–1193.

35. Leonard, B.E. 1994. Serotonin receptors—where are they going? *Int. Clin. Psychophamacol.* **9**(Suppl 1): 7–17.

36. Golder, F.J., L. Ranganathan, I. Satriotomo, *et al.* 2008. Spinal adenosine A2a receptor activation elicits long-lasting phrenic motor facilitation. *J. Neurosci.* **28:** 2033–2042.

37. Hoffman, M.S. & G.S. Mitchell. 2011. Spinal 5-HT7 receptor activation induces long-lasting phrenic motor facilitation. *J. Physiol. (Lond.)* **589**(Pt 6): 1397–1407.

38. Hoffman, M.S., F.J. Golder, S. Mahamed & G.S. Mitchell. 2010. Spinal adenosine A2(A) receptor inhibition enhances phrenic long term facilitation following acute intermittent hypoxia. *J. Physiol. (Lond.)* **588**(Pt 1): 255–266.

39. Nichols, N.L., E.A. Dale & G.S. Mitchell. 2012. Severe acute intermittent hypoxia elicits phrenic long-term facilitation by a novel adenosine-dependent mechanism. *J. Appl. Physiol.* **112:** 1678–1688.

40. Dale-Nagle, E.A., I. Satriotomo & G.S. Mitchell. 2011. Spinal vascular endothelial growth factor induces phrenic motor facilitation via extracellular signal-regulated kinase and Akt signaling. *J. Neurosci.* **31:** 7682–7690.

41. Dale, E.A., I. Satriotomo & G.S. Mitchell. 2012. Cervical spinal erythropoietin induces phrenic motor facilitation via extracellular signal-regulated protein kinase and Akt signaling. *J. Neurosci.* **32:** 5973–5983.

42. Mahamed, S., K.A. Strey, G.S. Mitchell & T.L. Baker-Herman. 2011. Reduced respiratory neural activity elicits phrenic motor facilitation. *Respir. Physiol. Neurobiol.* **175:** 303–309.

43. Strey, K.A., N.L. Nichols, N.A. Baertsch, *et al.* 2012. Spinal atypical protein kinase C activity is necessary to stabilize inactivity-induced phrenic motor facilitation. *J. Neurosci.* **32:** 16510–16520.

44. Baker-Herman, T.L. & K.A. Strey. 2011. Similarities and differences in mechanisms of phrenic and hypoglossal motor facilitation. *Respir. Physiol. Neurobiol.* **179:** 48–56.

45. Wilkerson, J.E.R., I. Satriotomo, T.L. Baker-Herman, *et al.* 2008. Okadaic acid-sensitive protein phosphatases constrain phrenic long-term facilitation after sustained hypoxia. *J. Neurosci.* **28:** 2949–2958.

46. Turner, D.L. & G.S. Mitchell. 1997. Long-term facilitation of ventilation following repeated hypoxic episodes in awake goats. *J. Physiol. (Lond.)* **499**(Pt 2): 543–550.

47. Dwinell, M.R., P.L. Janssen & G.E. Bisgard. 1997. Lack of long-term facilitation of ventilation after exposure to hypoxia in goats. *Respir. Physiol.* **108:** 1–9.

48. Olson, E.B., Jr., C.J. Bohne, M.R. Dwinell, *et al.* 2001. Ventilatory long-term facilitation in unanesthetized rats. *J. Appl. Physiol.* **91:** 709–716.

49. Nakamura, A., E.B. Olson, J. Terada, *et al.* 2010. Sleep state dependence of ventilatory long-term facilitation following

acute intermittent hypoxia in Lewis rats. *J. Appl. Physiol.* **109:** 323–331.

50. Morgan, B.J., D.C. Crabtree, M. Palta & J.B. Skatrud. 1995. Combined hypoxia and hypercapnia evokes long-lasting sympathetic activation in humans. *J. Appl. Physiol.* **79:** 205–213.

51. McEvoy, R.D., R.M. Popovic, N.A. Saunders & D.P. White. 1996. Effects of sustained and repetitive isocapnic hypoxia on ventilation and genioglossal and diaphragmatic EMGs. *J. Appl. Physiol.* **81:** 866–875.

52. McEvoy, R.D., R.M. Popovic, N.A. Saunders & D.P. White. 1997. Ventilatory responses to sustained eucapnic hypoxia in healthy males during wakefulness and NREM sleep. *Sleep* **20:** 1008–1011.

53. Xie, A., J.B. Skatrud, D.S. Puleo & B.J. Morgan. 2001. Exposure to hypoxia produces long-lasting sympathetic activation in humans. *J. Appl. Physiol.* **91:** 1555–1562.

54. Tamisier, R., A. Anand, L.M. Nieto, *et al.* 2005. Arterial pressure and muscle sympathetic nerve activity are increased after two hours of sustained but not cyclic hypoxia in healthy humans. *J. Appl. Physiol.* **98:** 343–349.

55. Querido, J.S., P.M. Kennedy & A.W. Sheel. 2010. Hyperoxia attenuates muscle sympathetic nerve activity following isocapnic hypoxia in humans. *J. Appl. Physiol.* **108:** 906–912.

56. Harris, D.P., A. Balasubramaniam, M.S. Badr & J.H. Mateika. 2006. Long-term facilitation of ventilation and genioglossus muscle activity is evident in the presence of elevated levels of carbon dioxide in awake humans. *Am. J. Physiol. Regul. Integr. Comp. Physiol.* **291:** R1111–R1119.

57. Griffin, H.S., P. Kumar, K. Pugh & G.M. Balanos. 2012. Long-term facilitation of ventilation following acute continuous hypoxia in awake humans during sustained hypercapnia. *J. Physiol. (Lond.)* **590**(Pt 20): 5151–5165.

58. Bach, K.B., R. Kinkead & G.S. Mitchell. 1999. Post-hypoxia frequency decline in rats: sensitivity to repeated hypoxia and alpha2-adrenoreceptor antagonism. *Brain Res.* **817:** 25–33.

59. Fuller, D.D., K.B. Bach, T.L. Baker, *et al.* 2000. Long term facilitation of phrenic motor output. *Respir. Physiol.* **121:** 135–146.

60. Baker-Herman, T.L. & G.S. Mitchell. 2008. Determinants of frequency long-term facilitation following acute intermittent hypoxia in vagotomized rats. *Respir. Physiol. Neurobiol.* **162:** 8–17.

61. Mahamed, S. & G.S. Mitchell. 2008. Simulated apnoeas induce serotonin-dependent respiratory long-term facilitation in rats. *J. Physiol. (Lond.)* **586:** 2171–2181.

62. Lovett-Barr, M.R., G.S. Mitchell, I. Satriotomo & S.M. Johnson. 2006. Serotonin-induced in vitro long-term facilitation exhibits differential pattern sensitivity in cervical and thoracic inspiratory motor output. *Neuroscience* **142:** 885–892.

63. Felder, C.C., R.Y. Kanterman, A.L. Ma & J. Axelrod. 1990. Serotonin stimulates phospholipase A2 and the release of arachidonic acid in hippocampal neurons by a type 2 serotonin receptor that is independent of inositolphospholipid hydrolysis. *Proc. Natl. Acad. Sci. U.S.A.* **87:** 2187–2191.

64. Murray, K.C., A. Nakae, M.J. Stephens, *et al.* 2010. Recovery of motoneuron and locomotor function after spinal cord injury depends on constitutive activity in 5-HT2C receptors. *Nat. Med.* **16:** 694–700.

65. Rahman, S. & R.S. Neuman. 1993. Multiple mechanisms of serotonin 5-HT2 receptor desensitization. *Eur. J. Pharmacol.* **238:** 173–180.

66. Zhang, J., S.S. Ferguson, L.S. Barak, *et al.* 1997. Molecular mechanisms of G protein-coupled receptor signaling: role of G protein-coupled receptor kinases and arrestins in receptor desensitization and resensitization. *Recept. Channels* **5:** 193–199.

67. Berry, S.A., M.C. Shah, N. Khan & B.L. Roth. 1996. Rapid agonist-induced internalization of the 5-hydroxytryptamine2A receptor occurs via the endosome pathway in vitro. *Mol. Pharmacol.* **50:** 306–313.

68. Hanley, N.R.S. & J.G. Hensler. 2002. Mechanisms of ligand-induced desensitization of the 5-hydroxytryptamine(2A) receptor. *J. Pharmacol. Exp. Ther.* **300:** 468–477.

69. Winder, D.G. & J.D. Sweatt. 2001. Roles of serine/threonine phosphatases in hippocampal synaptic plasticity. *Nat. Rev. Neurosci.* **2:** 461–474.

70. Colbran, R.J. 2004. Protein phosphatases and calcium/calmodulin-dependent protein kinase II-dependent synaptic plasticity. *J. Neurosci.* **24:** 8404–8409.

71. Massaad, C.A. & E. Klann. 2011. Reactive oxygen species in the regulation of synaptic plasticity and memory. *Antioxid. Redox. Signal.* **14:** 2013–2054.

72. MacFarlane, P.M., I. Satriotomo, J.A. Windelborn & G.S. Mitchell. 2009. NADPH oxidase activity is necessary for acute intermittent hypoxia-induced phrenic long-term facilitation. *J. Physiol. (Lond.)* **587**(Pt 9): 1931–1942.

73. MacFarlane, P.M., J.E.R. Wilkerson, M.R. Lovett-Barr & G.S. Mitchell. 2008. Reactive oxygen species and respiratory plasticity following intermittent hypoxia. *Respir. Physiol. Neurobiol.* **164:** 263–271.

74. Fabian, R.H., J.R. Perez-Polo & T.A. Kent. 2004. Extracellular superoxide concentration increases following cerebral hypoxia but does not affect cerebral blood flow. *Int. J. Dev. Neurosci.* **22:** 225–230.

75. Wilkerson, J.E.R., P.M. Macfarlane, M.S. Hoffman & G.S. Mitchell. 2007. Respiratory plasticity following intermittent hypoxia: roles of protein phosphatases and reactive oxygen species. *Biochem. Soc. Trans.* **35**(Pt 5): 1269–1272.

76. Conde, S.V. & E.C. Monteiro. 2004. Hypoxia induces adenosine release from the rat carotid body. *J. Neurochem.* **89:** 1148–1156.

77. Abraham, W.C. & M.F. Bear. 1996. Metaplasticity: the plasticity of synaptic plasticity. *Trends Neurosci.* **19:** 126–130.

78. MacFarlane, P.M., S. Vinit, A. Roopra & G.S. Mitchell. 2010. Enhanced phrenic long-term facilitation following repetitive acute intermittent hypoxia: role of glycolytic flux. *FASEB J.* **24** (Meeting Abstract Supplement): 799.15.

79. Vinit, S., P.M. MacFarlane, I. Satriotomo & G.S. Mitchell. 2010. Enhanced phrenic long-term facilitation (pLTF) following repetitive acute intermittent hypoxia. *FASEB J.* **24** (Meeting Abstract Supplement): 799.14.

80. Reeves, S.R., G.S. Mitchell & D. Gozal. 2006. Early postnatal chronic intermittent hypoxia modifies hypoxic respiratory responses and long-term phrenic facilitation in adult rats. *Am. J. Physiol. Regul. Integr. Comp. Physiol.* **290:** R1664–R1671.

81. Peng, Y.-J. & N.R. Prabhakar. 2003. Reactive oxygen species in the plasticity of respiratory behavior elicited by chronic intermittent hypoxia. *J. Appl. Physiol.* **94:** 2342–2349.

82. Lesske, J., E.C. Fletcher, G. Bao & T. Unger. 1997. Hypertension caused by chronic intermittent hypoxia—influence of chemoreceptors and sympathetic nervous system. *J. Hypertens.* **15**(Pt 2): 1593–1603.

83. Row, B.W., L. Kheirandish, J.J. Neville & D. Gozal. 2002. Impaired spatial learning and hyperactivity in developing rats exposed to intermittent hypoxia. *Pediatr. Res.* **52:** 449–453.

84. Yan, B., L. Li, S.W. Harden, *et al.* 2009. Chronic intermittent hypoxia impairs heart rate responses to AMPA and NMDA and induces loss of glutamate receptor neurons in nucleus ambiguous of F344 rats. *Am. J. Physiol. Regul. Integr. Comp. Physiol.* **296:** R299–R308.

85. Wilkerson, J.E.R. & G.S. Mitchell. 2009. Daily intermittent hypoxia augments spinal BDNF levels, ERK phosphorylation and respiratory long-term facilitation. *Exp. Neurol.* **217:** 116–123.

86. Satriotomo, I., E.A. Dale, J.M. Dahlberg & G.S. Mitchell. 2012. Repetitive acute intermittent hypoxia increases expression of proteins associated with plasticity in the phrenic motor nucleus. *Exp. Neurol.* **237:** 103–115.

87. Abraham, W.C. 2008. Metaplasticity: tuning synapses and networks for plasticity. *Nat. Rev. Neurosci.* **9:** 387–387.

88. Tenorio, G., S.A. Connor, D. Guévremont, *et al.* 2010. "Silent" priming of translation-dependent LTP by β-adrenergic receptors involves phosphorylation and recruitment of AMPA receptors. *Learn. Mem.* **17:** 627–638.

89. Kim, J.J. & K.S. Yoon. 1998. Stress: metaplastic effects in the hippocampus. *Trends Neurosci.* **21:** 505–509.

90. Powell, F.L., K.A. Huey & M.R. Dwinell. 2000. Central nervous system mechanisms of ventilatory acclimatization to hypoxia. *Respir. Physiol.* **121:** 223–236.

91. Dwinell, M.R. & F.L. Powell. 1999. Chronic hypoxia enhances the phrenic nerve response to arterial chemoreceptor stimulation in anesthetized rats. *J. Appl. Physiol.* **87:** 817–823.

Ann. N.Y. Acad. Sci. ISSN 0077-8923

ANNALS OF THE NEW YORK ACADEMY OF SCIENCES

Issue: *Neurons, Circuitry, and Plasticity in the Spinal Cord and Brainstem*

Axon regeneration and exercise-dependent plasticity after spinal cord injury

John D. Houle and Marie-Pascale Côté

Department of Neurobiology and Anatomy, Spinal Cord Research Center, Drexel University College of Medicine, Philadelphia, Pennsylvania

Address for correspondence: Dr. John D. Houle, Department of Neurobiology and Anatomy, Drexel University College of Medicine, 2900 Queen Lane, Philadelphia, PA 19129. jhoule@drexelmed.edu

Current dogma states that meaningful recovery of function after spinal cord injury (SCI) will likely require a combination of therapeutic interventions comprised of regenerative/neuroprotective transplants, addition of neurotrophic factors, elimination of inhibitory molecules, functional sensorimotor training, and/or stimulation of paralyzed muscles or spinal circuits. We routinely use (1) peripheral nerve grafts to support and direct axonal regeneration across an incomplete cervical or complete thoracic transection injury, (2) matrix modulation with chondroitinase (ChABC) to facilitate axonal extension beyond the distal graft–spinal cord interface, and (3) exercise, such as forced wheel walking, bicycling, or step training on a treadmill. We and others have demonstrated an increase in spinal cord levels of endogenous neurotrophic factors with exercise, which may be useful in facilitating elongation and/or synaptic activity of regenerating axons and plasticity of spinal neurons below the level of injury.

Keywords: rehabilitation; neurotrophic factors; neurotransplantation; cFos

Introduction

Spinal cord injury (SCI) initiates a progressive state of degeneration, with structural, biochemical, and physiological changes occurring over weeks to months postinjury. Secondary responses to the initial trauma include inflammation, excitotoxicity, apoptosis of neurons and glia, axon retraction, glial scarring, demyelination and exposure of myelin-associated inhibitory molecules, changes in electrophysiological properties of neurons, and aberrant sprouting/plasticity of spared nerve fibers.[1–4] When viewed as a whole, clinicians and researchers are faced with multiple changes ranging from molecular to gross anatomical disturbances that need to be addressed, while having the knowledge that no single treatment approach to date has been effective in promoting recovery of volitional movement. Different therapeutic strategies have, in the past, focused on a single aspect of the injury response, be it neuroprotection, neuroregeneration, or neurorehabilitation. In contrast, many recent studies indicate that a greater effect may be realized when a combination of interventions is implemented. This short review will describe the need to balance the effort to promote recovery through use of neural tissue transplantation with the attempt to diminish growth inhibitory factors, such as chondroitin sulfate proteoglycans that prevent regeneration of axons, while incorporating exercise as a treatment strategy to help shape cell and molecular plasticity distal to an injury site. Ongoing research will establish whether exercise has beneficial effects on the regenerative efforts of neurons directly injured by trauma to the spinal cord.

Results

Spinal cord injury and transplantation models

One approach used to promote the regeneration and integration of injured axons across a spinal cord lesion is neural tissue transplantation, in which the intent is to provide a substrate that will support and guide axons toward a specified target. Much has been written about the sources of these tissues, with strong evidence of axonal growth into and through transplants of embryonic neural tissue, Schwann cells, segments of peripheral nerve,

doi: 10.1111/nyas.12052

Ann. N.Y. Acad. Sci. 1279 (2013) 154–163 © 2013 New York Academy of Sciences.

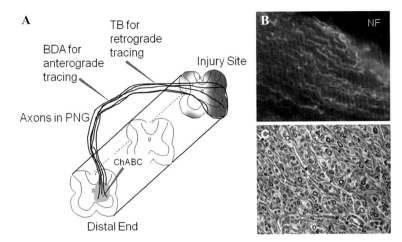

Figure 1. Axon regeneration into peripheral nerve grafts. (A) Diagrammatic representation of a PNG bridging a spinal cord lesion with a demonstration of axonal growth into and through the distal end of the graft, where ChABC was delivered to degrade inhibitory matrix molecules. This illustration shows the flexibility of the PNG, in that grafts of a specified length can be created, the distal end can be apposed to a specific region or level of the spinal cord, and the graft is external to the spinal canal that facilitates isolation for electrophysiological testing and microinjection of tracers into the graft. (B) Regenerating axons immunolabeled with antibody to neurofilament are oriented in a longitudinal array within the PNG. (C) Evaluation of successful grafting includes quantitation of the number of myelinated axons observed in a transverse section through the PNG. Here, myelinated axons of various diameters are seen surrounded by fibrous-like connective tissue.

genetically modified fibroblasts, and stem cells.[5–9] Other reports suggest that biomaterial scaffolds effectively maintain continuity between injured surfaces of the spinal cord to facilitate axon growth.[10–12] It is important to recognize that most of these transplant approaches have been applied to multiple forms of SCI, including hemisection and complete transection aspirations, penetrating knife cuts, and contusion injuries. Survival and integration of the transplant in a variety of injury situations suggests that an injury model can be chosen to address a specific set of questions rather than having the need to mimic a human injury condition. This provides for greater variability of research design in collecting a broad range of information about, for example, transplant survival, support of axonal growth, integration with host tissues, and the possibility of immunorejection.

There has been considerable success with the use of peripheral nerve grafts (PNGs) following the groundbreaking demonstration by Aguayo *et al.* of axon regeneration by adult mammalian neurons.[13,14] There are several advantages to the use of PNGs for promoting axon regeneration, which have been highlighted in a recent review.[15] Briefly, it is possible to direct growing axons toward a specific target, which in our case is the intermediate gray

of spinal cord distal to an injury; the target site can be treated to make it either more attractive or less inhibitory for the extension of axons beyond the distal end of the PNG; the PNG contains a pure population of regenerating axons that provides a valuable resource for the assessment of molecular signaling during regrowth; axons coursing through the PNG and the path of axonal outgrowth from the PNG can be mapped out by application of tracer compounds directly to the PNG; and behavioral recovery due to regenerated axons can be evaluated by cutting the PNG and subsequently determining whether there is a loss of functional activity. Several of these features are highlighted in Figure 1A. The basic techniques of creating a lesion cavity, of transplanting a PNG to form a bridge across the lesion, and intraspinal microinjection of matrix-modifying enzymes are available for video viewing (see Ref. 16).

Over the last 25 years, PNGs have been used to define the regenerative effort of supraspinal and propriospinal neurons following a variety of spinal cord injuries. From this body of work we know that acutely and chronically injured neurons exhibit the capacity for regeneration when provided with an appropriate substrate, and that these newly growing axonal processes will extend to the distal end of the PNG.[17,18] Both descending motor axons and

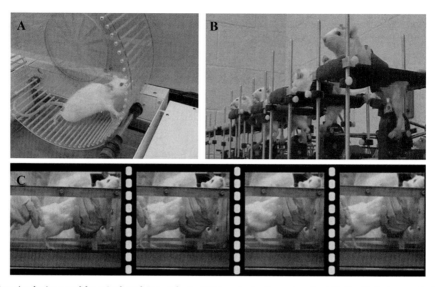

Figure 2. Exercise devices used for spinal cord–injured rats. (A) A motorized exercise wheel (Lafayette Instruments, Lafayette, IN) is useful for rats that have a unilateral injury so they can maintain a slow walking speed using three limbs initially, until the impaired limb regains some measure of function. Changes in speed and duration of exercise are routinely implemented to encourage rats to perform at a higher level. (B) Motorized cycles (custom-made) for spinalized rats provide rhythmic sensory input to the lumbar spinal cord as they move the hindlimbs through a complete range of motion. As little as 15 min of cycling per day is sufficient to increase intraspinal NTF levels. (C) Step training of spinalized rats on a motorized treadmill provides loading of hindlimb muscles that does not occur with cycling. Similar to treadmill training of cats,[37] spinalized rats require assistance with balance and perineal stimulation to initiate locomotion.

ascending sensory axons exhibit a pattern of linear growth within the graft (Fig. 1B), and axons of small to large caliber were well myelinated (Fig. 1C). We have observed similar results when grafting into spinal cord–injured adult rats or cats.[19,20] After treatment with chondroitinase to degrade chondroitin sulfate proteoglycans of the glial scar, it is observed that approximately 10–15% of the axons in the PNG extend beyond the distal graft to integrate with adjacent spinal neurons.[17] These axons do not extend very far from the PNG (<1 mm), but many form synaptic contacts with spinal neurons, and the observed increase in cFos protein in spinal cord neurons after electrical stimulation of the PNG is indicative of the formation of functional synapses between regenerated axons and their target. Functional recovery due to synaptic connections made by axons growing into and out of the grafts is directly tested by evaluating kinematic data, EMG activity, and recording of spinal cord field potentials after electrical stimulation of the PNG.[20–25]

On the positive side, it is encouraging to see a robust regenerative response following a traumatic spinal cord injury. After a cervical level injury, we have traced axonal growth into PNGs by all major descending spinal pathways except the corticospinal tract.[26] After a lower thoracic injury, significant regeneration by thoracic propriospinal neurons was observed, but few supraspinal neurons responded to an injury at this distant level by regenerating their axon.[27] Ascending sensory fibers exhibit a strong regenerative effort without regard to the level of injury as dorsal root ganglion neurons (both small and large) and thoracic propriospinal neurons have been identified as sources of axons regenerating into a PNG that was apposed to the caudal surface of the injury site. In adult rats, the mean number of myelinated axons within PNGs is about 1,200 regardless of whether they arise from descending or ascending spinal pathways. In adult cats, we have observed more than 2,400 myelinated axons in our grafts, which likely reflects the larger diameter of the nerve used for grafting and the ability to have room for more axonal ingrowth. Despite demonstration of anatomical and behavioral recovery by axons extending across a lesion, the overall benefit of chondroitinase treatment is not very compelling as the majority of axons remain in the transplant (no matter what the tissue source) with only a small number reaching target areas, and there is no convincing evidence of restoration of volitional control of movement. It is our goal to develop better strategies to encourage outgrowth by a greater percentage of those axons that have regenerated into a PNG, with the rationale that greater axonal integration with spinal cord neurons will result in greater functional reorganization.

Exercise promotes neural plasticity

The capacity for CNS reorganization[28,29] is a well-known property that may underlie many examples of functional recovery.[30] SCI leads to rapid changes in motoneuron (MN) activity patterns because of the interruption of descending supraspinal and propriospinal pathways. Removal of descending presynaptic inhibition of group Ia axons may contribute to the decrease in reflex threshold and loss of habituation by MNs.[31–33] Changes in reflex circuitry properties likely underlie the progressive development of a hyperreflexive state that eventually leads to spasticity. After SCI, the H-reflex exhibits lower negative modulation with increasing stimulation frequency compared to intact animals; whereas in exercised animals, there is a restoration of frequency-dependent depression as a measure of H-reflex excitability[34] and prevention of atrophy of hindlimb muscles.[35] Thus, a component of the attenuating effects of SCI on motor systems with passive exercise may reside in the spinal cord and may include the afferent connections to MNs or properties of the MNs themselves. Rats with a complete lower thoracic transection injury that received cycling exercise demonstrated retention of MN resting membrane potential and spiked trigger levels compared to unexercised animals,[36] consistent with influences on ion conductances at or near the initial segment. This study supports therapeutic roles for exercise in slowing the deleterious responses of MNs to SCI and supporting structural and physiological plasticity in components of neuromuscular junctions associated with MNs located below the level of injury.

The lumbosacral spinal cord apparently can learn several tasks in which specific supraspinal and sensory information associated with movement influence and shape the activity of spinal networks. Even after complete transection injuries, cats can be trained to stand and step on a treadmill.[37] They are capable of adjusting the movement to change speed or weight support, and relearn a task after an interruption or change in training type.[38–41] Sen-

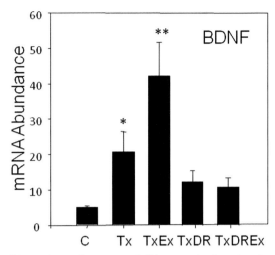

Figure 3. Deafferentation abolishes exercise-dependent increase in NTF mRNA. Rats with a complete thoracic transection injury show significantly increased levels of BDNF mRNA after cycling exercise. When dorsal root axons caudal to the injury are removed bilaterally, there is no benefit from exercise of these rats. This indicates the necessity of sensory information being transmitted to the spinal cord during exercise. C, control, uninjured rats; Tx, complete thoracic level-10 transection; Ex, cycling exercise for 10 days; DR, bilateral dorsal rhizotomy, T10–S2; *, significant increase from C; **, significant increase from Tx.

sory feedback is critical to coordinate limb function in spinalized cats during treadmill training, suggesting that spinal pathways can integrate dynamic sensory inputs provided by repetitive activation to shape motor output.[42,43] We have been applying training procedures to SCI rats[44,45] (Fig. 2) and preventing afferent feedback to drive spinal networks during exercise through either T12–S2 dorsal rhizotomy or pyridoxine toxicity.[46] Either of these approaches prevented rate-dependent reflex modulation, and there was no exercise-related increase in brain-derived neurotrophic factor (BDNF) mRNA (Fig. 3), indicating that input arising from large-diameter proprioceptors is required to realize the beneficial effects of exercise after SCI.

Change in neurotrophic factor levels with exercise

Neurotrophic factors (NTFs) are a class of polypeptide growth factors that promote neuron survival, modulate neuronal function, and enhance plasticity. For example, BDNF[47–50] and neurotrophin-3 (NT3)[51,52] have been shown to enhance survival of damaged neurons and contribute to the recovery of function following injury. With regard to SCI, exogenous NTFs rescue neurons after injury,[53,54] and in combination with a neural tissue transplant, promote the regenerative effort of acutely and chronically injured neurons.[6,49,55–57] Our studies with PNGs demonstrate that exogenous delivery of several NTFs, such as BDNF, glial cell line–derived neurotrophic factor (GDNF), or NT3, significantly increases the number of chronically injured neurons that regenerate an axon into the PNG.[17,49,58]

There is also evidence that NTFs play a role in maintaining and promoting function at the peripheral neuromuscular junction (NMJ). Indeed, recent studies demonstrate that NMJs in hindlimb muscles innervated by MNs caudal to spinal lesions are not stable, but extensively disassembled, resulting in substantial loss of nerve-muscle connectivity as early as two weeks after SCI,[59] when vulnerable NMJs lose synaptic organization by either excessive sprouting of their nerve terminals or failing to maintain muscle acetylcholine receptors (AChRs). On the other hand, cycling exercise that increased neuromuscular activity in muscles caudal to a complete transection suppressed sprouting and enhanced receptor maintenance, thereby stabilizing synaptic connectivity in hindlimb muscles.

Exercise and subsequent activation of neural networks influences transcription of neurotrophic factors, both centrally and peripherally, in skeletal muscle. In the developing CNS, activity stabilizes synapses[60] and disuse leads to synaptic elimination (i.e., pruning).[61,62] In intact adult rats, treadmill training increases BDNF and NT3 expression (protein and mRNA) in both spinal MNs and skeletal muscle.[63–65] BDNF and NT3 are localized by immunohistochemistry to the ventral horn, and NT3 staining is also dense in the substantia gelatinosa. Elimination of hindlimb EMG activity by complete thoracic and sacral cord transections, and bilateral lumbosacral dorsal rhizotomy (i.e., spinal isolation)[66] results in greatly diminished activity and reduced transcription of neurotrophins. Levels of neurotrophins (BDNF, NT3) and the downstream effector (synapsin I) were significantly diminished in the lumbar cord. BDNF and NT3 expression has also been shown to be downregulated after moderate spinal cord contusion, and increased with treadmill training,[67] cycle training,[68] or voluntary locomotion, in freely moving running wheels.[69]

We examined the effects of nonrobotic step training on a treadmill or motorized cycling on the ex-

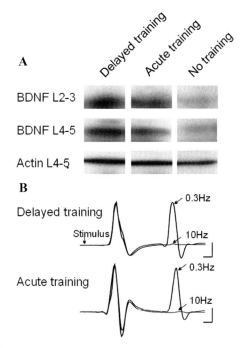

Figure 4. Neurotrophin expression with acute or delayed exercise after SCI. (A) Western blot data demonstrate the presence of comparable levels of BDNF in the lumbar spinal cord when cycling exercise was initiated 5 days after injury (acute) or 28 days after injury (delayed). (B) Frequency-dependent depression of the H reflex was restored in spinalized rats that received acute or delayed cycling exercise. Together, these observations demonstrate the potential for exercise-dependent plasticity in the injured spinal cord at short and long intervals after SCI.

pression of NTFs in the spinal cord.[45] Each of these exercise paradigms led to comparable increases in BDNF, NT3, NT4, and GDNF protein levels in the lumbar enlargement (L4–L6), as well as in upper lumbar (L2–3) and thoracic (T12–L1) regions just caudal to the lesion site. Interestingly, there was also a significant increase in NT3 levels in the thoracic spinal cord proximal to the level of complete transection injury. The increase in spinal NTFs and recovery of H-reflex frequency-dependent depression was positively correlated and required the afferent feedback provided by exercise,[46] suggesting that normalization of reflexes in the injured spinal cord is effectively triggered by afferent activity and local increase in NTFs. Motor recovery after SCI has been linked directly to the increase in BDNF levels within the lumbar spinal cord,[70] and a recent review of BDNF treatments after SCI[71] suggests that exercise is the safest intervention to achieve beneficial effects without some of the undesired effects often seen with exogenous delivery of BDNF over a prolonged period.

One issue that we and others have faced over the years is whether the effect of exercise is due to the latest bout or to cumulative effects over the entire period of exercise. Also it was not clear how a delay in exercise might affect levels of NTFs in the spinal cord and recovery of MN reflex activity. To address this, we prepared three groups of adult rats with complete lower thoracic transection injuries. Animals in group one began exercising five days after injury for five consecutive days. Animals in group two were subjected to only one bout of exercise (on the last day of exercise for group one animals), and those in group three were subjected to five days of exercise beginning four weeks after the transection injury. Western blot analysis of protein from the lumbar enlargement of these animals indicated that the effect of exercise on NTF production was indeed cumulative, as there was a significant increase in BDNF with five days of exercise (group one) over animals in group two, which received only a single bout of exercise (data not shown). Perhaps of greater importance was the observation of comparable levels of BDNF in animals of group one and group three, indicating the potential for exercise to increase NTFs with acute or delayed application (Fig. 4A). In these same groups, there was no significant difference in frequency-dependent depression of the H-reflex when training was provided

acutely or after a four-week delay (Fig. 4B). Whether this is related to the expression of NTFs with training is not known; these results nevertheless indicate a capacity for considerable neuroplasticity that remains long after the initial injury to the spinal cord.

Cell and molecular changes with exercise

Downstream effectors of BDNF, including TrkB, synapsin I, and CREB, also increase with exercise and are conversely diminished by botulinum-induced paralysis.[64] The effect on synapsin is particularly interesting because it is postulated to play an important role in neurotransmitter release and maintenance of synaptic contacts.[72,73] In a series of studies, we found that exercise after SCI led to a significant increase in the distribution of immuno-labeled synapsin around neurons in the intermediate gray and ventral horn of the lumbar spinal cord (data not shown). Whether this is indicative of an increase in synaptic sites per neuron or in the density of synaptic vesicles per synapse is unknown, but the results clearly demonstrate an effect of exercise. Correlating with this is the observation of an increase in the number of neurons expressing cFos in dorsal horn and intermediate gray regions of the lumbar spinal cord following hindlimb exercise of spinalized rats (Fig. 5). An increase in cFos is an indication of increased neuronal activity resulting from synaptic activity,[74,75] in this case during

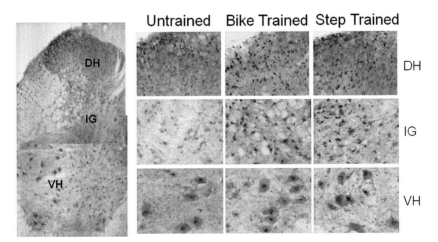

Figure 5. Immunolabeling for cFos after SCI and exercise. With bike or step training, there is an increase in the number of cFos-labeled neurons in the L4 dorsal horn (DH) and intermediate gray (IG) compared to neurons in untrained animals. The base level of cFos in MNs of the ventral horn (VH) does not appear to be changed with exercise. The expression of cFos is used as an indicator of neuronal activity following hindlimb exercise of spinalized rats.

Table 1. Cellular and molecular targets of exercise[a]

Cellular and molecular plasticity with exercise

Outcome	Location	Effects	What's next
Increased expression of neurotrophic factors	Below and above injury site	Neuroprotective Reduce axon dieback Rescue injured neurons	Examine the potential of exercise to promote axon growth into and out of a PNG
Upregulation of cFos expression	Multiple neuronal cell types below injury site	Neuroplasticity Enhanced neuronal activity	
Increased expression of phosphorylated S6	Interneurons below injury site	Neuroregenerative S6 is a molecule downstream of AKT/mTOR; suggests a strong regenerative potential	
Retention of MN properties, muscle size, and modulation of H-reflex	Below injury site	Neurorehabilitative Promote interneuronplasticity Modulate MN excitability Reduce muscle atrophy	

[a]As shown in the summary, known effects of exercise in the injured spinal cord demonstrate a wide range of action, including local and distant increase in NTF; upregulation of cFos expression in multiple neuron cell types; expression of downstream molecules of the AKT/mTOR pathway that may reflect a strong regenerative potential, and retention of muscle size, MN membrane properties, and reflex activity. In future experiments, exercise will be employed to attempt to increase the growth of regenerating axons into and, subsequently, out of a PNG.

exercise. It is not clear whether long-term consequences may arise from this transient increase in neuron activity, but it is tempting to put forward the notion that neurons in a heightened state of activity may be an attractive target for regenerating axons entering the area from a nerve graft. Continuing the theme of cellular plasticity after exercise, we identified an increase in the number of interneurons expressing phosphorylated-S6 (P-S6) and an increase in dendritic branching and content of P-S6 by MNs after cycling of SCI rats.[76] The absence of P-S6 expression in exercised rats concurrently treated with rapamycin is an indication of the role of mTOR in this example of exercise-dependent plasticity.

In other experiments, we examined the expression of microRNAs (miRs) in spinal cord tissue caudal to a transection injury with or without hindlimb exercise. Significant changes in response to cycling exercise were identified in genes associated with apoptosis and cell growth signaling pathways.[77] Upregulation of miR21 was associated with decreased expression of mRNA for PTEN and PDCD4, which

are upstream of caspase expression. Also, miR15b was downregulated in association with upregulation of Bcl2 mRNA (another signaling molecule upstream of caspase expression), and cycling exercise led to a decrease in protein levels of caspase 7. We also measured changes in miRs that affect expression of PTEN, AKT, and mTOR and S6 mRNA, which have been recently implicated in axon regeneration after optic nerve or spinal cord injury.[78,79] The increased expression of miR21 after exercise was associated with a decrease in PTEN message and protein, and an increase in AKT message and protein. A decrease in miR199a-3p with exercise was associated with an increase in mTOR and S6 message and protein. Thus it is tempting to suggest that modulation of the PTEN/AKT/mTOR pathway is one mechanism by which exercise might influence axon growth by neurons affected by SCI.

Summary

Several thousand axons regenerate into PNGs, yet only 10–15% of these appear to exit the graft when inhibitory proteoglycans in the spinal cord are digested before apposition of the distal graft end.[17,19] We have identified functional changes attributable to this relatively modest axonal outgrowth, but it is important to know whether the number of axons extending beyond the graft can be increased and whether such an increase would result in more dramatic functional recovery. Our previous work[58] demonstrated that short-term delivery of exogenous NTFs enhanced the regenerative effort of chronically injured neurons, and several recent studies demonstrate that creation of neurotrophic gradients effectively increase axonal regeneration beyond a transplant of neural precursor cells.[80,81] As a noninvasive treatment strategy, hindlimb exercise of spinalized rats significantly increases the level of NTFs in spinal cord tissue both proximal and distal to an injury site,[45] increases the expression of cFos in dorsal horn and intermediate gray neurons, promotes expression of molecules involved in regulation of protein synthesis, and restores reflex activity of MNs (Table 1). These results advocate the potential utility of this intervention for increasing the growth response of injured neurons, both in acute and chronic injury situations. As a first step toward enhancing the possibility of reconnection of axons across an injury, it is important to determine whether we can increase the number of axons available at the distal end of the PNG to form new contacts once they reenter the spinal cord. An increase in the number of outgrowing axons could result in more synaptic contacts being made with spinal cord neurons, possibly leading to an enhanced behavioral and/or functional outcome. A key feature of this work will be to test whether repetitive stimulation of spinal cord circuitry will be beneficial in attracting axons to areas of the spinal cord affected by exercise. In effect, exercise may be useful in creating suitable target areas within the spinal cord according to some of the plasticity observed so far. There remain several fundamental areas of exercise-dependent plasticity to explore; studies in these areas will influence our views of how to move forward with combination treatment strategies for functional recovery after SCI.

Acknowledgments

This work was supported by grants from the NIH (NS26380, NS055976, and NS074406 to J.D.H.) and the Craig H. Neilsen Foundation (to M.P.C.).

Conflicts of interest

The authors declare no conflicts of interest.

References

1. Crowe, M.J., J.C. Bresnahan, S.L. Shuman, *et al.* 1997. Apoptosis and delayed degeneration after spinal cord injury in rats and monkeys. *Nat. Med.* **3:** 73–76.
2. Bareyre, F.M., M. Kerschensteiner, O. Raineteau, *et al.* 2004. The injured spinal cord forms a new intraspinal circuit in adult rats. *Nat. Neurosci.* **7:** 269–277.
3. Jones, T.B., E.E. McDaniel & P.G. Popovich. 2005. Inflammatory-mediated injury and repair in the traumatically injured spinal cord. *Curr. Pharm. Des.* **11:** 1223–1236.
4. Fitch, M.T. & J. Silver. 2008. CNS injury, glial scars, and inflammation: Inhibitory extracellular matrices and regeneration failure. *Exp. Neurol.* **209:** 294–301.
5. Cao, Q., R.L. Benton & S.R. Whittemore. 2002. Stem cell repair of central nervous system injury. *J. Neurosci. Res.* **68:** 501–510.
6. Xu, X.M., V. Guenard, N. Kleitman & M.B. Bunge. 1995. Axonal regeneration into Schwann cell seeded guidance channels grafted into transected adult rat spinal cord. *J. Comp. Neurol.* **351:** 145–160.
7. Murray, M., D. Kim, Y. Liu, *et al.* 2002. Transplantation of genetically modified cells contributes to repair and recovery from spinal injury. *Brain Res. Brain Res. Rev.* **40:** 292–300.
8. Lee, Y.S., C.Y. Lin, R.T. Robertson, *et al.* 2004. Motor recovery and anatomical evidence of axonal regrowth in spinal cord-repaired adult rats. *J. Neuropathol. Exp. Neurol.* **63:** 233–245.
9. Reier, P.J. 2004. Cellular transplantation strategies for spinal cord injury and translational neurobiology. *NeuroRx* **1:** 424–451.
10. Gros, T., J.S. Sakamoto, A. Blesch, *et al.* 2010. Regeneration of long-tract axons through sites of spinal cord injury using template agarose scaffolds. *Biomaterials* **31:** 6719–6729.
11. Wang, M., P, Zhai, X. Chen, *et al.* 2011. Bioengineered scaffolds for spinal cord repair. *Tissue Eng. B* **17:** 177–194.
12. Liu, T., J.D. Houle, J. Xu, *et al.* 2012. Nanofibrous collagen nerve conduits for spinal cord repair. *Tissue Eng. A* **18:** 1057–1066.
13. David, S. & A.J. Aguayo. 1981. Axonal elongation into peripheral nervous system "bridges" after central nervous system injury in adult rats. *Science* **214:** 931–933.
14. Richardson, P.M., U.M. McGuinness & A.J. Aguayo. 1980. Axons from CNS neurons regenerate into PNS grafts. *Nature* **284:** 264–265.
15. Côté, M.-P., A.A. Amin, V.T. Tom & J.D. Houle. 2011. Peripheral nerve grafts support regeneration after spinal cord injury. *Neurotherapeutics* **8:** 294–303.
16. Houle, J.D., A. Amin, M.P. Cote, *et al.* 2009. Combining peripheral nerve grafting and matrix modulation to repair the injured rat spinal cord. *J. Visual. Exp.* **33:** 1324.
17. Tom, V.J., H.R. Sandrow-Feinberg, K. Miller, *et al.* 2009. Combining peripheral nerve grafts and chondroitinase promotes functional axonal regeneration in the chronically injured spinal cord. *J. Neurosci.* **29:** 14881–14890.

18. Houle, J.D. & A. Tessler. 2003. Repair of chronic spinal cord injury. *Exp. Neurol.* **182:** 247–260.

19. Houle, J.D., V.J. Tom, D. Mayes, *et al.* 2006. Combining an autologous peripheral nervous system "bridge" and matrix modification by chondroitinase allows robust, functional regeneration beyond a hemisection lesion of the adult rat spinal cord. *J. Neurosci.* **26:** 7405–7415.

20. Cote, M.P., A. Hanna, M.A. Lemay, *et al.* 2010. Peripheral nerve grafts after cervical spinal cord injury in adult cats. *Exp. Neurol.* **225:** 173–82.

21. Salame, C.G. & R.P. Dum. 1985. Central nervous system axonal regeneration into sciatic nerve grafts: physiological properties of the grafts and histologic findings in the neuraxis. *Exp. Neurol.* **90:** 322–340.

22. Pinzon, A., B. Calancie, M. Oudega & B.R. Noga. 2001. Conduction of impulses by axons regenerated in a Schwann cell graft in the transected adult rat thoracic spinal cord. *J. Neurosci. Res.* **64:** 533–541.

23. Gauthier, P., P. Réga, N. Lammari-Barreault & J. Polentes. 2002. Functional reconnections established by central respiratory neurons regenerating axons into a nerve graft bridging the respiratory centers to the cervical spinal cord. *J. Neurosci. Res.* **70:** 65–81.

24. Nordblom, J., J.K. Persson, M. Svensson & P. Mattsson. 2009. Peripheral nerve grafts in a spinal cord prosthesis result in regeneration and motor evoked potentials following spinal cord resection. *Restor. Neurol. Neurosci.* **27:** 285–295.

25. Lee, Y.S., I. Hsiao & V.W. Lin. 2002. Peripheral nerve grafts and aFGF restore partial hindlimb function in adult paraplegic rats. *J. Neurotrauma* **19:** 1203–1216.

26. Sandrow, H.R., J.S. Shumsky, A. Amin & J.D. Houle. 2008. Aspiration of a cervical spinal contusion injury in preparation for delayed peripheral nerve grafting does not impair forelimb behavior or axon regeneration. *Exp. Neurol.* **210:** 489–500.

27. Amin, A.A. & J.D. Houle. 2010. The role of neurotrophic factors and their receptors in ascending and descending axon regeneration through intraspinal peripheral nerve grafts (PNGs). *Soc. Neurosci. Abstr.* Program #588.8.

28. Edgerton, V.R., R.D. de Leon, N. Tillakaratne, *et al.* 1997. Use dependent plasticity in spinal stepping and standing. *Adv. Neurol.* **72:** 233–247.

29. de Leon, R.D., J.A. Hodgson, R.R. Roy & V.R. Edgerton. 1999. Retention of hindlimb stepping ability in adult spinal cats after the cessation of step training. *J. Neurophysiol.* **81:** 85–94.

30. Dobkin, B.H. 1998. Activity-dependent learning contributes to motor recovery. *Ann. Neurol.* **44:** 158–160.

31. Gustafsson, B., R. Katz & J. Malmsten. 1982. Effects of chronic partial deafferentation on the electrical properties of lumbar alpha-motoneurones in the cat. *Brain Res.* **246:** 23–33.

32. Hochman, S. & D.A. McCrea. 1994a. Changes of chronic spinalization on ankle extensor motoneurons. I: composite monosynaptic Ia EPSPs in four motoneuron pools. *J. Neurophysiol.* **71:** 1452–1467.

33. Hochman, S. & D.A. McCrea. 1994b. Effects of chronic spinalization on ankle extensor motoneurons. II: motoneuron electrical properties. *J. Neurophysiol.* **71:** 1468–1479.

34. Skinner, R.D., J.D. Houle, N.B. Reese, *et al.* 1996. Effects of exercise and fetal spinal cord implants on the H-reflex in chronically spinalized adult rats. *Brain Res.* **729:** 127–131.

35. Murphy, R.J.L., E.E. Dupont-Versteegden, C.A. Peterson & J.D. Houle. 1999. Two experimental strategies to restore muscle mass in adult rats following spinal cord injury. *Neurorehab. Neural Repair* **13:** 125–134.

36. Beaumont, E., J.D. Houle, C.A. Peterson & P.F. Gardiner. 2004. Fetal spinal cord transplant and passive exercise help to restore motoneuronal properties after spinal cord transection in rats. *Muscle Nerve* **29:** 234–242.

37. Martinez, M. & S. Rossignol. 2013. A dual spinal cord lesion paradigm to study spinal locomotor plasticity in the cat. *Ann. N.Y. Acad. Sci.* **1279:** 127–134. This volume.

38. Barbeau, H. & S. Rossignol. 1987. Recovery of locomotion after chronic spinalization in the adult cat. *Brain Res.* **412:** 84–95.

39. Hodgson, J., R. Roy, B. Dobkin & V.R. Edgerton. 1994. Can the mammalian spinal cord learn a motor task? *Med. Sci. Sports Exer.* **26:** 1491–1497.

40. Lovely, R., R. Gregor, R. Roy & V.R. Edgerton. 1986. Effects of training on the recovery of full-weight-bearing stepping in the adult spinal cat. *Exp. Neurol.* **92:** 421–435.

41. Lovely, R., R. Gregor, R. Roy & V. Edgerton. 1990. Weight-bearing hindlimb stepping in treadmill-exercised adult spinal cats. *Brain Res.* **514:** 206–218.

42. Barriere, G., A. Frigon, H. Leblond, *et al.* 2010. Dual spinal lesion paradigm in the cat: evolution of the kinematic locomotor pattern. *J. Neurophysiol.* **104:** 1119–1133.

43. Martinez, M., H. Delivet-Mongrain, H. Leblond & S. Rossignol. 2012. Incomplete spinal cord injury promotes durable functional changes within the spinal locomotor circuitry. *J. Neurophysiol.* **108:** 124–134.

44. Sandrow-Feinberg, H.R., J. Izzi, J.S. Shumsky, *et al.* 2009. Forced exercise as a rehabilitation strategy after unilateral cervical spinal cord contusion injury. *J. Neurotrauma* **26:** 1–11.

45. Côté, M.-P., G.A. Azzam, M.A. Lemay, *et al.* 2011. Activity-dependent increase in neurotrophic factors is associated with an enhanced modulation of spinal reflexes after spinal cord injury. *J. Neurotrauma* **28:** 299–309.

46. Ollivier-Lanvin, K., B.E. Keeler, R. Siegfried, *et al.* 2010. Proprioceptive neuropathy affects normalization of the H-reflex by exercise after spinal cord injury. *Exp. Neurol.* **221:** 198–205.

47. Koliatsos, V.E., R.E. Clatterbuck, J.W. Winslow, *et al.* 1993. Evidence that brain-derived neurotrophic factor is a trophic factor for motor neurons in vivo. *Neuron* **10:** 359–367.

48. Friedman, B., D. Kleinfeld, N.Y. Ip, *et al.* 1995. BDNF and NT4/5 exert neurotrophic influences on injured adult spinal motor neurons. *J. Neurosci.* **15:** 1044–1056.

49. Ye, J.H. & J.D. Houle. 1997. Treatment of the chronically injured spinal cord with neurotrophic factors can promote axonal regeneration from supraspinal neurons. *Exp. Neurol.* **143:** 70–81.

50. Gomez-Pinilla, F., Z. Ying, R.R. Roy, *et al.* 2002. Voluntary exercise induces a BDNF-mediated mechanism that promotes neuroplasticity. *J. Neurophysiol.* **88:** 2187–2195.

51. Koliatsos, V.E., R.E. Clatterbuck, J.W. Winslow, *et al.* 1993. Evidence that brain-derived neurotrophic factor is a trophic factor for motor neurons in vivo. *Neuron* **10**: 359–367.

52. Grill, R., K. Murai, A. Blesch, *et al.* 1997. Cellular delivery of neurotrophin-3 promotes corticospinal axonal growth and partial functional recovery after spinal cord injury. *J. Neurosci.* **17**: 5560–5572.

53. Giehl, K.M. & W. Tetzlaff. 1996. BDNF and NT-3 but not NGF, prevent axotomy-induced death of rat corticospinal neurons in vivo. *Eur. J. Neurosci.* **8**: 1167–1175.

54. Novikova, L.N., L.N. Novikov & I.-O. Kellerth. 2000. Survival effects of BDNF and NT-3 on axotomized rubrospinal neurons depend on the temporal pattern of neurotrophin administration. *Eur. J. Neurosci.* **12**: 776–780.

55. Bregman, B.S., J.-V. Coumans, H.N. Dai, *et al.* 2002. Transplants and neurotrophic factors increase regeneration and recovery of function after spinal cord injury. In *Progress in Brain Research*. L. McKerracher, G. Doucet & S. Rossignol, Eds.: 258–273. Elsevier. Montreal.

56. Coumans, J.V., T.T.-S. Lin, H.N. Dai, *et al.* 2001. Axonal regeneration and functional recovery after complete spinal cord transection in rats by delayed treatment with transplants and neurotrophins. *J. Neurosci.* **21**: 9334–9344.

57. Kadoya, K., S. Tsukada, P. Lu, *et al.* 2009. Combined intrinsic and extrinsic neuronal mechanisms facilitate bridging axonal regeneration one year after spinal cord injury. *Neuron* **64**: 165–172.

58. Dolbeare, D. & J.D. Houle. 2003. Restriction of axonal retraction and promotion of axonal regeneration by chronically injured neurons after intraspinal treatment with glial cell line-derived neurotrophic factor (GDNF). *J. Neurotrauma* **20**: 1251–1261.

59. Burns, A.S., S. Jawaid, H. Zhong, *et al.* 2007. Paralysis elicited by spinal cord injury evokes selective disassembly of neuromuscular synapses with and without terminal sprouting in ankle flexors of the adult rat. *J. Comp. Neurol.* **500**: 116–33.

60. Shatz, C.J. 1990. Impulse activity and the patterning of connections during CNS development. *Neuron* **5**: 745–756.

61. Constantine-Paton, M., H.T. Cline & E. Debski. 1990. Patterned activity, synaptic convergence, and the NMDA receptor in developing visual pathways. *Ann. Rev. Neurosci.* **13**: 129–154.

62. Constantine-Paton, M. & H.T. Cline. 1998. LTP and activity-dependent synaptogenesis: the more alike they are, the more different they become. *Curr. Opin. Neurobiol.* **8**: 139–148.

63. Gomez-Pinilla, F., Z. Ying, P. Opazo, *et al.* 2001. Differential regulation by exercise of BDNF and NT3 in rat spinal cord and skeletal muscle. *Eur. J. Neurosci.* **13**: 1078–1084.

64. Gomez-Pinilla, F., Z. Ying, R.R. Roy, *et al.* 2002. Voluntary exercise induces a BDNF-mediated mechanism that promotes neuroplasticity. *J. Neurophysiol.* **88**: 2187–2195.

65. Ying, Z., R.R. Roy, V.R. Edgerton & F. Gomez-Pinilla. 2003. Voluntary exercise increases neurotrophin-3 and its receptor TrkC in the spinal cord. *Brain Res.* **987**: 93–99.

66. Gomez-Pinilla, F., Z. Ying, R.R. Roy, *et al.* 2004. Afferent input modulates neurotrophins and synaptic plasticity in the spinal cord. *J. Neurophysiol.* **92**: 3423–3432.

67. Hutchinson, K.J., F. Gomez-Pinilla, M.J. Crowe, *et al.* 2004. Three exercise paradigms differentially improve sensory recovery after spinal cord contusion in rats. *Brain* **127**: 1403–1414.

68. Dupont-Versteegden, E.E., J.D. Houle, R.A. Dennis, *et al.* 2004. Exercise-induced gene expression in soleus muscle is dependent on time after spinal cord injury in rats. *Muscle Nerve* **29**: 73–81.

69. Ying, Z., R.R. Roy, V.R. Edgerton & F. Gomez-Pinilla. 2005. Exercise restores levels of neurotrophins and synaptic plasticity following spinal cord injury. *Exp. Neurol.* **193**: 411–419.

70. Ying, Z., R.R. Roy, H. Zhong, *et al.* 2008. BDNF-exercise interactions in the recovery of symmetrical stepping after a cervical hemisection in rats. *Neuroscience* **155**: 1070–1078.

71. Weishaupt, N., A. Blesch & K. Fouad. 2012. BDNF: the career of a multifaceted neurotrophin in spinal cord injury. *Exp. Neurol.* **238**: 254–264.

72. Brock, T.O. & J.P. O'Callaghan. 1987. Quantitative changes in the synaptic vesicle proteins synapsin I and p38 and the astrocyte-specific protein glial fibrillary acidic protein are associated with chemical-induced injury to the rat central nervous system. *J. Neurosci.* **7**: 931–942.

73. Wang, S.D., M.E. Goldberger & M. Murray. 1991. Plasticity of spinal systems after unilateral lumbosacral dorsal rhizotomy in the adult rat. *J. Comp. Neurol.* **304**: 555–568.

74. Ahn, S.N., J.J. Guu, A.J. Tobin, *et al.* 2006. Use of c-fos to identify activity-dependent spinal neurons after stepping in intact adult rats. *Spinal Cord* **44**: 547–559.

75. Morgan, J.I. & T. Curran. 1991. Stimulus-transcription coupling in the nervous system: involvement of the inducible proto-oncogenes fos and jun. *Ann. Rev. Neurosci.* **14**: 412–452.

76. Liu, G., M.R. Detloff, K.N. Miller, *et al.* 2011. Exercise modulates microRNAs that affect the PTEN/mTOR pathway in rats after spinal cord injury. *Exp. Neurol.* **233**: 447–456.

77. Liu, G., B.E. Keeler, V. Zhukareva & J.D. Houle. 2010. Cycling exercise affects the expression of apoptosis-associated microRNAs after spinal cord injury in rats. *Exp. Neurol.* **226**: 200–206.

78. Park, K.K., K. Liu, Y. Hu, *et al.* 2008. Promoting axon regeneration in the adult CNS by modulation of the PTEN/mTOR pathway. *Science* **322**: 963–966.

79. Liu, K., Y. Lu, J.K. Lee, *et al.* 2010. PTEN deletion enhances the regenerative ability of adult corticospinal neurons. *Nat. Neurosci.* **13**: 1075–1081.

80. Bonner, J.F., T.M. Connors, W.F. Silverman, *et al.* 2011. Grafted neural progenitors integrate and restore synaptic connectivity across the injured spinal. *J. Neurosci.* **31**: 4675–4686.

81. Liu, P., Y. Wang, L. Graham, *et al.* 2012. Long-distance growth and connectivity of neural stem cells after severe spinal cord injury. *Cell* **150**: 1264–1273.

Ann. N.Y. Acad. Sci. ISSN 0077-8923

Accelerating locomotor recovery after incomplete spinal injury

Brian K. Hillen,[1] James J. Abbas,[2] and Ranu Jung[1]

[1]Department of Biomedical Engineering, Florida International University, Miami, Florida. [2]Center for Adaptive Neural Systems, Arizona State University, Tempe, Arizona

Address for correspondence: Ranu Jung, Ph.D., Professor and Chair, Department of Biomedical Engineering, Wallace H. Coulter Eminent Scholars Chair in Biomedical Engineering, Florida International University, 10555 W. Flagler Street, EC2602, Miami, FL 33174. rjung@fiu.edu

A traumatic spinal injury can destroy cells, irreparably damage axons, and trigger a cascade of biochemical responses that increase the extent of injury. Although damaged central nervous system axons do not regrow well naturally, the distributed nature of the nervous system and its capacity to adapt provide opportunities for recovery of function. It is apparent that activity-dependent plasticity plays a role in this recovery and that the endogenous response to injury heightens the capacity for recovery for at least several weeks postinjury. To restore locomotor function, researchers have investigated the use of treadmill-based training, robots, and electrical stimulation to tap into adaptive activity-dependent processes. The current challenge is to maximize the degree of functional recovery. This manuscript reviews the endogenous neural system response to injury, and reviews data and presents novel analyses of these from a rat model of contusion injury that demonstrates how a targeted intervention can accelerate recovery, presumably by engaging processes that underlie activity-dependent plasticity.

Keywords: spinal cord injury; locomotion; electrical stimulation; plasticity; kinematics; coordination

The critical role of the spinal cord in a broad range of physiological functions is clearly demonstrated by the deficits observed after spinal cord injury (SCI) and by the medical conditions that develop in the acute and chronic phases after injury.[1–3] Locomotor deficits have been extensively studied, and numerous techniques designed to promote and accelerate recovery of locomotor function have been developed.[4] This emphasis on locomotor function is based on several considerations: its clinical relevance, its repetitive nature, and the fact that it can be readily observed and measured.

Locomotion is clinically relevant because mobility has a direct effect on quality of life by enabling individuals to participate in a range of activities and has an indirect effect on health by enabling exercise and physical activity. These secondary health benefits can reduce the likelihood and/or severity of conditions such as diabetes, cardiovascular disease, and soft tissue breakdown—all of which are of great concern for individuals with SCI.[5] Restoration of locomotor ability can slow or reverse the downward spiral that can occur when injury leads to reduced activity.[6]

Since locomotion involves repetitive activation of neural circuits, it is well suited for the study of the processes of activity-dependent plasticity and their potential role in providing functional benefits. In many spinal cord injuries[7] (and in many animal models[8]), a critical subset of the spinal circuitry responsible for generating the oscillatory patterns that drive locomotion may retain substantial capabilities and may be particularly responsive to therapeutic strategies that activate them in a repetitive manner.[9]

Finally, locomotion can be directly observed, can be rated by widely used clinical measures,[1] and can be quantitatively characterized using a combination of well-established biomechanical measures[10,11] as well as some novel dynamical systems approaches.[12] This observability facilitates the use of locomotion as a behavior that is a biomarker of neural function

doi: 10.1111/nyas.12061

as well as deficit, and of subsequent recovery after injury.

In developing and evaluating therapies for promoting recovery after SCI, it is important to recognize that the degree of success is often highly dependent on the time (postinjury) at which they were administered. Many clinical trials are performed using subjects in the chronic postinjury phase, which provides a stable baseline from which to assess the therapeutic intervention. Furthermore, it facilitates participation in the experimental protocol, since potential subjects would have worked through many of the lifestyle adjustments brought on by the injury. While there is evidence that therapy administered even long after the injury can have positive effects,[13] administering therapy early has the potential to accelerate recovery.[14] This window of opportunity may primarily be the result of factors involved in the physiological response to trauma, but they may also benefit from an upward spiral effect in which capabilities enable activity, which in turn enhances capabilities.

This review provides a brief overview of spinal cord injury, its pathophysiology, and its endogenous recovery, and then describes current methods of rehabilitation with an emphasis on strategies that leverage the endogenous neural response to repetitive stimuli. We present results from a therapy that uses electrical stimulation in a rat model to accelerate locomotor recovery following contusion injury.

Physiological response to spinal cord injury

Although spinal cord injury can result from tumor growth or other disorders, it typically occurs because of mechanical trauma to the spine that breaks vertebrae and/or displaces adjacent vertebrae.[15] The two primary factors that determine the clinical impact of an injury are the spinal level and the severity. Most injuries are incomplete (iSCI) in that some connections across the injury site remain intact,[16] thus sparing some supraspinal influence on function and, perhaps more importantly, providing a substrate for plasticity that may improve communication across the injury level, thereby affording some degree of functional recovery. In instances of complete SCI, plasticity in circuits below the level of injury may enhance function.[17–19]

Directly following the traumatic injury is a complex cascade of events, which, if unaltered, tends to cause additional damage to the nervous system. Knowing the details of this sequence may help determine appropriate timing for rehabilitation strategies. The following description of the stages of injury summarizes a detailed explanation provided by Bareyre and Schwab.[20] Beginning with a mechanical lesion that causes immediate damage to the cord, the acute phase of injury continues until a few days after injury. The mechanical trauma and initial response may result in reduced blood flow, which can cause ischemic necrosis as well as subsequent edema in the spinal cord. These events can produce changes in extracellular ion concentrations, which, over time (days in animal models, weeks in humans), can lead to neural failure and spinal shock. The secondary phase of injury includes the time period from minutes to weeks following injury. It is marked by excessive release of neurotransmitters, inflammatory reactions, and apoptosis. During this period, the size of the astrocytes increases in a process termed *reactive gliosis*, creating a glilal scar at the site of injury. Neutrophils, macrophages, and T cells migrate toward the injury site. Although they perform important functions in the body's response to injury, they can also have detrimental effects on central nervous system tissue. In the chronic phase of injury (days to years), apoptosis can continue and surviving cells can be further affected by impairments in channel and receptor functions. Finally, there can be significant scar formation along with demyelination of some fibers and Wallerian degeneration of axons that no longer have a target due to cell loss that immediately followed the injury or occurred in subsequent phases of the response.

During all phases following SCI, the central nervous system undergoes substantial reorganization. These changes include synaptic plasticity, axonal sprouting, and cellular proliferation throughout the central nervous system, not just caudal to the injury.[21] Some of the changes may involve the formation of new pathways to bypass the damaged regions, while other changes may involve reorganization in undamaged regions to take on new or modified functions, as occurs in the reorganization of cortical maps.[22] Although expression of neurotrophic factors increases in the days following injury, neurotransmitter receptors and transporters and ion channels are downregulated.[20] This is likely due to decreased firing of the neurons both from the lack of supraspinal input following the injury as well as the typical immobilization following injury. Investigations of the dendrites caudal to the injury demonstrate a significant loss of dendritic branching[23] as well as changes at the synaptic level that result in substantial alterations in electrophysiological properties.[24] Of particular clinical importance are changes in electrophysiological properties and connectivity that result in hyperreflexia and autonomic dysreflexia.[25–27]

Despite the loss of cells due to the initial trauma and the potentially detrimental effects of the physiological response, there is clinical evidence that at least some degree of functional recovery is possible. Most importantly, there is growing clinical evidence that some interventions can enhance and/or accelerate functional recovery, and the scientific community is developing an improved understanding of the mechanisms that mediate functional recovery.

From our perspective, a combination of four critical observations motivate the investigation of interventions delivered at specific times postinjury that use electrical stimulation: (1) the downregulation in the production of proteins that form neurotransmitter receptors, transporters, and ion channels may be due to reduced activity levels, which would suggest that the downregulation could be reversed by enhancing neural activity; (2) many of the processes involved in plasticity at the subcellular, cellular, and circuit levels are upregulated after injury, thus making the system highly responsive to inputs that could help to direct plasticity toward functional reorganization; (3) the complex interactions among the various components of the physiological response to injury are time dependent and are likely to affect the efficacy of interventions designed to promote functional recovery; and (4) electrical stimulation provides a means to directly activate specific neurons and/or to indirectly activate diverse sets of neurons in a coordinated manner by producing movements through activation of motoneurons to generate muscle contractions. Cyclic movements, such as locomotion, can elicit repetitive patterns of activation that may maximize the benefits to be derived from activity-dependent plasticity.

By delivering an appropriately timed intervention that utilizes electrical stimulation to produce cyclic movements, it may be possible to maximize the therapeutic benefits by reversing the downregulation of receptors and other proteins, and directing plasticity to produce functional reorganization.

Assessment of locomotor recovery

When evaluating new therapies, animal models are often used for SCI research since techniques have been developed to induce injuries with less variability than is observed in human patients. In both animal models and humans, sensitive and reliable assessment tools are needed to determine the effectiveness of SCI therapies on any timescale. Assessments of movement utilize outcome measures that reflect the varying extent to which recovery can occur, from increasing range of motion of a joint to increasing speed. Locomotion is often used, as it is a repetitive, reproducible movement that relies on coordination of multiple neurological and biomechanical levels. Locomotor assessments range from clinical measures[1] to 3D kinematic analysis.[10] Locomotion can be tested over ground or on a treadmill (Fig. 1A). While the clinical measures do capture the level of recovery, they are not very sensitive[28] and rely on subjective assessment by individual experimenters. However, these types of measures are relatively easy to assess, so they are included in most experiments. In the rat model, the Basso Beatie and Breshnehan locomotor score[29] fulfills a similar role. However, to view smaller changes in recovery without relying on experimenter subjectivity, 3D gait analysis is often used (e.g., Fig. 1C).

Quantitative measures available from 3D gait analysis include phase lag, symmetry, complexity, and range of motion. Left–right symmetry is perhaps the most easily identifiable measurement. Uninjured individuals tend to have gait patterns that exhibit left–right symmetry.[10] Following injury, this symmetry can be disrupted, although it is often regained to some extent during recovery. Intralimb joint angle phase relationships help identify coordination within the limb. As the phase values approach 1, the joints move in sync with one another, as can be seen following various injuries. Following SCI, range of motion in the ankle increases while it decreases within the knee. Additionally, the ankle trajectory loses a prestance extension phase. Complexity measures may characterize the changes in neural control following iSCI.

Although characterizing overground locomotion can be informative, more challenging locomotor activities may provide a more sensitive assessment of locomotor capabilities. One possibility is to use a more complex task that requires a higher level of supraspinal input. The horizontal ladder task

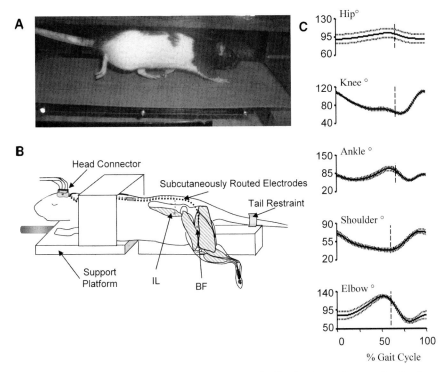

Figure 1. Rat locomotor testing and therapy. (A) Rat walking on a treadmill; reflective markers are placed on bony prominences for 3D visualization to collect kinematic data. (B) Stimulation therapy setup (figure modified from Ref. 64). Electrodes for neuromuscular stimulation are implanted in the hip flexors (IL-iliacus) and extensors (BF$_h$: biceps remoralis anterior head). Electrode leads are routed to a head connector for connection to a stimulator. The rat is mounted on a platform for stimulation training. (C) Sham injured kinematic data from a rat walking on a treadmill (B.K. Hillen, Arizona State University, Tempe, unpublished results). Note the biphasic nature of the knee and ankle trajectories.

assesses more fine motor control than the gross locomotor skill needed for overground walking; fine motor control recovery is much more minimal than gross locomotor recovery following injury.[30] Accordingly, fine motor rehabilitative tasks may be more efficacious. For instance, in a horizontal ladder task (for training and assessment), SCI rats showed significantly more improvement when the training was administered immediately (three to four days) following injury as opposed to three months later.[31] The early time period may be very specific. In animals trained on a ladder starting one week postinjury, performance was not improved.[32] Determining the correct window of opportunity and measurement technique for therapy may enhance rehabilitation following SCI. Some other specific tasks may also suffer when training is applied early. When rats are trained to swim, early training (three days postinjury) is actually worse that delayed training (two weeks postinjury) in the swimming assessment

and neither time point provides overground locomotor improvement.[33] Finally, when using reaching training following cervical injury, early training has a negative effect on untrained tasks (horizontal ladder).[34] This may point to overground locomotion (including ladder locomotion) as one of the few tasks that should be trained early. These differential responses may be due to the different tracts involved in these tasks.[35] The corticospinal tract is involved with skilled forelimb movements in rats[36] while treadmill locomotion can persist even following complete transection in neonatal rats.[37] As described in the example later on, even short-term therapy applied at the right time point has the potential to significantly accelerate locomotor recovery.

Accelerating locomotor recovery

While the body does undergo some recovery on its own, targeted interventions have the possibility of accelerating or enhancing recovery following iSCI. Many existing therapies utilize the body's own adaptive processes. The degree of recovery is dependent on a large number of factors, including location of injury, severity of injury, and immediate and long-term care.

In addition to the repetitive motion therapies discussed below, a large number of different types of therapy have been attempted for spinal cord injury. These include the application of transplants and neurotrophic factors,[38] viral vectors,[39] neural bypasses,[40] and combinations of treatments.[41,42]

Passive exercise. One of the most basic types of rehabilitation following iSCI is passive exercise. Limbs are moved with the help of physical therapists, motorized bikes, or robots. Much of the control of movement is organized around reflex circuits. These circuits coordinate movement around a joint and between joints.[43] Thus, when reflexes are disrupted, as seen following iSCI, movement coordination may suffer as well. Passive movement therapy has been shown to normalize some of these improper reflex actions[44] and enhance general reflex modulation following SCI.[45] For instance, passive cycling exercise has been shown to affect a pathway involved in regulation of protein synthesis, cell growth and proliferation,[46] and upregulate neurotrophic factors while minimizing muscle mass loss following spinal cord injury.[47] The primary limitation of passive exercise is that it docs not directly activate motoneurons, thus possibly limiting (though not eliminating) the activation of processes that would promote activity-dependent plasticity.

Active exercise. Active exercise in which the individual generates volitional contractions to perform specific, repetitive movements may be the most intuitive rehabilitative strategy following iSCI. One of the most common is treadmill-based locomotor retraining therapy in which the individual walks on a treadmill with a harness providing some body weight support and physical therapists helping to move the legs. Newer versions of this therapy include robot-assisted movements.[48] The addition of volitional drive to the passive exercise regime may help to activate a network of supraspinal and spinal circuits that may engage activity-dependent processes. If the volitional drive is successful in producing muscle contractions, it may enhance sensory feedback by activating muscle spindles and Golgi tendon or-

gans. Active locomotor training following SCI in rodent models has led to an increase in expression of neurotrophic factors[49] mitigating the effects of some negative adaptations. The effect of active exercise on increased expression of neuotrophic factors and synaptic plasticity in the spinal cord is particularly strong.[50,51] Repetitive active training has also been shown to promote cortical remapping[52–54] and to improve corticospinal drive.[55,56] The primary limitation of active exercise is that it requires some level of function in the individual to work. The results achieved by the individual are often relative to the amount of function they have before starting therapy.

Electrical stimulation. A number of clinical and basic science studies suggest that therapy is most likely to efficiently promote functional reorganization if it produces rich, appropriately timed patterns of activity that are associated with functional activities.[57,58] While performing passive movements or active (volitional) activities, the patterns of activity produced in the spinal cord and supraspinal circuits are likely to be deficient relative to the patterns in an uninjured individual. Most notably, paralyzed or paretic muscles may not receive sufficient excitation to generate contractions of sufficient strength. This insufficiency in motor drive would reduce the activity of sensors in the muscle and tendon and would limit or alter the activity in other somatic sensors by impairing the production of a suitable movement pattern. If delivered appropriately, it may be possible to use electrical stimulation of motoneurons to produce muscle contractions in an appropriately timed sequence to activate sensors in the muscles and to generate functional movements that activate somatic sensors. This rich pattern of activity in sensorimotor circuits may prove to be more effective than passive or volitional exercise in activating activity-dependent processes to provide functional gains.[6] Furthermore, it may be possible to initiate an electrical stimulation intervention earlier in the postinjury phase than protocols that depend upon volitional activation. In a number of studies it has been demonstrated that electrical stimulation can affect the properties of the spinal pattern generator,[59] it can increase neurotransmitter expression,[60] and it has been shown to partially restore spinal reflex action.[61] While the goal of electrical stimulation is generally to elicit appropriate movement, even stimulation of the sensory affer-ents alone may increase recovery following iSCI.[62] Thus, electrical stimulation addresses much of the pathophysiology of SCI as well as the sites of neu-ral plasticity identified during the recovery phase. The properties described previously apply to elec-trical stimulation in general (including open-loop stimulation, as described in our example below), but further gains may be obtained with more ad-vanced controllers. One important challenge is to deliver patterns of electrical stimulation that are ap-propriately timed to generate the desired movement pattern in a specific individual. Given the complex, time-dependent interactions of the processes of the endogenous response to injury, a second key chal-lenge is to identify the appropriate time-window after injury to deliver electrical stimulation-based therapy.

Adaptive control of electrical stimulation. For any given movement, although there is some consis-tency in muscle activity patterns across individuals, there is substantial variability due to individual dif-ferences in neural and musculoskeletal properties, and such differences are likely to be exacerbated after injury.[28,63] Predefined, or open-loop, stimu-lation patterns are therefore limited in their abil-ity to produce desired movement patterns and the limitations are magnified by the high rate of fatigue of electrically stimulated muscle.[64–66]

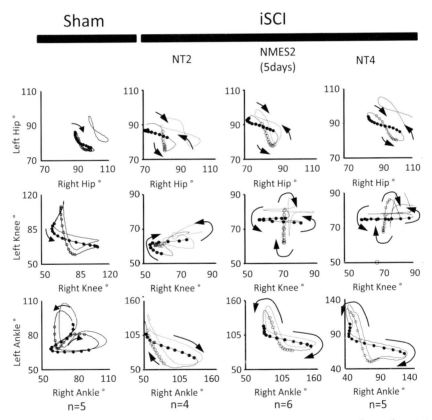

Figure 2. Sham and incomplete spinal cord injury (iSCI) intralimb coordination patterns. Averaged traces from 4–6 animals, 5- to 12-step cycles each. Sham, sham injured rats; NT2 and NT4, iSCI rats with no training tested two weeks and four weeks postinjury, respectively; NMES2, iSCI rats with five days of electrical stimulation therapy tested two weeks postinjury. Sham and NT4 data derived from B.K. Hillen, Arizona State University, Tempe (unpublished results); NT2 and NMES2 derived from Ref. 64. Left joint angle is plotted versus right joint angle. Symmetry in movement is indicated by symmetry about the diagonal. Symmetry is lost in the knee and ankle two weeks following injury. Symmetry is maintained if NMES therapy is added or an additional two weeks have elapsed. Note the simplification of the ankle–ankle pattern. In both the NMES and the NT4 animals, the knee–knee pattern shows a cruciform shape, with one joint moving while the opposite one remains stationary. Although the NMES2 and NT4 animals do not show the same trajectories as the sham animals, both groups were able to walk proficiently.

Adaptive control strategies have been developed to help account for the variability observed across individuals and for muscle fatigue.[65, 67–70] If successful, this type of control accurately produces the desired motion in a highly repeatable manner[66] and may therefore maximize the therapeutic benefits of electrical stimulation.

Early application of neuromuscular electrical stimulation. Neuromuscular electrical stimulation (NMES) (open-loop) applied during the second week postinjury may accelerate recovery following iSCI. As previously mentioned, the timing of rehabilitation techniques may contribute to their effectiveness. Rats reach the chronic phase of their recovery approximately four to six weeks following iSCI, so immediate intervention is needed to assess effectiveness of therapies within the window of recovery. In Jung *et al.*,[64] five days of electrical stimulation to hip flexors and extensors (Fig. 1B) was applied one week following a mild–moderate T9 spinal contusion injury (tested two weeks following the injury). Animals with the same level of injury but without therapy were assessed at two weeks and four weeks postinjury (B.K. Hillen, Arizona State University, Tempe, unpublished results). Animals were tested using 3D gait analysis while walking on

a treadmill.[10] This analysis yielded trajectories of individual joint angles of the hip, knee, and ankle over multiple cycles of movement. Inter- and intralimb angle–angle trajectories reflect the coordination among the different joints and limbs. Quantitative measures of symmetry between left and right joint angles can be obtained. Similarly, phase delays between particular events during the gait cycles (e.g., touch-down of each hindlimb or maximum joint angle during the gait cycle) can be determined. The stimulation training protocol caused the animals to improve on a number of measures and adopt gait characteristics similar to those of animals four weeks postinjury. This is evident in the left–right interlimb coordination plots (Fig. 2), particularly at the knee. When the right joint angle trajectory is plotted versus that of the left, symmetry about the diagonal indicates symmetry in locomotion. The pattern is clearly more symmetric in the two-week trained and four-week animals compared to the two-week untrained animals. It, however, is still very different from the normal (sham) animals. Quantitative symmetry analysis also indicated that the NMES trained animals have more symmetric gait than the untrained animals.[64] Both two-week NMES animals and four-week untrained animals showed a cruciform pattern in the interlimb coordination

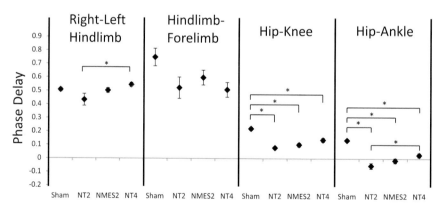

Figure 3. Phase delays for kinematic events (+/- SE). Sham and NT4 data derived from B.K. Hillen, Arizona State University, Tempe (unpublished results); NT2 and NMES2 data derived from Ref. 64. Hindlimb–forelimb and left–right hindlimb phase delay: relative phase of each limb touchdown with respect to another within a gait cycle. Phase is defined at discrete moments within a gait cycle and varies between 0 and 1 (two consecutive touchdown events for each limb). Hip–knee and hip–ankle: relative phase calculated from time between maximum joint angles for each joint during a gait cycle, within a limb. Injury causes a change in forelimb–hindlimb coordination, shifting it from approximately 0.75 to approximately 0.5, or completely out of phase. Left–right phase delay is altered following injury; however, NMES training or an additional two-week recovery period returns left–right coordination to symmetry. Intralimb phase delays (hip–knee and hip–ankle) are significantly decreased following injury, with a progressive increase from stimulation therapy or additional time for recovery. Sham, sham injured rats; NT2 and NT4, iSCI rats with no training tested two weeks and four weeks postinjury, respectively; NMES2, iSCI rats with five days of electrical stimulation therapy tested two weeks postinjury.

plots. This pattern indicates that one joint moves while the other is held constant. Although this pattern is different from normal, it is a further time point in the natural recovery. Because the pattern for the two-week trained animals is similar to that of the four-week untrained animals, the data suggest that NMES accelerated recovery following injury. This pattern of accelerated recovery is also reflected by the phase delay analysis (Fig. 3). Again, while not the same as the sham animals, the phase delay for the two-week stimulation-trained animals is similar to the four-week untrained animals. While the left–right phase in the two-week untrained animals is different from the four-week untrained animals, it is not different from the two-week NMES animals. Both intralimb phase delay measures (hip–knee and hip–ankle) show differences following injury, with a trend toward recovery seen. Particularly, the NMES trained animals have values between the untrained two-week and untrained four-week animals. One method for quantitative analysis of gait parameters is permutation entropy (PE).[71] Permutation entropy quantifies the likelihood of a signal to change direction. This new analysis of the data from Jung *et al.*[64] uses the code for PE from Olofsen *et al.*[72] for tau = 1.

The level of complexity at the hip, as measured by permutation entropy, decreased due to the stimulation (Fig. 4), presumably because the intervention was a simple alternating flexor/extensor pattern of stimulation at the hip. Most notably, the ankle lost significant complexity, mostly due to the loss of the second local maxima (prestance extension) in the ankle trajectory. Overall, there was a trend for increasing complexity from untrained two-week to NMES two-week to untrained four-week.

In this example, applying a very small dose of stimulation to a small number of muscles very soon following injury was able to accelerate recovery. Although the intervention had an effect, a more complex stimulation pattern may be able to increase the effect. The course of recovery was not followed beyond this time point; however, future studies will be needed to determine whether accelerated recovery leads to improved outcomes that are persistently improved following therapy. An adaptive stimulation paradigm that can produce a movement pattern similar to that in intact animals[66] may be able to provide a greater degree of recovery of the complexity of movement at each joint. It is also important to

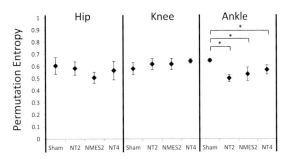

Figure 4. Permutation entropy complexity measures. Sham and NT4 data derived from B.K. Hillen, Arizona State University, Tempe (unpublished results), and new analysis based on data of NT2 and NMES2; data from Ref. 64. Permutation entropy quantifies the complexity of a time series signal, indicating how often it changes direction. Higher values indicate higher complexity. Following stimulation therapy, hip complexity decreased due to entrainment to the simple stimulation pattern. Following injury, ankle complexity decreased with a concomitant increase in knee complexity. Sham, sham injured rats; NT2 and NT4, iSCI rats with no training tested two weeks and four weeks postinjury, respectively; NMES2, iSCI rats with five days of electrical stimulation therapy tested two weeks postinjury.

note that accelerating recovery may have a positive feedback effect on clinical outcomes because faster recovery can lead to higher activity levels, which would in turn speed recovery.

Conclusions

The complex time course of the various components of the endogenous response to injury provides the substrate for rehabilitation interventions. Results from clinical and basic science studies indicate that repetitive movements may promote functional recovery by tapping into the processes that underlie activity-dependent plasticity. The efficacy of any rehabilitation strategy may be determined, or at least strongly influenced, by the extent to which it can produce functional, coordinated patterns of activity across spinal and supraspinal circuits. Electrical stimulation therapy can accelerate recovery of function and can be readily implemented in the clinic. It can be incorporated into movement therapies of varying complexity that meet clinical constraints and are appropriate at the various stages of progression after injury. Current goals are to utilize the technique for producing complex movement patterns and to determine the appropriate windows of opportunity to maximize functional benefits. If successful, electrical stimulation may play a critical role in slowing or reversing the downward spiral that can occur when injury leads to reduction in activity levels.

Acknowledgment

This work was supported by NIH R01-NS054282.

Conflicts of interest

The authors declare no conflicts of interest.

References

1. Marino, R.J. & D.E. Graves. 2004. Metric properties of the ASIA motor score: subscales improve correlation with functional activities. *Arch. Phys. Med. Rehabil.* **85:** 1804–1810.
2. McKinley, W.O. *et al.* 1999. Long-term medical complications after traumatic spinal cord injury: a regional model systems analysis. *Arch. Phys. Med. Rehabil.* **80:** 1402–1410.
3. Weaver, L.C. *et al.* 2006. Autonomic dysreflexia after spinal cord injury: central mechanisms and strategies for prevention. *Progr. Brain Res.* **152:** 245–263.
4. Ramer, M.S., G.P. Harper & E.J. Bradbury. 2000. Progress in spinal cord research—a refined strategy for the International Spinal Research Trust. *Spinal Cord* **38:** 449–472.
5. Bajotto, G. & Y. Shimomura. 2006. Determinants of disuse-induced skeletal muscle atrophy: exercise and nutrition countermeasures to prevent protein loss. *J. Nutr. Sci. Vitaminol.* **52:** 233–247.
6. Baldi, J.C. *et al.* 1998. Muscle atrophy is prevented in patients with acute spinal cord injury using functional electrical stimulation. *Spinal Cord* **36:** 463–469.
7. Kiehn, O. 2006. Locomotor circuits in the mammalian spinal cord. *Annu. Rev. Neurosci.* **29:** 279–306.
8. Ichiyama, R.M. *et al.* 2008. Step training reinforces specific spinal locomotor circuitry in adult spinal rats. *J. Neurosci.* **28:** 7370–7375.
9. Minassian, K. *et al.* 2012. Neuromodulation of lower limb motor control in restorative neurology. *Clin. Neurol. Neurosurg.* **114:** 489–497.
10. Thota, A.K. *et al.* 2005. Neuromechanical control of locomotion in the rat. *J. Neurotrauma* **22:** 442–465.
11. Johnson, W.L. *et al.* 2012. Quantitative metrics of spinal cord injury recovery in the rat using motion capture, electromyography and ground reaction force measurement. *J. Neurosci. Methods* **206:** 65–72.
12. Spardy, L.E. *et al.* 2011. A dynamical systems analysis of afferent control in a neuromechanical model of locomotion: I. Rhythm generation. *J. Neural. Eng.* **8:** 065003.
13. Harkema, S.J. *et al.* 2012. Balance and ambulation improvements in individuals with chronic incomplete spinal cord injury using locomotor training-based rehabilitation. *Arch. Phys. Med. Rehabil.* **93:** 1508–1517.
14. Wernig, A. *et al.* 1995. Laufband therapy based on 'rules of spinal locomotion' is effective in spinal cord injured persons. *Eur. J. Neurosci.* **7:** 823–829.
15. Devivo, M.J. 2012. Epidemiology of traumatic spinal cord injury: trends and future implications. *Spinal Cord* **50:** 365–372.
16. Sekhon, L.H. & M.G. Fehlings. 2001. Epidemiology, demographics, and pathophysiology of acute spinal cord injury. *Spine* **26:** S2–12.
17. Ferguson, A.R. *et al.* 2012. Central nociceptive sensitization vs. spinal cord training: opposing forms of plasticity that dictate function after complete spinal cord injury. *Front. Physiol.* **3:** 396.
18. Raineteau, O. & M.E. Schwab. 2001. Plasticity of motor systems after incomplete spinal cord injury. *Nat. Rev. Neurosci.* **2:** 263–273.
19. Roy, R.R., S.J. Harkema & V.R. Edgerton. 2012. Basic concepts of activity-based interventions for improved recovery of motor function after spinal cord injury. *Arch. Phys. Med. Rehabil.* **93:** 1487–1497.
20. Bareyre, F.M. & M.E. Schwab. 2003. Inflammation, degeneration and regeneration in the injured spinal cord: insights from DNA microarrays. *Trends Neurosci.* **26:** 555–563.
21. Lynskey, J.V., A. Belanger & R. Jung. 2008. Activity-dependent plasticity in spinal cord injury. *J. Rehabil. Res. Dev.* **45:** 229–240.
22. Bareyre, F.M. *et al.* 2004. The injured spinal cord spontaneously forms a new intraspinal circuit in adult rats. *Nat. Neurosci.* **7:** 269–277.
23. Bose, P. *et al.* 2005. Morphological changes of the soleus motoneuron pool in chronic midthoracic contused rats. *Exp. Neurol.* **191:** 13–23.
24. Thompson, F.J., R. Parmer & P.J. Reier. 1998. Alteration in rate modulation of reflexes to lumbar motoneurons after midthoracic spinal cord injury in the rat. I. Contusion injury. *J. Neurotrauma* **15:** 495–508.
25. Weaver, L.C. *et al.* 2001. Autonomic dysreflexia and primary afferent sprouting after clip-compression injury of the rat spinal cord. *J. Neurotrauma* **18:** 1107–1119.
26. Krenz, N.R. & L.C. Weaver. 1998. Sprouting of primary afferent fibers after spinal cord transection in the rat. *Neuroscience* **85:** 443–458.
27. Wong, S.T., B.A. Atkinson & L.C. Weaver. 2000. Confocal microscopic analysis reveals sprouting of primary afferent fibres in rat dorsal horn after spinal cord injury. *Neurosci. Lett.* **296:** 65–68.
28. McKay, W.B. *et al.* 2011. Neurophysiological characterization of motor recovery in acute spinal cord injury. *Spinal Cord* **49:** 421–429.
29. Basso, D.M., M.S. Beattie & J.C. Bresnahan. 1995. A sensitive and reliable locomotor rating scale for open field testing in rats. *J. Neurotrauma* **12:** 1–21.
30. McEwen, M.L. & J.E. Springer. 2006. Quantification of locomotor recovery following spinal cord contusion in adult rats. *J. Neurotrauma* **23:** 1632–1653.
31. Norrie, B.A., J.M. Nevett-Duchcherer & M.A. Gorassini. 2005. Reduced functional recovery by delaying motor training after spinal cord injury. *J. Neurophysiol.* **94:** 255–264.
32. Onifer, S.M. *et al.* 2011. Horizontal ladder task-specific retraining in adult rats with contusive thoracic spinal cord injury. *Restor. Neurol. Neurosci.* **29:** 275–286.
33. Smith, R.R. *et al.* 2009. Swim training initiated acutely after spinal cord injury is ineffective and induces extravasation in and around the epicenter. *J. Neurotrauma* **26:** 1017–1027.

34. Krajacic, A. *et al.* 2009. Advantages of delaying the onset of rehabilitative reaching training in rats with incomplete spinal cord injury. *Eur. J. Neurosci.* **29:** 641–651.

35. Kanagal, S.G. & G.D. Muir. 2009. Task-dependent compensation after pyramidal tract and dorsolateral spinal lesions in rats. *Exp. Neurol.* **216:** 193–206.

36. Anderson, K.D., A. Gunawan & O. Steward. 2005. Quantitative assessment of forelimb motor function after cervical spinal cord injury in rats: relationship to the corticospinal tract. *Exp. Neurol.* **194:** 161–174.

37. Giszter, S.F. *et al.* 2010. How spinalized rats can walk: biomechanics, cortex, and hindlimb muscle scaling–implications for rehabilitation. *Ann. N.Y. Acad. Sci.* **1198:** 279–293.

38. Bregman, B.S. *et al.* 2002. Transplants and neurotrophic factors increase regeneration and recovery of function after spinal cord injury. *Progr. Brain Res.* **137:** 257–273.

39. Wu, H.F. *et al.* 2013. The promotion of functional recovery and nerve regeneration after spinal cord injury by lentiviral vectors encoding Lingo-1 shRNA delivered by Pluronic F-127. *Biomaterials* **34:** 1686–1700.

40. Campos, L.W. *et al.* 2008. Regenerating motor bridge axons refine connections and synapse on lumbar motoneurons to bypass chronic spinal cord injury. *J. Comp. Neurol.* **506:** 838–850.

41. van den Brand, R. *et al.* 2012. Restoring voluntary control of locomotion after paralyzing spinal cord injury. *Science* **336:** 1182–1185.

42. Dose, F. & G. Taccola. 2012. Coapplication of noisy patterned electrical stimuli and NMDA plus serotonin facilitates fictive locomotion in the rat spinal cord. *J. Neurophysiol.* **108:** 2977–2990.

43. Kandel, E.R., J.H. Schwartz & T.M. Jessell. 2000. *Principles of Neural Science.* McGraw-Hill, Health Professions Division. New York.

44. Skinner, R.D. *et al.* 1996. Effects of exercise and fetal spinal cord implants on the H-reflex in chronically spinalized adult rats. *Brain Res.* **729:** 127–131.

45. Cote, M.P. *et al.* 2011. Activity-dependent increase in neurotrophic factors is associated with an enhanced modulation of spinal reflexes after spinal cord injury. *J. Neurotrauma* **28:** 299–309.

46. Liu, G. *et al.* 2012. Exercise modulates microRNAs that affect the PTEN/mTOR pathway in rats after spinal cord injury. *Exp. Neurol.* **233:** 447–456.

47. Dupont-Versteegden, E.E. *et al.* 2004. Exercise-induced gene expression in soleus muscle is dependent on time after spinal cord injury in rats. *Muscle Nerve* **29:** 73–81.

48. Hornby, T.G., D.H. Zemon & D. Campbell. 2005. Robotic-assisted, body-weight-supported treadmill training in individuals following motor incomplete spinal cord injury. *Phys. Ther.* **85:** 52–66.

49. Hutchinson, K.J. *et al.* 2004. Three exercise paradigms differentially improve sensory recovery after spinal cord contusion in rats. *Brain* **127:** 1403–1414.

50. Ying, Z. *et al.* 2005. Exercise restores levels of neurotrophins and synaptic plasticity following spinal cord injury. *Exp. Neurol.* **193:** 411–419.

51. Ying, Z. *et al.* 2008. BDNF-exercise interactions in the recovery of symmetrical stepping after a cervical hemisection in rats. *Neuroscience* **155:** 1070–1078.

52. Beekhuizen, K.S. & E.C. Field-Fote. 2005. Massed practice versus massed practice with stimulation: effects on upper extremity function and cortical plasticity in individuals with incomplete cervical spinal cord injury. *Neurorehabil. Neural Repair* **19:** 33–45.

53. Hoffman, L.R. & E.C. Field-Fote. 2007. Cortical reorganization following bimanual training and somatosensory stimulation in cervical spinal cord injury: a case report. *Phys. Ther.* **87:** 208–223.

54. Winchester, P. *et al.* 2005. Changes in supraspinal activation patterns following robotic locomotor therapy in motor-incomplete spinal cord injury. *Neurorehabil. Neural Repair* **19:** 313–324.

55. Thomas, S.L. & M.A. Gorassini. 2005. Increases in corticospinal tract function by treadmill training after incomplete spinal cord injury. *J. Neurophysiol.* **94:** 2844–2855.

56. Norton, J.A. & M.A. Gorassini. 2006. Changes in cortically related intermuscular coherence accompanying improvements in locomotor skills in incomplete spinal cord injury. *J. Neurophysiol.* **95:** 2580–2589.

57. Perez, M.A., E.C. Field-Fote & M.K. Floeter. 2003. Patterned sensory stimulation induces plasticity in reciprocal ia inhibition in humans. *J. Neurosci.* **23:** 2014–2018.

58. Knikou, M. 2010. Neural control of locomotion and training-induced plasticity after spinal and cerebral lesions. *Clin. Neurophysiol.* **121:** 1655–1668.

59. Lavrov, I. *et al.* 2006. Plasticity of spinal cord reflexes after a complete transection in adult rats: relationship to stepping ability. *J. Neurophysiol.* **96:** 1699–1710.

60. Al-Majed, A.A., S.L. Tam & T. Gordon. 2004. Electrical stimulation accelerates and enhances expression of regeneration-associated genes in regenerating rat femoral motoneurons. *Cell. Mol. Neurobiol.* **24:** 379–402.

61. Knikou, M. & B.A. Conway. 2005. Effects of electrically induced muscle contraction on flexion reflex in human spinal cord injury. *Spinal Cord* **43:** 640–648.

62. Dobkin, B.H. 2003. Do electrically stimulated sensory inputs and movements lead to long-term plasticity and rehabilitation gains? *Curr. Opin. Neurol.* **16:** 685–691.

63. Leroux, A., J. Fung & H. Barbeau. 1999. Adaptation of the walking pattern to uphill walking in normal and spinal-cord injured subjects. *Exp. Brain Res.* **126:** 359–368.

64. Jung, R. *et al.* 2009. Neuromuscular stimulation therapy after incomplete spinal cord injury promotes recovery of interlimb coordination during locomotion. *J. Neural Eng.* **6:** 55010.

65. Kim, S.J. *et al.* 2009. Adaptive control of movement for neuromuscular stimulation-assisted therapy in a rodent model. *IEEE Trans. Biomed. Eng.* **56:** 452–461.

66. Fairchild, M.D. *et al.* 2010. Repetitive hindlimb movement using intermittent adaptive neuromuscular electrical stimulation in an incomplete spinal cord injury rodent model. *Exp. Neurol.* **223:** 623–633.

67. Abbas, J.J. & H.J. Chizeck. 1995. Neural network control of functional neuromuscular stimulation systems: computer simulation studies. *IEEE Trans. Biomed. Eng.* **42:** 1117–1127.

68. Basu, I. *et al.* 2011. Adaptive control of deep brain stimulator for essential tremor: entropy-based tremor prediction using surface-EMG. Conference proceedings: IEEE

Engineering in Medicine and Biology Society Conference. 2011: 7711–7714.

69. Emborg, J. *et al.* 2011. Design and test of a novel closed-loop system that exploits the nociceptive withdrawal reflex for swing-phase support of the hemiparetic gait. *IEEE Trans. Biomed. Eng.* **58:** 960–970.

70. Ferrante, S. *et al.* 2004. Functional electrical stimulation controlled by artificial neural networks: pilot experiments with simple movements are promising for rehabilitation applications. *Funct. Neurol.* **19:** 243–252.

71. Bandt, C. & B. Pompe. 2002. Permutation entropy: a natural complexity measure for time series. *Phys. Rev. Lett.* **88:** 174102.

72. Olofsen, E., J.W. Sleigh & A. Dahan. 2008. Permutation entropy of the electroencephalogram: a measure of anaesthetic drug effect. *Br. J. Anaesth.* **101:** 810–821.